Eberhard Teuscher, Ulrike Lindequist
**Natural Poisons and Venoms**

## Also in this Series

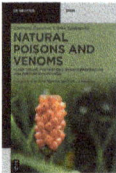

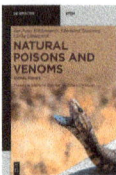

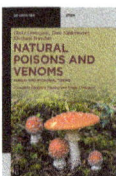

Eberhard Teuscher, Ulrike Lindequist

# Natural Poisons and Venoms

Volume 1
Plant Toxins: Terpenes and Steroids

Founded by
Eberhard Teuscher and Ulrike Lindequist

**DE GRUYTER**

**Authors**
Prof. Dr. Eberhard Teuscher
Goethestr. 9
07950 Zeulenroda-Triebes
Germany
teuscher-triebes@t-online.de

Prof. Dr. Ulrike Lindequist
Lise-Meitner-Str. 6a
17491 Greifswald
Germany
lindequi@uni-greifswald.de

ISBN 978-3-11-072472-1
e-ISBN (PDF) 978-3-11-072473-8
e-ISBN (EPUB) 978-3-11-072486-8

**Library of Congress Control Number: 2023930659**

**Bibliographic information published by the Deutsche Nationalbibliothek**
The Deutsche Nationalbibliothek lists this publication in the Deutsche Nationalbibliografie;
detailed bibliographic data are available on the internet at http://dnb.dnb.de.

www.degruyter.com

# Preface

Living beings produce pharmacological active substances to ensure their survival. In many cases, such substances can harm or kill other living creatures or can trigger allergies. Some of these compounds pose a danger not only to human beings but also to animals.

Our concern is to try to acquaint the readers – especially physicians, veterinarians, pharmacists, biologists, chemists, food chemists, biochemists, students of these disciplines, but interested laymen also – with poisonous microorganisms, fungi, plants, and animals, in words and pictures.

We want to inform the reader about the biology of the poisonous living beings and about the structure, the effects and the mode of action of their toxic ingredients.

We report on the numerous possibilities to poison, which are, e.g., misapplication of drugs and natural substances as therapeutics in medicine, folk medicine and folk customs, misuse of plants and biogenic substances as narcotics, confusion of poisonous plants with edible ones, testing the edibility of unknown plants, attempted suicides, homicides, poison arrows, poison baits, animal attacks or infections.

We inform about the symptoms of the poisonings, in many cases shown on case reports, and we give hints on poisoning prevention and about measures to treat the poisonings. However, the books are not intended as a guide to medical practice in cases of poisoning.

For poisonings of animals, especially of farm and domestic animals, by plants we characterize the preconditions of poisonings and show the symptoms and the possible treatment of these poisonings, as well as their prevention.

We inform not only about poisonous drugs and their active ingredients which in earlier and current times are used as therapeutics, but also about drugs which may be starting points for the development of new medicines.

Over the decades, our knowledge in the field of natural poisons has expanded rapidly. Reasons for this include, e.g., modern molecular biological methods available today. The easier accessibility to exotic sources through globalization and the easier accessibility of living beings from the marine habitat have fostered the interest in investigations of these sources.

We have always tried to see the toxins not only as a danger to humans and animals but also as an admirable variety of substances that have been evolved in the course of evolution to protect the living products of evolution and thus enable them to continue to exist.

The books are based on our book *Biogene Gifte*, third edition, published in 2010 by Wissenschaftliche Verlagsgesellschaft Stuttgart in German language. Because of the enormous increase in knowledge in the field of natural poisons and venoms, to reach a larger readership and to fit into the STEM series of the De Gruyter publishing house, the topic is now covered in five volumes in English language.

https://doi.org/10.1515/9783110724738-202

After a fundamental chapter on toxins (definitions of terms used in toxicology, history, chemistry, biology, first aid, and general and clinical toxicology) Volumes 1 to 3 are devoted to poisonous plants and their active substances. Volume 4 summarizes the current knowledge about poisonous animals, their poisons, their attacks, and the measures after an attack. Volume 5 deals with poisonous mushrooms and cyanobacteria and their ingredients and selected microbial toxins. The volumes are staggered in time, so that the highest possible topicality is guaranteed. The volumes can be purchased independently, but at least the joint purchase of the three plant volumes is recommended.

In terms of nomenclature, we have adhered for plants to 'The World Flora Online, WFO 2022'. We have taken the English language trivial names predominantly from the 'Der große Zander, Enzyklopädie der Pflanzennamen, 2008'.

Despite the extensive bibliographies at the end of each chapter, the cited literature represents only a selection. For reasons of space, the information is as short as possible. Preference was given to review papers. Further literature is summarized at the end of the volumes.

We would like to thank everyone who helped us with the creation of the books. We would especially like to thank Mr. Dipl.-Phys. Karl-Heinz Lichtnow, Greifswald (Germany), for drawing the drafts of structural formulas and schemes and his readily available help with critical computer problems of any kind. We would like to thank all image authors, and Getty images for the provision of the photos. Our thanks go to De Gruyter publishing house, especially Mrs. Karin Sora, for making these books possible, Mrs. Dr. Bettina Noto for her always pleasant and helpful cooperation in the completion of the volumes and her careful editing work and Mrs. Anne Stroka and her staffs for the preparation of the manuscript to the printable form.

We are always grateful for critical information on errors and suggestions for improvement.

Triebes and Greifswald, January 2023
Eberhard Teuscher and Ulrike Lindequist, editors

# A Short Glance on This Volume

In the first chapter of this book, one will find the definitions of toxicological terms and a very interesting section about the history of natural poisons. Basic information about the characterization, isolation, distribution, and structural elucidation of natural poisons and about their importance in biological systems is given. Besides, it gives an overview about pharmacological and toxicological effects and mode of actions of the toxins as well as about their dangers for living beings. It informs about precautions to avoid poisonings of human beings, especially of children, and of pets. Measures in the case of poisoning, inclusive proposals for first aid and diagnostics, are suggested. The section about the usefulness of natural poisons for humans shows the great importance and enormous potential of natural poisons for drug research and development.

This volume focuses on the toxic representatives of two important biogenetically related groups of natural products: terpenes and steroids. They include, for example, poisonous ingredients of essential oils, pyrethrins, highly effective diterpenes from plants of the heather family and the spurge family and other species, terpenes from seaweeds, limonoids, cucurbitacins, cardioactive steroid glycosides, withanolides, petunia sterones, and triterpene and steroid saponins.

In each case, chemistry, biogenesis, distribution, pharmacology, and toxicology of the poisonous compounds and biology and distribution of the toxin-producing plants are described. Selected poisoning cases illuminate possible causes of poisoning, the typical symptoms, and possible treatment measures.

The many new findings in the field of biogenic poisons in recent years are taken into the account. For example, we now know much more than we did just a few years ago about the diverse effects of cardiac glycosides or saponins and the mechanisms behind them; about the possible danger of licorice, especially for pregnant women, and other components of our nutrition; about the presence of vitamin D in numerous plants and potential consequences; or about the threats from exotic plants such as *Hippomane mancinella*, manchineel, or *Cerbera odollam*, Madagascar ordeal bean.

Each chapter contains an extensive bibliography, which also allows recourse to older publications.

In the appendix you will find a list of Poison Information Centers of the whole world and their contact data, which can be helpful in the case of suspected poisoning.

The authors

https://doi.org/10.1515/9783110724738-203

# Contents

# Abbreviations and Icons

| | |
|---|---|
| ADI | acceptable daily intake |
| Approx. | approximately |
| B.C. | before Christ |
| B.P. | boiling point |
| BW | body weight |
| CNS | central nerve system |
| CYP | cytochrom P |
| DW | dry weight |
| EFSA | European Food Safety Authority |
| EMA | European Medicines Agency |
| esp. | especially |
| FDA | U.S. Food and Drug Administration |
| FW | fresh weight |
| GABA | γ-aminobutyric acid |
| H.I. | hemolytic index |
| HMPC | Herbal Medicinal Products Committee of the EMA |
| i.a. | intraarterial |
| IL | interleukin |
| i.m. | intramuscular |
| i.p. | intraperitoneal |
| i.v. | intravenous |
| LD | letal dose |
| LOAEL | lowest observed adverse effect level |
| M.P. | melting point |
| Mr | molecular weight |
| NF-kB | nuclear factor kB |
| NOAEL | no observed adverse effect level |
| OECD | Organisation for Economic Co-operation and Development |
| p.e. | parenteral |
| P-gp | P-glycoprotein |
| p.i. | per inhalationem |
| p.o. | peroral |
| resp. | respectively |
| s.c. | subcutaneous |
| ssp. | subspecies |
| TCM | traditional Chinese medicine |
| TRP | transient receptor potential cation channels |
| WHO | World Health Organization |

| | |
|---|---|
| ☠ | Symptoms of acute poisoning |
| (☠) | Low risk of poisoning |
| ⚕ | Case description |
| 🐾 | Poisoning of animals |
| ✳ | Allergy |
| ☹ | Chronic poisoning |
| 🚑 | First aid and potential treatment |

https://doi.org/10.1515/9783110724738-205

# 1 Natural Poisons

## 1.1 Definitions

**Poisons** are substances that, under certain circumstances, can cause temporary or permanent harm to the body, organs, tissues, cells, or DNA, or can lead to death when introduced or absorbed.

**Toxins**, also called 'natural poisons', 'biological poisons', or 'biotoxins', are poisons produced by living organisms. The term 'toxin' was first used in 1890 by the physician Ludwig Brieger (1849–1919), professor at the Humboldt University in Berlin, during his work with bacterial poisons. Initially, the term was used to describe animal or bacterial compounds that trigger the formation of antibodies in animals and humans.

Today, the term 'toxin' is used with different meanings:
- as a general expression for poisons produced by living organisms, or
- as an expression for natural or synthetic poisons that target specific organs, e.g., **nephrotoxins** (damaging the kidneys), **hepatotoxins** (harming the liver), **cardiotoxins** (harming the heart), or **hemotoxins** (damaging red blood cells), regardless of whether they are of natural or synthetic origin.

**Toxicant** is a poorly defined term. Mostly, it is used to describe a toxic substance capable of causing deleterious effects in living organisms. Some authors include in this category only man-made synthetic substances showing such effects.

**Toxicology** is the scientific discipline describing poisons, poisonings, kind of action of poisons and potential treatments of poisoned organisms.

**Toxinology** is a branch of toxicology. It comprises the knowledge about the chemistry of natural poisons, the biology of their producers, their biosynthesis, their pharmacokinetic behavior (**toxicokinetics**), their mechanism of action (**toxicodynamics**), their effect pattern on the affected living being (**toxicography**), and the knowledge about the treatment of poisonings (**clinical toxicology**). The term 'toxinology' is not generally accepted.

**Venoms** are mixtures of toxins produced and used by animals for their defense. They are specifically designed for the poisoning of individuals of other species. Venoms are purposefully applied to target organisms using specific tools like teeth, other mouthparts, or stings.

**Poisoning** is the process in which an organism becomes intoxicated by an exogenous substance.

**Toxicosis** is any kind of disease condition caused by poisoning.

**Envenomation** is the process of venom application by venomous animals to target organisms, mostly by biting or by stinging.

https://doi.org/10.1515/9783110724738-001

The clinical pictures of poisoning in humans are also called **intoxications** or **toxicoses** (e.g., mycotoxicoses). If they are due to the exposure to mixtures of different toxins, they are also referred to as **polyintoxications**.

**Toxicity** is the extent of the poisonous effect. It is determined by
- the type of substance;
- the amount of the substance administered in relation to the body weight (BW) of the affected living being;
- the mode of application into the organism: peroral, percutaneous, intramuscular, intravenous, intraarterial, rectal, by inhalation;
- in case of local application, the duration of exposure and the sensitivity of the affected individual (there are significant differences in toxin sensitivities between children and adults, females and males, and humans and different animal species);
- the frequency and time course of toxin intake designated as
  - **acute toxicity:** adverse effects occurring following a single exposure or multiple exposures in a short period of time (<24 h) manifested within 14 days upon administration;
  - **subacute toxicity:** adverse effects following multiple toxin exposures (13–40 doses over a period of 14–28 days);
  - **chronic toxicity:** adverse effects following long-term exposure to multiple doses of toxins that have per se no acute toxicity;
  - **subchronic toxicity:** adverse effects occurring during a treatment regime of 90 daily doses that have per se no acute toxicity;
  - **long-term toxicity:** adverse effects that become manifest only after a long latency period, mainly mutagenic and carcinogenic effects (see later).

Thus, the designation of a substance as a poison or toxin depends on many variables and is not to be understood as an *absolutum*.

**Habituation** is the development of tolerance for a poison upon frequent exposure to doses of a poison that has per se no acute toxicity (also designed as mithridatism).

**Addiction** is the overwhelming desire or need to continue the use of a substance after habituation.

**Withdrawal syndrome** is a term to describe a group of symptoms occurring upon abrupt discontinuation of intake of a medicine, a narcotic substance or a toxin after development of addiction.

**Toxicological relevance** is a term which describes the likeliness of exposure of humans or animals to a certain poison.

**Toxification** means that a substance, nontoxic in its native form, is converted into a poison by biotransformation in an organism.

**Detoxification** means the conversion of a poison to a nontoxic substance by biotransformation in an organism.

**Mutagenicity** is the ability of a substance, a **mutagen**, to promote the occurrence of mutations, i.e., to induce changes in the nucleotide sequence of the DNA in cells or organisms.

**Carcinogenicity** is the potential of a substance, a **carcinogen**, to provoke the development of cancer. **Primary carcinogens** have a carcinogenic effect without metabolic activation. **Secondary carcinogens** are noncarcinogenic precursors which must be transformed into effective carcinogens in the organism's metabolism. Most natural carcinogens, e.g., aflatoxins or pyrrolizidine alkaloids, are secondary carcinogens.

**Tumor promoters** are substances that promote tumor growth but are not carcinogens themselves.

**Teratogenicity** is the capacity of a substance, a **teratogen**, to induce irregularities in the embryonic development causing malformations in the structure or function in the offspring or fetal or embryonic mortality.

**Phototoxins** are substances that cause deleterious photosensitivity of the skin or other external organs.

**Phytotoxins** are toxins produced by parasitic microorganisms that are directed against plants and cause chlorosis (bleaching), wilting, malformations of the plant, or its death. Substances that are covered by this definition are not considered in this book series. However, the term **phytotoxins** is sometimes used to describe toxic peptides and proteins produced by plants and directed against other organisms, e.g., as defensive chemicals. Such toxins are considered in this book series.

In the present series of our books, only those natural poisons are considered which are capable to cause adverse effects in animals and humans. Interesting toxins that are harmful to microorganisms, fungi, or plants are mentioned but not comprehensively discussed.

## 1.2  History of Natural Poisons

Our knowledge about poisonous plants, mushrooms, and animals is as old as mankind itself. It was acquired in the course of exploration of the environment through accidental discoveries at the cost of many sacrifices. Some of them have already been made by our animal ancestors.

Every adult wild animal 'knows' dangerous plants and poisonous animals in its familiar environment. Poisoning occurs in animals only when they encounter poisonous plants in an unfamiliar form, e.g., in hay, when animals encounter 'unknown' poisonous plants, e.g., exotic ornamental plants, or plants whose poisonous effects become apparent only after a long period of time. Animals will also sometimes eat otherwise avoided plants in deficiency situations, e.g., overgrazing. Warning colors (aposematism) of poisonous animals are recognized by other organisms and used to avoid such creatures.

Humans discovered the poisonous effects of plants especially when testing plant parts for edibility or for use in curing diseases. Observing which plants were avoided by animals as food also contributed to the recognition of poisonous plants by humans.

The use of natural poisons already began in primitive societies. Thus, poisons were used to kill prey (poisoned arrows, poisoned spears, poisoned bait, and fish poisons), as insecticides or in small doses as medicines, but also as abortifacients, suicide poisons, means for executions, as weapons, as intoxicants, witch poisons, for the preparation of love-potions, or in a criminal manner as murder poisons.

The oldest evidence of poisonous plant exploration comes from China, India, Persia, and Egypt.

Pharaoh Menes (ruled around 3000 B.C.) is reported to have cultivated and studied poisonous and medicinal plants. In the Egyptian Papyrus Ebers (written around 1550 B.C.), information is given about the effects of aconite, castor beans, hemlock, and opium, among others. The preparation of homicidal poisons is also described there.

The Greek philosopher Theophrastus of Eresos (371–287 B.C.), a student of Aristotle, and the Greek physicians Nicander of Colophon (197–133 B.C.), Dioscorides (40–90 B.C.), and Claudius Galenus (129–199) described mineral and plant poisons known at that time, their effects, and the treatment of poisonings caused by them.

Muslim physicians such as Ibn Wahshiyya (ninth century) and Ibn Moses Maimonides (1135–1204) summarized and expanded the knowledge of the Greeks in their books.

In ancient Rome, Pliny the Elder (23/24–79) and Pliny the Younger (61–113) listed 7,000 poisons, most of them natural, in their multivolume book *Natural History*. Roman rulers stockpiled poisons for potential use in political assassinations. In particular, the poison collections of the Emperors Caligula (12–41) and Caracalla (188–217) are famous and notorious.

The European literature of the Middle Ages on poisons is based primarily on classical Greek, Roman, and Arabic publications. Supposed poisonous ingredients were summarized in long lists. In his writings, the Italian physician Peter of Abano (1257–1315) listed arsenic sublimate, lead, and mercury, as well as Aristolochia, bryony, colocynth, cherry laurel, coriander, fool's parsley, hellebore, Strychnos, and cantharides, among others. The Italian toxicologist Sante Arduino (1430–?) describes in his book bryony, cherry laurel, hellebore, mandrake, opium, and cantharides.

From late antiquity to modern times, especially in the Victorian era (1837–1901), arsenic salts, particularly the white, odorless, and tasteless arsenic(III) oxide, were often used as murder poisons. They were sold in pharmacies as rat and mice poisons and were thus easily accessible. Moreover, they were not detectable in the victims at that time. Among the natural poisons, especially the aconitine-containing monkshood was used for murders. The Italian physician Giovanni Battista della Porta (1535–1615) published in his book *Neapolitani Magiae Naturalis* (1589) instructions for poison mixers, e.g., for the preparation of a concoction, called Venenum Lupinum, which contained extracts of bitter almond, monkshood, yew, and toxic minerals.

In the early history of mankind, experiments with poisons to test their possible use as murder poisons or as weapons have been performed. At the same time, investigations to find antidotes have been done. Lewin [87, 88] reported about experiments with animals, but also with humans, at Pontic, Pergamic, Alexandrian, and Roman courts. For example, Attalus III Philometor (last King of Pergamon, 170–132 B.C.), Mithridates VI Eupator (king of Pontos, 135–63 BC), and the Roman emperor Nero (ruled from 54–68) are known as notorious experimenters. Cleopatra (queen of Egypt, 69–30 B.C.) is said to have tested the toxic effects of belladonna, henbane, and Strychnos on her slaves. In the Middle Ages, it was Catherine de Medici (1519–1589), queen of France, who systematically tested the action of poisonous concoctions on poor and sick people.

Already in ancient times, efforts were made to protect oneself from poisonous attacks. This was done, among others, by ingesting poisonous mixtures in harmless quantities at first and then carefully increasing the dose to achieve habituation to the poisons. This method was practiced among others by Mithridates VI (therefore called mithridatization). Furthermore, Mithridates developed a universal antidote (called mithridaticum) with 54 ingredients. The personal physician of Nero, Andromachus the Elder, extended the mixture to 80 components. Later, the mixture, sometimes containing up to 300 components, was called Theriac and was used as a medicine until the eighteenth century. Its effectiveness seems at least doubtful.

Beginning in the fifteenth century, attempts were made to explore the nature of poisons and their sources. The merit of Paracelsus (1493–1541) is to explore the active ingredients of medicines. He promoted the development of the so-called iatrochemistry. Distillation was used to enrich and to isolate substances. A variety of essential oils and volatile substances, such as succinic acid, camphor, or thymol, could be obtained and purified. The statement of Paracelsus is, at least partly, still valid today: 'What is there, that is not poison? All things are poison and nothing is without poison. Solely the dose determines that a thing is not a poison' or shortly 'The dose makes the poison.' Today we know that not only the dose but also other factors determine toxicity.

Since the beginning of the nineteenth century, the inclusion of chemistry, biochemistry, pharmacology, and other branches of natural sciences in toxicological investigations allowed a considerable extension of the knowledge about natural poisons and their producers. The chemistry was developed far enough to enable isolation of pure active ingredients from a biological material.

A breakthrough was the extraction of morphine in 1806 by the pharmacist F.W. Sertürner (1783–1841) from opium. This was quickly followed by other discoveries of active plant substances.

In the twentieth and twenty-first centuries, research into all aspects of natural poisons took off considerably, thanks to modern methods in biochemistry, chemistry, biotechnology, pharmacology, toxicology, and molecular biology.

Books and reviews: [3, 16, 18, 19, 28, 34, 38, 45, 49, 54, 65, 66, **87, 88, 112, 123, 167**].

## 1.3 Isolation and Characterization of Natural Poisons

### 1.3.1 Isolation of Natural Poisons

Initially, differences in solubility in various solvents, in partitioning behavior between two immiscible liquid phases at different pH values by the so-called shaking out, in volatility during distillation, and in chemical reactivity, e.g., with precipitants, were used to separate the active ingredients from the accompanying substances.

The development of chromatographic methods in 1903 by the Russian botanist M.S. Zwet (also written Tswett, 1872–1919), at the beginning of the twentieth century, enabled a very strong progress. This method is based on the different abilities of substances to adsorb to solid particles (adsorption chromatography) or on the different distribution of substances between two immiscible liquid phases, the stationary phase adsorbed on solid particles and the mobile phase (distribution chromatography). The mixture of substances to be separated is dissolved in the mobile phase, which flows through a particulate or monolithic (compact) stationary phase in a column. At the end of the column, the substances, retained to a lesser or greater extent, appear sequentially in the outflow (column chromatography and liquid chromatography (LC)).

**Column chromatography** allows rapid separation of substances and yields quantities of pure substances sufficient for pharmacological testing and structure elucidation. It is usually performed using high pressure systems, HPLC (**H**igh **P**erformance **L**iquid **C**hromatography), or UHPLC (**U**ltra **H**igh **P**erformance **L**iquid **C**hromatography) with stationary phase of spherical, particularly small particles, or as medium pressure MPLC (**M**edium **P**ressure **L**iquid **C**hromatography). Distribution between two immiscible liquids is used in FCPC (**F**ast **C**entrifugal **P**artion **C**hromatography), DCCC (**D**roplet **C**ounter-**C**urrent **C**hromatography) or RLCC (**R**otation **L**ocular **C**ounter-**C**urrent **C**hromatography). If the stationary phase is polar (e.g., in hydrous silica gel), it is referred to as NP-HPCL (**NP** = normal phase) or simply HPLC, if the stationary phase is apolar (e.g., compounds with a silicone backbone linked to long-chain hydrocarbons), it is referred to as RP-HPLC (**RP** = reversed phase).

**Electrophoretic separation methods**, e.g., CE (**C**apillary **E**lectrophoresis), CEC (**C**apillary **E**lectro**C**hromatography), and MEKC (**M**icellar **E**lectro**K**inetic **C**hromatography) are based on the different migration speeds of substances with different electrical charges in an electric field. SDS-PAGE (**S**odium **D**odecyl-**S**ulfate-**P**olyacryl**A**mide-**G**el **E**lectrophoresis) is particularly useful for the separation of peptides and proteins.

**Molecular sieve chromatography** or **gel filtration**, e.g., performed using dextran–epichlorohydrin copolymers (molecular sieves), is based on the different molecular sizes (more precisely the hydrodynamic volumes) of the substances to be separated. These molecular sieves, available as small pearls with different pore sizes, are placed in a column. The mixtures of substances dissolved in water are allowed to flow through the column. In this process, small molecules can penetrate the pores and thus have a larger diffusion volume, while large molecules cannot do it. Depending on their size, the smaller molecules travel a greater distance than the larger molecules and therefore appear later in the effluent. This method is particularly suitable for the separation of macromolecules.

**Ion exchange chromatography** is based on the electrical charge-dependent binding of ions to exchange resins. Exchange resins are polymers that carry charged groups that can reversibly bind cations or anions.

**Affinity chromatography** is based on the affinity of a ligand, e.g., an antigen or a lectin, to proteins, e.g., antibodies or glycoproteins.

**Fractional precipitation** is used to separate proteins. It is carried out by precipitation with concentrated salt solutions, e.g., highly soluble ammonium sulfate, by changing the pH of the solution, or with water-miscible organic solvents, e.g., acetone. When precipitating with organic solvents, one must work at temperatures close to the freezing point in order to prevent denaturation of the proteins.

These isolation processes are usually performed as bioactivity-guided fractionations, i.e., each fraction of the eluate of a column is tested for its biological effects. The effective fractions are then further separated to obtain the active ingredient in its pure form.

## 1.3.2 Elucidation of the Chemical Structure of Natural Poisons

Elucidation of the chemical structure was initially performed only with the aid of chemical methods, especially by quantitative elemental analysis, by investigation of chemical behavior to determine the functional groups, and by partial degradation to already known compounds. This was often followed by structure confirmation by synthesis. This type of structure elucidation was very time consuming. In many cases, more than half a century elapsed between the isolation of a compound and the elucidation of its structure.

Digitoxin, for example, was isolated from the leaves of the red foxglove in 1869. The sum formula of the aglycone digitoxigenin was determined in 1927, its structural formula was established in 1935, and its configuration was elucidated in 1945. It had already been recognized in 1925 that the aglycone digitoxigenin in digitoxin is linked to three molecules of digitoxose. The full structure was not determined until 1962. Thus, after almost 100 years of research and through the efforts of several generations of researchers, the structural formula of digitoxin was completely known.

Since the middle of the twentieth century, the methods of structure elucidation underwent a tremendous development through the possibility to use instrumental methods. The development of ultraviolet (UV) spectroscopy, infrared (IR) spectroscopy, mass spectrometry (MS), nuclear magnetic resonance (NMR) spectroscopy, Raman spectroscopy, X-ray structural analysis, and neutron scattering meant that now the isolation of a substance and its structural formula with all stereochemical details are generally published together at the same time.

Today, methods for separation and structure clarification are frequently combined by using combination instruments (hyphenated techniques), e.g., GC-MS (**Gas**

Chromatography/Mass Spectrometry), HPLC/UV, HPLC/MS, HPLC/NMR, GC-IRMS (GC-Isotope Ratio Mass Spectrometry), GC-FTIR (GC-Fourier Transform Infrared Spectrometry) and HPLC-SPE-NMR (SPE = Solid Phase Extraction).

The progress of separation and structure elucidation techniques leads to the isolation and structural characterization of several thousand new natural products every year.

These new methods in combination with modern so-called omics technologies (genomics, transcriptomics, proteomics, metabolomics) and, if possible, cultivation techniques allow the investigation of toxin sources that are not or only insufficiently accessible: from noncultivable microorganisms, protected organisms, organisms living in environments difficult to access, e.g., deep sea, rain forests, Arctic or Antarctic territories, or organisms that cannot be collected in larger quantities. Metagenomics allows the study of the genetic material from a mixed community of organisms, e.g., a sponge with microorganisms living in it, and to explore its potential for synthesis of natural compounds without separation of the organisms [56].

Reviews: [23, 26, 50, 53, 58].

### 1.3.3 Pharmacological Investigations of Natural Poisons

To isolate pure substances from natural sources and to elucidate their chemical structure is relatively easy today. It is more difficult to clarify their effects on living organisms. Therefore, many chemically characterized natural substances are known today, whose physiological effects are not or only incompletely characterized.

The classical method was to test first the substances on animals, e.g., on mice, rats, guinea pigs, or rabbits. Already the pharmacist Friedrich Sertürner (1783–1831), at that time still a pharmacist's assistant, who isolated morphine, the main active ingredient of opium, tested its effect in experiments on a little dog and later in an almost fatal self-experiment.

These methods are partially replaced today by in vitro pretests on organ preparations, isolated and cultivated cells, cell components, isolated enzymes, cell-free systems, or bacteria. In vitro assays allow a high throughput, are ethically more defensible than animal assays, and need lower amounts of test samples [6, 13, 35, 52].

Alternative methods to obtain indications of potential effects are studies in silico without using organisms by virtual screening. This is done by examining the possible interaction of three-dimensional, computer-rendered structural models of potential poisons with three-dimensional models of pharmacophores (groups responsible for the pharmacological action of a molecule), e.g., of enzymes or receptors [1, 32].

Up to now, it is not possible to completely replace animal testing. Reasons for this are that in vitro and in silico studies do not reflect pharmacokinetic processes including possible toxification or detoxification, pleiotropic effects of a possible toxin, the influence of microbiota, and the uncertainty whether isolated cells react in the same way as the whole organism. Effects on central and peripheral nerves and organotypic effects, e.g., hormonal, cardiotoxic, hepatotoxic, nephrotoxic, spastic, spasmolytic, and skin irritant effects, are often not considered in in vitro testing. Therefore, extrapolation of toxicity evaluations from in vitro studies to live animals or humans is possible only to a very limited extent. The preliminary in vitro tests must still be followed by in vivo tests, i.e., on live animals, despite the ethical concerns [6, 73]. In compliance with the principles of 3R (replace, reduce, and refine), it is intended to increasingly replace animal experiments and still provide good risk assessments [29, 31].

Nevertheless, there can be considerable differences in sensitivity against certain substances between different animal species and between animals and humans. For example, coumarin is highly toxic to rats, but relatively nontoxic to humans. Theobromine is toxic to dogs and cats at low doses, but not to humans. Thus, doses of certain chemicals used in animal testing may or may not match with the effective doses in humans. Causes for these differences are mostly genetically determined and manifested, for example, in different metabolic processes. The development of transgenic and knockout mice provides models to investigate the role of specific genes and their products targeted by poisonous compounds [9].

## 1.4 Sources of Natural Poisons

### 1.4.1 Definitions

Substances produced by living organisms can be primary or secondary metabolites.

**Primary metabolites** are substances involved in the energy or the anabolic metabolism of the organism or only temporarily excreted from these processes. They occur in all living organisms and are vital for the producer.

**Secondary metabolites** are formed by living organisms as end products, i.e., they are not further used in metabolic pathways but instead for nonmetabolic purposes or as excretory substances. They are not directly involved in energy and building metabolism, but some of them may be involved in controlling these metabolic processes. The generation, accumulation, and the use of secondary metabolites provide the organism with selective advantages.

Toxins are secondary metabolites. They confer primary or secondary toxicity to their producers.

**Primary toxic organisms** produce their toxins themselves. In order not to endanger their producer, these toxins are often stored as nontoxic precursors and activated only at the time of use or are strongly compartmentalized to avoid interference with the own metabolism.

**Secondary toxic organisms** owe their toxicity to ingested and stored toxins that have been synthetized by other organisms and taken up via the food web. A prerequisite for storage is that the ingested substances are tolerated by the creature that stores them. For example, the occurrence of toxins from cyanobacteria, eubacteria, or dinoflagellates in mussels, crabs, or fish is well known. Some insects as well as poison dart frogs use plant toxins or their derivatives to protect themselves from predators.

Plant as well as bacterial toxins and mycotoxins may occur in animal products (milk, eggs, and meat) after ingestion of poisonous plants or moldy feed. Of toxicological significance is also the formation of toxins in spoiled food by bacteria or fungi.

**Active poisonous organisms** (or **venomous organisms**) like snakes, spiders, or many insects have a venom apparatus with which they can deliver their venom to or into target organisms for the purposes to defend themselves or to catch prey.

**Passive poisonous organisms** (or **poisonous organisms**) like amphibia use their toxins to impregnate skin or inner organs to become toxic to potential predators or secrete their toxins through defense glands to deter potential predators.

### 1.4.2 Biosynthesis of Natural Poisons

Elucidations of the reaction steps by which toxins arise from primary metabolites in their producers became possible to a greater extent only in the middle of the twentieth century, when easily detectable, specifically radioactively labeled, potential precursors of compounds became available, e.g., possible precursors containing radioactive $^{14}$C atoms instead of natural $^{12}$C atoms. Thus, the pathways of biosynthesis could be followed by capturing intermediates from radioactively labeled precursors.

Another method was to use defect mutants, organisms that were unable to complete the biosynthesis of a toxin due to a lack in one or more of the required enzymes. Such mutants accumulate intermediates of toxin biosynthesis pathways which are easily recognizable as such by their accumulation in high concentrations.

Considerable progress has been made in the field of isolation and characterization of enzymes involved in toxin biogenesis. Presently, the genes coding for these enzymes are characterized, and their expression is studied using transcriptomics or proteomics.

The use of these techniques has substantially advanced investigations on toxin synthesis pathways. We know now that monosaccharides serve as precursors of phenylpropane derivatives and as building blocks of glycosides and glycopeptides, that amino acids are precursors and building blocks of cyanogenic glycosides, alkaloids, peptides, and proteins, and that carboxylic acids are precursors and building blocks of polyketides, terpenes, or steroids. Also, the intermediate reaction steps from these primary substances to toxins are mostly known.

These reactions are catalyzed by specific enzymes which are expressed in specialized cells, e.g., in epithelial cells of venom glands of snakes. While the enzymes building the basic molecular structure often have high specificities, the enzymes that generate the secondary metabolites in their final form, e.g., methyl transferases and glycosyl transferases, are often less specific. This seems to be tolerable because these enzymes underlie tissue-specific expression patterns.

It is not easy to answer the question how the biogenetic pathways of toxins exactly originated from the pathways of basic metabolism. The genes encoding the enzymes of secondary metabolism have presumably emerged in the course of evolution through mutations from duplicates of genes of primary metabolism.

Relatively easy to explain is the origin of toxic peptides and proteotoxins from snake venoms. They originate from the digestive enzymes of the salivary glands, which were transformed into venom glands. Sequence analysis of the toxins and DNA of the snake venoms demonstrate the origin of the toxins from phospholipase $A_2$, ribonucleases, or proteases. The genes for these enzymes were multiplied and subject to evolutionary adaptation. Selection pressure ensured that at least one gene of such isoenzymes formed remained stable because its product was useful in its original purpose while all other genes were allowed to change freely. They were stabilized in their current state only when their products brought a selective advantage to their carriers, e.g., when useful as toxins.

More difficult to interpret is the evolution of genes for the extensive enzyme sets necessary for the biogenesis of, for example, an alkaloid. Certainly, the great flexibility of the genome in providing many isoforms of genes, of which some may be useful isoenzymes, plays a decisive role in this context. Probably, the 'ancient deadly nightshade', *Atropa belladonna*, million years ago, did not immediately start with the formation of hyoscyamine as a protective factor. Presumably, its precursor hygrine, formed from the ubiquitous amino acid L-arginine, already brought a modest selection advantage, which was then extended further, resulting in the biogenesis of the very toxic hyoscyamine. Besides point mutations in the genome, as occurred in the formation of proteotoxins by the snakes, major changes in genome arrangements (mobile DNA elements like transposons) may have facilitated the evolutionary development of metabolic networks favoring toxin production. Their significance is not yet understood. The future will provide new insights into the relations between primary and secondary metabolisms and the biochemical pathways involved.

Reviews: [57, 69, **96**].

### 1.4.3 Distribution of Natural Poisons and Conditions for Their Production

Natural poisons are secondary metabolites, i.e., they are neither involved in energy metabolism nor in further biosynthetic processes. As metabolites, they are not directly vital for the producer. This is shown by their tissue distribution. While the amino acid L-arginine, for example, occurs as a primary metabolite in all living organisms and in all cells of a multicellular organism, secondary metabolites derived from L-arginine, e.g., hyoscyamine, may or may not occur in a particular organism or may just occur in specialized cells of the producer.

Toxin-forming and/or toxin-storing creatures occur in almost all taxa of living organisms. The kinds and quantities of toxins present in a certain species are genetically determined (chemical races and chemotypes), may be determined by the developmental state of the organism or by environmental factors (e.g., nutrition in animals, or soil quality in plants). In the case of secondary toxic organisms, the differences are mainly determined by the amounts of toxins ingested and stored.

In microorganisms, the ability to synthesize certain substances is often encoded in extranuclear genetic elements, e.g., plasmids. Copies of such genetic elements can be transferred to other organisms, usually of the same species or across species boundaries (horizontal gene transfer). For example, strains of the cyanobacterium *Anabaena flos-aquae* can transfer a plasmid to other strains enabling those to biosynthesize anatoxin-A.

The ability of an organism to produce a certain toxin may be lost in the course of evolution when its selective function is no longer relevant. Poison-free races of originally toxic higher plants have often been elevated to the rank of edible plants, e.g., cucurbitacin-free cucumbers. This is also true of the appearance of chemical races of rowan that have lost the ability to produce parasorboside, the precursor of the mucosa-irritating parasorbic acid. The two chemotypes of the almond tree that yields seeds containing cyanogenic glycosides (bitter almond) or no cyanogenic glycosides (sweet almond) are another example.

Chemotypes often have a specific regional distribution so that studies of the spectrum of active ingredients of the same species in one region do not necessarily allow conclusions about the active ingredient composition of individuals of the same species in another region [24, 68]. These differences may be based on genetic traits (local adaptation) or on environmental effects (phenotypic plasticity).

Toxins from berry fruits, drupes, or fleshy seed coats, which are intended to protect the immature fruits from predators, are broken down as the fruit ripens. If the seeds of the fruit are to be dispersed by animals, the flesh of the fruit must be nontoxic to animals suitable for dispersal and provide a 'reward', such as sugar for the spreader. 'Harvest maturity' is indicated by a ripening signal, e.g., red, yellow, or blue coloration.

The ripening signal sometimes applies only to specialists, e.g., the blue color of belladonna berries applies only to wild boars, rabbits, and some birds that can break down the toxic hyoscyamine of the plant. For humans, the berries of belladonna are poisonous. Humans seem not to be the right spreaders of the plant. After ripening, the toxicity of the seeds of such fruits is maintained or enhanced. The seeds, usually additionally protected by a hard shell, should not be destroyed by the spreaders to avoid mobilization of the toxins.

Examples of the influence of ripening state on the toxicity of a plant organ include the presence of toxic steroid alkaloid glycosides in immature tomatoes and their absence in the ripe fruits and the presence of the toxic L-hypoglycine in the arils of the immature fruits and its absence in the palatable arils of the ripe fruits of ackee plum.

Environmental factors that influence the toxin content (elicitors) are unusual climatic conditions or stress like plant pests, diseases, or physical factors. Some elicitors, e.g., plant pathogenic microorganisms, induce the increased formation of bioactive secondary substances, so-called **phytoalexins**, as a defense reaction. Physical stress factors, e.g., injury, and exposure of underground tubers, e.g., potatoes, to light or unusual temperatures may also induce the formation of phytoalexins. Phytoalexins are secondary plant substances that do not or only in traces occur in healthy plants. Their unexpected occurrence in plants can lead to poisonings in plant eating animals and in humans. The chemical nature of phytoalexins is mostly specific to plant families. For example, Fabaceae produce isoflavonoids, Solanaceae sesquiterpenes, Orchidaceae dihydrophenanthrenes, Vitaceae and Pinaceae stilbenes, and Asteraceae polyynes. Of special toxicological interest are, e.g., the occurrence of hepatotoxic furanosesquiterpenes in the tubers of sweet potato, batate, following infections of the plant with fusaria, and the stress-induced increase in the content of steroid glycosides in potato tubers.

Reviews: [2, 8, 27].

The toxin content of potentially secondary toxic organisms varies widely. It is determined primarily by the type and quantity of food they ingest or the extent of toxin production by their symbionts living in or on them. Often, potentially secondary poisonous organisms may be nontoxic and are used as food when raised in an environment free of toxin-producing microorganisms. In other cases, poisons are deposited only in some parts of plants or organs of animals. Then only the consumption of these parts of the organisms is dangerous. In puffer fish, for example, the highly toxic tetrodotoxin is found in high concentrations only in internal organs while muscle tissue contains only low amounts and may be consumed in limited amounts.

## 1.4.4 Importance of Natural Poisons for Biological Systems

Toxins as secondary metabolites are not vital for the producer. Therefore, breeding lines of originally toxic species that are free of the toxins are not restricted in their viability. Nicotine-free tobacco, caffeine-free coffee seeds, and cucurbitacin-free cucumbers are fully viable in a controlled environment.

In other cases, secondary metabolites may have great importance for the producing organism as they may control the course of other metabolic processes.

An important function of secondary metabolites is that they can provide their producer with advantages in dealing with external threats. The poisonous organisms may be better suited than nontoxic ones to fight infections when producing antimicrobial substances. Others may be able to repel predators by producing substances with repulsive smell or taste.

The repellent effect of the skin secretions in some fish species protects them from shark attacks. Secretions of the skin glands of toads and salamanders, for instance, even combine the two functions. Here, a repulsive taste of the poisonous components of the skin secretions is probably coupled with an antimicrobial effect that prevents colonization of the moist skin with microorganisms.

Secondary metabolites of actively poisonous animals may serve them for catching prey or to protect them from attacks by predators. For defense purposes, the poisons are the more effective the faster they trigger a defensive effect in the attacker. Relatively slow-acting toxin ingredients, e.g., peptides of bee venom, are therefore frequently accompanied by immediately (!) pain-triggering substances, e.g., serotonin or kinins.

The aposematic coloration of insects, e.g., the conspicuous warning costume of a wasp or a ladybug, plays an essential role in protecting poisonous animals from predators. It supports the learning process of potential attackers. The effectiveness of this measure is shown by the survival advantage of animals ('impostors') imitating on the coloration of truly toxic species. For example, the harmless hoverflies (Syrphidae), which have the appearance of a dangerous wasp with their yellow/black banded abdomen, deter not only insectivores but even uninformed humans.

Plants also use signals to support the avoidance learning process of the predators. Such signals are the bitter or pungent tastes of some plant species. Although these taste characteristics are not essential for pharmacological action, almost all toxic plants taste bitter or pungent. In animals and humans, the taste sensation 'bitter' is instinctively associated with the impulse of rejection. Here, too, some plants are 'impostors' in that they produce badly tasting ingredients without being toxic. For example, the nontoxic yellow gentian (*Gentiana lutea*) resembles the white hellebore (*Veratrum album*) in appearance and bitter taste.

In the course of evolution, some organisms evolved resistance to certain toxins. For example, the caterpillar of the cabbage white butterfly (*Pieris rapae*) is adapted to the cleavage products of the glucosinolate/myrosinase system of cabbage species. Not only does cabbage not harm the caterpillars, but the defensive chemicals of the plant have even become signals of attraction for caterpillars, indicating rich food sources that have not to be shared with other herbivores [71].

Even humans have inherited such resistances from their animal ancestors or have acquired them in the course of evolution. We possess enzyme sets that detoxify toxins frequently found in our food. Oxygen-transferring enzymes, especially cytochrome P450-dependent monooxygenases, help to detoxify toxins. About 20 different CYP450-dependent isozymes are known. CYP3A4, CYP2D6, and CYP2C9 are particularly important for the degradation of toxins. The conjugation of OH groups present or newly formed in this way with glucose, glucuronic acid, or sulfuric acid leads to the conversion of lipophilic toxins into hydrophilic metabolites that may be effectively excreted with urine formation. Further mechanisms of resistance in humans are resorption barriers that prevent the absorption of toxins from the intestine or molecular evolution processes that render previously toxin-sensitive target molecules of these animals into toxin-insensitive.

About 0.5 million different secondary compounds of other organisms are present in our food. All of them can be detoxified or are more or less well tolerated. This is one of the reasons why substances that have been shown to be toxic in in vitro systems, e.g., when tested on bacteria and cell cultures, are harmless to humans. For example, flavonoids, which are inhibitors of a variety of enzymes and are mutagenic in bacterial systems, are tolerated by our bodies without harm. In this way, humans have tapped into a pallet of foods larger in scope than that of almost any animal species.

Of course, substances with which humans rarely come into contact are not covered by these detoxyfying mechanisms. Failures occur. Sometimes substances, e.g., the pyrrolizidine alkaloids, even become toxic in the human organism due to enzyme activities in our detoxification system. Here the selection has failed because the poisonous effects appear only after a long time, and the toxicity of the plants containing them is not immediately recognizable.

Our intestinal flora assists in detoxification. For example, it detoxifies saponins by detachment of their sugar components and flavonoids by transforming them into nontoxic metabolites. An interesting example are the babies of koalas (*Phascolarctos cinereus*) who eat from the feces of their mothers to acquire the intestinal flora capable of detoxifying the 1,8-cineol of Eucalyptus leaves. But also toxification happens in the intestine, caused by bacteria. For example, anthraquinone derivatives acquire their laxative effect only after chemical transformation by the intestinal microflora.

The evolution of toxins and venoms required the parallel development of mechanisms which protect the producer from its own toxins and venoms and prevent the producer from harming itself. Actively poisonous animals usually store their poisons in poison glands well shielded from other tissues. In passively poisonous plants, the toxins must always be present throughout the whole organism in case of attack. To protect the plant itself from the own poison it is deposited in compartments within the cell separated from the cytoplasm or storage forms are physiologically inactive 'prodrugs' (precursors), activated precisely when the plant is attacked by an aggressor. The latter is only useful in the case of slow 'working' attackers, e.g., in the case of attack by phytopathogenic fungi.

Storage of toxins in cells outside the cytoplasm may occur in plants also, for example, in the case of alkaloids, by uptake into the vacuoles of the cells, or by secretion into milk juices in extracellular milk tubes. Inactive precursors that are 'unlocked' only when the plant tissue is injured by an attacker are the cyanogenic glycosides, the glucosinolates, the alliins, the cucurbitaceous glycosides, the bisdesmosidic saponins, ranunculin, the phytoalexins, and parasorboside.

Reviews: [15, 22, 44, 60, 61].

## 1.5 Structure–Activity Relationships of Natural Poisons

Natural poisons exhibit a very large structural diversity. Many toxins contain additionally to C, O, H, N heteroatoms (B, S, Se, P, Cl, Br, I). Almost all types of organic compounds are represented, and almost all known functional groups are present as well. Most of the toxins are polyfunctional compounds, i.e., their molecules carry several functional groups that may interact with different partner molecules.

Natural poisons include aliphatic alcohols (e.g., cicutoxin), aldehydes (e.g., citral), carboxylic acids (e.g., oxalic acid and monofluoroacetic acid), and lactones (e.g., protoanemonin, butanolides, and sesquiterpene lactones); alicyclic hydrocarbons (e.g., sabinene), alcohols (e.g., sabinol), ketones (e.g., thujone), and hydroperoxides (e.g., crispolide); aromatic compounds such as phenols (e.g., safrole), quinones (e.g., *p*-benzoquinone), aldehydes (e.g., salicylaldehyde), and carboxylic acids (e.g., lichen acids). Polycyclic hydrocarbons and their derivatives (e.g., steroids, many terpenes), heterocyclic compounds (e.g., alkaloids), amino acids (e.g., coprin), peptides (e.g., amanitins), proteins (e.g., ricin), amines (e.g., muscarin), amides (e.g., palytoxin), hydrazines (e.g., gyromitrin), nitriles (e.g., linamarin), isothiocyanates (e.g., allyl mustard oil), azo compounds (e.g., cycasin), and halogen derivatives (e.g., surugatoxin) occur, frequently as glycosides. Moreover, epoxy groups, acetal and ketal formations, ester bonds (carboxylic acid esters, phosphates, and sulfates), SH groups, nitro groups, ether bridges, or intramolecular acid anhydride formations occur.

Despite our extensive knowledge about the occurrence of these diverse toxin structures, elucidation of the relationships between structure and function in target organisms is still in its infancy. Although in some cases the presence of specific structural features of a toxin may allow to predict some of the functions of this molecule (such molecular features are called pharmacophoric groups) we can rarely deduce possible effects from just analyzing the structures of newly discovered natural substances. However, in some cases, predictions about possible effects of a toxic substance by knowing its chemical structure can be done. For a chemist, the ability of a substance to alkylate a substrate or to split off HCN or to cleave hydrazines or isothiocyanates can usually be read from its structure. Likewise, we can predict whether a substance will interact with hydrophilic or rather with lipophilic partners, which may provide information about its resorption behavior and its influence on biomembranes.

The analysis of structure–activity relationships is complicated by the fact that, especially in the case of complex natural products, the pharmacophoric groups are hidden in a large number of molecular accessories. It is obvious that toxins are not evolved in a targeted manner, but by accidentally occurring minor structural changes in nontoxic precursors, triggered by undirected mutations in the enzymes involved in the synthesis pathways. In most cases, the effect of the toxin in a target organism does not require the entire molecular structure but could also be achieved with a smaller molecule just comprising the pharmacophore. The essential function of the pharmacophore is illustrated by the observation that even minor changes in the pharmacophore, be it only the conversion of one enantiomer into another, can drastically alter the effectiveness of the substance. (–)-Hyoscyamine, for example, is about 100 times more potent than the isomeric (+)-hyoscyamine. To make things even more complicated, modifications of molecular groups apart from the pharmacophore may decisively alter the quality of the toxin effect in a target organism, e.g., by altering the binding mode or affinity of the pharmacophore to its molecular receptor. Thus, acetylcholine (pharmacophores: $N^+$ and both O atoms) is a cholinergic agonist, physostigmine is an acetylcholinesterase inhibitor, and atropine is a cholinergic antagonist, although all these compounds contain the same pharmacophores as acetylcholine.

## 1.6  Natural Poisons as Sources of Danger for Humans and Animals

Accidental poisonings in humans or animals occur when microorganisms, mushrooms, animals, and plants or parts of them are touched, tasted, or used as food, spices, stimulants, or medicines in ignorance of their toxicity. Less frequently, accidental poisonings occur as a result of attacks by actively poisonous animals.

Children are at particularly high risk for accidental poisonings. In the course of exploration of their environment, children try conspicuous fruits, seeds, and also green plant parts, flowers, and roots, occurring in gardens, parks, at home, or in the field. In imitation of adults, they may pretend to smoke cigarettes made from plant materials found outside or they may chew on cigarette butts. Children are also in danger to consume medicines that are lying around at home. Typical for small children is hand-to-mouth activity. In the age of under 6 years, they are unable to read warning labels and to differentiate food from nonfood items. Children are attracted by colorful plants or packaging and enticing fragrances. The unpleasant taste of many poisonous plants and other materials does not keep children from swallowing them [64, **17, 72, 81, 161**]. Poisonings, especially of children, after licking or swallowing of highly toxic seeds, e.g., from *Abrus precatorius*, jequirity, *Ricinus communis*, castor bean, or *Cascabela thevetia*, yellow oleander, used to make jewelry necklaces, play presumably a minor role.

Attempts to play with poisonous animals, e.g., bees, can also hurt children. Poisonings of children are often more dangerous than those of adults because children, depending on their age, have a greater sensitivity to toxins. Due to their lower BW, higher tissue concentrations are achieved with the same amount of toxins.

The increasing alienation of humans from nature has led to the loss of the experience about potential sources of poisons in our natural environment that was general wisdom in previous generations. Confusion of edible plants or mushrooms with similar poisonous ones occurs frequently. This, together with growing efforts to consume plant-based diets and increasing travel to foreign geographic regions due to tourism or migration, which brings people in contact with unknown mushrooms, plants, or animals, resulted in increasing rates of accidental poisonings.

Consumption of natural products like cabbage or rhubarb over a long period of time and in excessive amounts, of improperly prepared food, e.g., cassava, beans, or tetrodotoxic fish, of improperly stored vegetables, e.g., potatoes or mushrooms, of contaminated or spoiled food or unripe fruits like unripe tomatoes can also lead to poisoning. Even some vitamins, consumed in excessive amounts in the form of dietary supplements, can cause damage in humans.

Another cause of accidental poisoning is the consumption of usually nontoxic animals or their products, which have become secondarily toxic due to toxin absorption from the environment. Thus, many hundreds of poisonings occur annually from shellfish, crabs, and fish that have absorbed and stored toxins. Cases of poisonings by honey obtained by bees from the nectar of poisonous plants, e.g., Rhododendron or Senecio species, have been reported. A much discussed but still largely unclear question is whether toxins occur in our daily food and if they are of any health concern. Do the D-amino acids or amines in fermented milk products harm humans? Are the conversion products of amino acids, formed during cooking of meat products, e.g., the known mutagen 2-amino-9$H$-pyrido[2,3-b]indole, which is formed from L-tryptophan, mutagenic for humans in vivo as in vitro? How dangerous are oxalates in food? Does the agaritine in mushrooms pose a hazard to humans? Are there hazards from substances in milk, derived from animal feed, e.g., aflatoxins, pyrrolizidine alkaloids, or ptaquilosides? Should cabbage and potatoes be avoided by pregnant women because of the possible teratogenicity of glucosinolates and steroid alkaloid glycosides, respectively? Do the lectins 'evolved' by plants for their protection, e.g., those in vegetables and fruits, damage our intestinal flora? How dangerous are fruits because of their content of fructose [30, **100**]?

Accidental poisoning is also possible due to the application of incorrectly dosed natural drugs, e.g., digitoxin, or due to the uncontrolled use of highly effective 'medicinal plants' by laypersons. Poisonings may also be caused by confusion of poisonous plants with medicinal plants, by self-experimentation with potential medicinal plants, by 'miracle cures' sold via the internet, by uptake of insufficiently prepared herbal drugs, e.g., from traditional Chinese medicine (TCM) and Ayurvedic medicine, or by longtime application of herbal medicines, which are harmless when used for short time periods, but may harm humans when used continuously over longer periods.

Acute illnesses or chronic damage in connection with use of narcotic drugs, doping agents, tobacco products, or excessive drinking of alcohol, are regarded as intentional poisonings. Here, the poisoned person is aware of the dangers of poisoning, at least partially. Narcotics (e.g., opium), stimulants (e.g., caffeine), hallucinogens (e.g., ayahuasco), as well as alcohol and tobacco may lead to dependence.

Of course, the uses of natural poisons for suicide or crime are special cases of intentional poisonings. The utilization of toxins in weapons and for bioterrorism is an especially disturbing matter [48].

Poisonings of animals by plants play an important role in agriculture and veterinary medicine. Pets such as dogs or cats can be poisoned by ingesting houseplants and garden plants or by being exposed to insecticides and other toxic materials used in households or on farms. Poisonings of grazing animals can be caused by ingestion of toxic plants growing on meadows. Typical symptoms are weight loss, infertility, abortions, or death. Ingested toxins may be deposited in meat and fat or transferred to animal products like milk or eggs. Humans may be exposed to such toxins via the food chain.

Reviews: [10, 47, **59, 78, 104, 108**].

Milk of dairy animals may contain, among others, pyrrolizidine alkaloids, piperidine alkaloids, quinolizidine alkaloids, glycoalkaloids, ptaquiloside, hypericin, and glucosinolates. Poisoning of humans through consumption of milk containing such toxins occurs especially when the milk comes from a few animals reared at the same location, e.g., a farm. When milk is collected from several farms at different locations, the toxin that may be contained in milk from one farm gets highly diluted, and the risk of poisoning consumers is negligible.

Human breast milk may also contain toxins if the nursing mother ingests medicines, narcotics, or plants. Such substances may be dangerous for the infant. The breast milk of female smokers contains the alkaloid nicotine which may cause irregularities in the ventilatory rhythm of the infant. Infants of drug-addicted mothers, who suffer from withdrawal symptoms after birth, may experience life-threatening conditions that require intensive treatment.

## 1.7  Usefulness of Natural Poisons for Humans

Many active ingredients of poisonous plants and animals can be used as medicines when therapeutic concentrations are employed, e.g., the cardioactive steroid glycosides, podophyllotoxins, anthracene derivatives, poppy alkaloids such as morphine and codeine, colchicine, ergot alkaloids, quinine, pilocarpine, atropine, and scopolamine [4, 5, 41, 46, **37, 109, 140, 149**].

In our book, we have provided references to the poisons for their medicinal use and often for dosage to make it clear which doses can be dangerous or fatal to humans.

Despite our limited insight into the structure–activity relationships of poisons to date, the active ingredients of poisonous plants and animals have often played a role as lead compounds in the synthesis of new drugs. Examples include synthetics derived from morphine (e.g., the benzomorphan derivatives and loperamide), (+)-tubocurarine (e.g., gallamine), khellin (e.g., disodium chromoglycate), salicin (e.g., acetylsalicylic acid), and peptide toxins, e.g., dolastatins (cematodin and soblidotin). These are the great reserves of drug research: elucidation of the pharmacophores of toxins, synthesis of model substances freed of all molecular accessories, and optimization of these substances with the aim to enhance desired effects and to avoid undesirable side effects, i.e., to increase their therapeutic index (relationship between the toxic dose of a drug and its therapeutic dose).

In medical research, natural poisons are important tools for the elucidation of pharmacological targets and mechanisms of action. Pharmacological receptors were often named after the toxin that specifically targets them as an agonist or antagonist. For example, we know muscarinic, nicotinic, or ryanodine receptors. With the help of labeled toxins, it is possible to investigate cellular processes in a targeted manner. Examples are the use of fluorescence-labeled phalloidin from *Amanita phalloides* to demonstrate cellular actin and of α-amanitin from the same mushroom species to investigate RNA polymerase II-mediated processes.

Some natural poisons serve as diagnostics. Colchicine stops cell division in metaphase and allows chromosomes to be visualized so that chromosomal abnormalities can be detected during pregnancy (amniocentesis). Lectins are used to study the functionality of the immune system because of their mitogenic properties and can be used for affinity chromatography because of their specific binding to glycoproteins.

Besides for medicinal use, toxins have long been used as pesticides (e.g., nicotine and strychnine), piscicides (e.g., rotenone), and arrow poisons (e.g., curare and Strophanthus glycosides). These applications began in the early days of mankind and are still practiced today. The same is true for the use of poisonous plants as hallucinogenic, narcotic, or stimulating preparations and for ritual acts.

## 1.8 Toxicology of Natural Poisons

### 1.8.1 Toxicity Determination

In determining the toxicity, a distinction must be made between acute and chronic toxicity and mutagenicity. In each case, the effect of a toxin determined by quantitative parameters is decisive.

The most important parameter for the evaluation of acute toxicity is the median lethal concentration, the **$LD_{50}$**, that is determined in animal assays. The $LD_{50}$ is the lowest single dose, usually related to the BW of animals (per kg BW), at which 50% of the test animals die. Sometimes lethal doses are also given in relation to a single animal.

The lethal dose depends to a large extent on the species of tested animal, its breed, age, sex, husbandry conditions, and the method of application of the toxin. The $LD_{50}$ values obtained with peroral application are usually higher than those obtained with parenteral application, i.e., application of the poisons bypassing the gastrointestinal tract. Ingested poisons only partially overcome the gastrointestinal tract: the attacks of gastric acid, the enzymes of the digestive tract, the intestinal flora, and the absorption barriers. During the passage of the liver, which immediately follows absorption as a 'first pass', the toxin can be completely or partially detoxified. The p.o. application in the experimental animal comes closest to the conditions of poisoning of humans by ingestion of the poison. Nevertheless, the transferability of the results obtained in the laboratory from animal to humans is limited due to the possibly deviating pharmacokinetics and differences in receptor equipment between laboratory animals and humans.

The lethal doses given in the literature for humans are based on empirical values obtained from documented cases of poisoning.

**LOAEL** (**l**owest **o**bserved **a**dverse **e**ffect **l**evel) is the lowest dose of a poison that causes harm when tested in animals.

**Chronic toxicity** is much more difficult to estimate than acute toxicity. It is usually evaluated in a 90-day experiment with daily oral administration of a potential toxin to rodents. Animals from several groups are given different doses of the toxin. Several body parameters of the test animals are evaluated. Chronic toxicity can be increased by accumulation of toxins, but can decrease also over time due to tolerance development. Particular attention must be given to the recording of possible chronic toxic effects in the case of compounds which are ingested as constituents of foodstuffs, teas, or medicinal preparations in small doses over a prolonged time period.

Standard tests for **mutagenicity** are the so-called reverse mutation test using bacteria (*Salmonella typhimurium*, referred to as Ames test) or mammalian tests such as the mouse lymphoma assay or micronucleus assay. In tests for **carcinogenicity**, animals receive the test sample over a long period of time. In future, toxicogenomics could possibly offer more effective mutagenicity and carcinogenicity tests [12].

Test methods are regulated by the OECD guidelines, e.g., OECD Guideline 452 for chronic toxicity tests of chemicals, OECD Guideline 471 for mutagenicity tests using bacteria (Ames test, reverse mutation test), OECD Guideline 473 for mutagenicity tests using mammalian cells (chromosome aberration), OECD Guideline 476 for detection of gene mutations, OECD Guideline 451 for studies of carcinogenicity, and OECD Guideline 453 for combined chronic toxicity/carcinogenicity studies.

The dangerousness of a toxic plant or animal to humans or animals does not depend solely on the toxicity of the constituents, but in practice it is determined by the degree of accessibility to humans or livestock and the incentive to consume or come into contact with it (**toxicological relevance**).

Information about the toxicity of poisonous substances including natural poisons is available in documents of the European Food Safety Authority, the Toxicology and Environmental Health Information Authority in the United States, the U.S. National Library of Medicine, the related Toxicology Data Network, several databases, e.g., Toxin and Toxin-Target Database [70], SuperToxic [55], and other sources.

## 1.8.2 Toxicokinetics

### 1.8.2.1 Pathways of Toxin Uptake
The intake of toxins can take place
- peroral (p.o.), by mouth, in animals also by gavage;
- parenteral (p.e.), i.e., bypassing the gastrointestinal tract:
    - intraperitoneal (i.p.) by injection into the peritoneum (abdominal and pelvic cavity),
    - intramuscular (i.m.), by injection into a muscle,
    - subcutaneous (s.c.) or percutaneous (p.c.), by application to or injection under the skin,
    - per inhalationem (p.i.), by inhalation,

- intraarterial (i.a.), by injection into an artery,
- intravenously (i.v.), by injection into a vein;
- transdermal, by application on the skin or transmucosal, by application on the mucosa.

### 1.8.2.2 Resorption

Prerequisite for the systemic effect of a toxic substance is its absorption (resorption) from the external environment or from spatially limited sites of the body interior into the tissues and into the bloodstream. The absorption takes place through the mucous membranes of the gastrointestinal tract after ingestion, the bronchia after inhalation, the abdominal cavity after intraperitoneal injection, the vagina after intravaginal application, the skin after topical application, and from the muscle after intramuscular injection or from the blood after intravenous or intraarterial injection.

In the case of p.o. ingestion, the rate of absorption of the toxins is initially limited by the rate of release from the toxin sources, e.g., from more or less shredded plant parts, and from binding to the accompanying substances, e.g., tannins. In addition, for some compounds, especially glycosides like anthraquinone glycosides and cyanogenic glycosides, cleavage by enzymes of the injured plant, the gastrointestinal tract, or the gastrointestinal microflora is a prerequisite for absorption.

The main site of absorption of p.o. ingested toxins is the small intestine. From there, the substances pass via the portal vein to the liver and then, after biotransformation, they enter via the superior vena cava the central bloodstream.

Absorption can be passive, by diffusion from sites of high concentration to sites of lower concentration, or active, involving energy consumption, by the body's own specific carrier contrary to a concentration gradient.

The extent of absorption is particularly determined by the molecular size and physicochemical parameters such as the lipophilicity of the substance. Lipophilic substances are better absorbed than hydrophilic substances. Lipophilicity is increased by a large number of $CH_2$, ether, and ester groups in the molecule, and reduced by phenolic hydroxy and carboxyl groups and quaternary N atoms in the alkaline environment of the intestine due to salt formation. Therefore, the formation of toxic alkaloids whose phenolic OH groups are partially methylated is nature's recipe for success in the evolution of poisons. Although lipophilic substances are well absorbed, they must have a low solubility in water for transport in the aqueous content of the gastrointestinal tract from the food pulp to the intestinal wall. Lipophilic substances in the intestinal contents, e.g., micelles of gallic acids, fatty acids, and cholesterol, or milk or alcohol mistakenly given in cases of poisoning, enhance the transport and thus the absorption of lipophilic toxins.

Very large molecules, in order to develop a systemic effect, must enter the bloodstream bypassing the gastrointestinal tract, e.g., by injection through the venom apparatus of an animal. They are therefore ineffective when given perorally.

### 1.8.2.3 Uptake into the Cells

The uptake of substances from the bloodstream or the tissue into cells occurs either by diffusion, following a concentration gradient, or by active uptake against a concentration gradient under energy consumption by specific carriers (over 400 are known) or by endocytosis. Toxins are objects of active uptake if they are able to compete with physiological compounds for the specific carriers.

Very large molecules, for example peptides, proteins, lectins, or some bacterial toxins, are taken up by endocytosis. It occurs after binding to the cell membrane, mostly receptor-mediated, in dissolved form (pinocytosis) or solid form (phagocytosis), enclosed in membrane vesicles.

### 1.8.2.4 Distribution in the Body

The distribution of toxins after absorption occurs through the blood and/or lymphatic stream, mostly adsorbed to transporter molecules, e.g., plasma proteins, in the case of lipophilic substances mainly integrated into transport proteins. Hydrophilic small molecules are sometimes really resolved.

When the toxins enter a blood vessel directly, e.g., from bites or stings by animals, the highest blood level is reached almost instantaneously. Toxins from venom depots in the skin or tissues, formed during intracutaneous, subcutaneous, or intramuscular application, e.g., following bites or stings by venomous animals, enter the blood and/or lymphatic pathways more slowly. The speed of release from these venom depots into the blood is influenced not only by the substance parameters but also by the condition of the application site (capillarization, blood flow, and condition of the intercellular cement substances).

The different distribution of substances in the organism is one reason why the effects of many toxins manifest themselves preferentially in certain organs. For example, cardioactive steroid glycosides are particularly concentrated in the heart muscle. Also noteworthy is the passage of many toxins into the placenta or breast milk, leading to impairment of the fetus or newborn child by toxins ingested by the mother. The different receptor equipment and other factors also contribute to the different sensitivity of individual organs.

### 1.8.2.5 Biotransformation

Biotransformation is the chemical modification of the absorbed toxin in the organism. In most cases, biotransformation diminishes the toxicity of the poison, but it can also increase its toxicity. For example, pyrrolizidine alkaloids are transformed into toxic metabolites by biotransformation.

Detoxification occurs primarily by increasing the water solubility of the toxin. It takes place in two phases:
- Phase I reactions:
    - introduction of polar groups, e.g., OH groups, mostly by more or less specific CYP450-dependent monooxygenases;

- – oxidative deamination;
- – modification of existing functional groups, e.g., by O- or N-demethylation.
- – Phase II reactions:
  conjugation of already existing or newly formed functional groups with hydrophilic molecules, e.g., with monosaccharides, glucuronic acid, sulfuric acid, mercapturic acid, glycine, glutamic acid, or cysteine.

The enzymes responsible for biotransformation are either already present in the body of the affected organism or their formation is induced by the toxins. Numerous receptors in the cell nucleus play a role in induction of such enzymes, e.g., the aryl hydrocarbon receptor and the orphan nuclear receptors, e.g., pregnane X receptor [72].

The speed of biotransformation of a toxin is, among others, determined by the way in which it enters the organism. Whereas active substances that are taken up perorally are transported after resorption to the liver via the portal vein and are rapidly metabolized there, substances entering the body parenterally can first spread throughout the body before reaching the liver.

Differences in the intensity and mode of biotransformation of a compound in different animal species, between individuals of one species or human beings, are genetically caused. They often explain the varying degrees of toxicity for different species. For example, hyoscyamine and its racemate atropine are highly toxic for humans. Rabbits, wild boars, and some birds possess special esterases that can cleave atropine very rapidly. Therefore, these animals can consume herbs and berries of atropine-containing plants like belladonna without any harm. Polymorphisms in the cytochrome system that result in different degradation products with potentially different effects are an example for differences between human individuals.

### 1.8.2.6 Elimination

Elimination (excretion) of unchanged toxins and their degradation products occurs, mostly mediated by inducible phase III transporters, predominantly via the kidneys after globular filtration or tubular secretion into the urine. Therefore, the kidneys are often among the organs most affected by toxins. Another part of the toxins is excreted with the bile from the liver, the most important organ of biotransformation, into the intestine. Due to reabsorption of these toxins in the lower intestinal segments, lipophilic substances in particular can enter the so-called enterohepatic circulation. This greatly delays their elimination and prolongs their toxic effect.

To a lesser extent, elimination may also occur via breast milk, salivary glands, lacrimal glands, and, in the case of volatile substances, through the lungs. Excretion into breast milk can lead to severe risk to the infant, whose liver and kidneys have little ability to transform and eliminate toxins.

### 1.8.3  Toxicodynamics

#### 1.8.3.1  General
The points of attack and the mechanisms of action of toxins in the organism are diverse and far from being clarified in all cases, although numerous new findings have been made in this area in recent years [1, 40, 51, 98, 103, 149, 168]. Only a few selected universal mechanisms of action can be mentioned here. Further mechanisms are described in the special chapters.

A distinction can be made between concentration poisons and summation poisons. In the case of concentration poisons, the effect at the target, e.g., at a receptor, increases with increasing concentration. The effect is reversible, i.e., when the toxin is removed, the effect recedes without damaging the target. In the case of summation toxins, the toxin causes an irreversible change of the target. With repeated administration of the poison, the effects can add up. For example, nicotine from tobacco smoke is a concentration poison, and benzo[a]pyrene from tobacco smoke, which is converted in the body to the carcinogen benzo[a]pyrene-7,8-dihydroxy-9,10-epoxide, is a summation poison.

#### 1.8.3.2  Attack on Receptors
Many poisons, especially alkaloids, attack pre- or postsynaptic receptors of the synapses. For example, nicotinergic-cholinergic receptors are targeted by nicotine, muscarinic-cholinergic receptors by pilocarpine or atropine, α-adrenergic receptors by ergot peptide alkaloids, GABA receptors by bicuculline, glycine receptors by strychnine, opioid receptors by morphine, purinergic receptors by caffeine, dopaminergic receptors by lysergic acid derivatives, glutaminergic receptors by acromelic acid, vanilloid receptors by capsaicin, cannabinoid receptors by tetrahydrocannabinol, and ryanodine receptors by boldine.

Depending on their structure, the toxins act as agonists or antagonists at the receptors. When attacking the presynaptic receptors, they can promote or inhibit the release of neurotransmitters into the synaptic cleft; when attacking the postsynaptic receptor, they can initiate an effectuation chain or prevent its start. For example, ephedrine promotes the release of norepinephrine in the central nervous system (CNS) and thus acts similarly to norepinephrine. Pilocarpine acts at muscarinic-cholinergic receptors as an agonist and exerts effects similar to acetylcholine. Atropine blocks these receptors as an antagonist and prevents acetylcholine from initiating its effectuation chain.

Receptor activities can also be influenced by inhibition of enzymes which, under physiological conditions, degrade neurotransmitters and thus limit their concentration in the synaptic cleft and the strength and duration of their action. The inhibition of these enzymes can lead to overshooting responses. For example, acetylcholinesterase inhibitors such as physostigmine cause severe intoxication by inducing an overreaction of the parasympathetic nervous system. By inhibiting monoamine oxidases (MAO), the concentration of neurotransmitters in the brain can be increased, causing a wake-up effect or, in high doses, a psychotomimetic effect. The effectuation chain, i.e., the chain of reactions between the attack on the receptor and the effect, is also interfered, for example, by suppressing the degradation of the second messenger cyclic adenosine monophosphate (cAMP) by inhibiting phosphodiesterase (e.g., by papaverine) or by promoting its formation by stimulating adenylate cyclase (e.g., by cholera toxin).

Chronic intake of some poisons, e.g., opiates, is responded by the body through reducing the number of receptors at which the poison attacks, in order to prevent persistent excessive reactions. Thus, with continuous use, the potency of the same doses of the substance decreases, and habituation sets in. To achieve the desired effect again, the dose must be increased. Withdrawal of the poison, in this case of morphine or heroin, leads to the fact that the endogenous neurotransmitters, in this case the endorphins, can no longer ensure the maintenance of the physiological state of the organism because of the small number of receptors. Severe withdrawal symptoms occur, which can be fatal. Only days to weeks after withdrawal, the number of the corresponding receptors is increased again to the physiological normal level.

### 1.8.3.3 Attack on Ion Channels and Other Membrane Components

Ion channels controlled by neurotransmitters (ligand-gated ion channels) and ion channels dependent on the membrane potential (voltage-gated ion channels) or the membrane itself are important targets for several natural poisons.

The selective permeability of cell membranes to ions, such as $Na^+$, $K^+$, and $Ca^{2+}$ ions, controlled by specific ion channels, is crucial for the function of irritable cells, such as nerve, muscle, and gland cells. Several toxins block ion channels or cause permanent opening by inhibiting their closure. Both processes lead to changes in membrane potential and thus to a lack of responsiveness of irritable cells. As a result, conduction in the nervous system and transmission of excitation from the nerve to the muscle can be interrupted. Consequences may include paralysis. Otherwise, the responsiveness of muscle cells to stimuli can also be increased by a lowered membrane potential. Consequences include convulsions. In both cases, death may result from respiratory paralysis. The change in the permeability of ion channels can also lead to the absence of certain ions in the cell that are needed for various processes, e.g., the absence of $Ca^{2+}$ ions for muscle contraction.

Ion channels for the same ions can have very different structures. For example, the proteins of a $K^+$ channel, which can be targeted by endogenous and exogenous ligands, are encoded by more than 80 genes. Therefore, the proteins can differ among the various isoforms of $K^+$ channels. Of the voltage-gated $Na^+$ channels, there are at least 10 isoforms that can be divided into the tetrodotoxin-sensitive and tetrodotoxin-insensitive groups. Each $Na^+$ channel has multiple targets (sites) for ligands, such as peptide toxins. In addition, different types of each type of ion channel exist with different tasks. Thus, there exist L-, N-, P-, Q-, R-, and T-type $Ca^{2+}$ channels with different control mechanisms and different functions. For example, L-type $Ca^{2+}$ channels are responsible for the inward flow of $Ca^{2+}$ ions necessary for contraction of muscle cells. They are the target for the $Ca^{2+}$ channel blockers used therapeutically. Their blockade prevents spasms when used as spasmolytics. N-type $Ca^{2+}$ channels are responsible for the transmission of pain stimuli in the CNS, and their blockade can suppress pain. Toxins usually attack only one type of ion channel, rarely several, in a very specific way.

The release of neurotransmitters from presynaptic vesicles of nerve terminals can also be promoted by modulating their $Na^+$, $K^+$, and $Ca^{2+}$ channels. Thus, nicotinergic or muscarinic (acetylcholine), adrenergic (epinephrine and norepinephrine), or nitrergic (NO) effects may occur upon exposure to ion channel-influencing toxins.

Tetrodotoxin and saxitoxin block $Na^+$ channels. Batrachotoxin, aconitine, and some Veratrum alkaloids, on the other hand, prevent their closure and thus the restoration of the membrane potential of the cell after a reaction. Other agents affect the permeability of $Ca^{2+}$ (e.g., maitotoxin, ω-conotoxin, and calciseptin) or $K^+$ channels (e.g., cicutoxin, κ-conotoxin, and apamin).

Inhibition of the $Na^+/K^+$-ATPase transport system, e.g., by cardioactive steroid glycosides, can also lead to changes in ion concentrations in cells and thus of their responsiveness. In toxic concentrations, they thereby trigger cardiac arrhythmias.

Some toxins change the ionic composition of the blood (hypokalemia and hypocalcemia) via damaging the kidneys and thus also lead to altered reactivity of irritable cells, e.g., the sinus node of the heart.

Lyobipolar compounds can become incorporated into membranes due to their surface activity, thus altering the permeability of the membranes even for noncharged molecules. In high concentrations, they destroy the membrane by lysis (cytolysis). Examples include the saponins and cytolysins of many animal toxins. In the latter, the cytolytic effect is often enhanced by their phospholipase activity. Membrane lysis in red blood cells occurs as hemolysis.

### 1.8.3.4 Attack on the Genetic Apparatus

Many toxic effects are based on interventions in nucleic acid metabolism, information transfer during protein synthesis, and cell division. They consist, for example, in the

- blockage of enzymes of DNA, RNA, or protein synthesis (e.g., the RNA-polymerase II of eukaryotic cells by amanitins);
- inhibition of topoisomerase I and/or II (e.g., by camptothecin and macaluvamine N);
- alkylation of nucleic acids (e.g., by sesquiterpene lactones);
- intercalation of toxins into the DNA resulting in alteration of the DNA structure (e.g., by furanocoumarins, berberine and palmatine) and mostly also in influence on replication;
- irreversible cross-linking of DNA strands or the reaction of DNA or RNA with proteins (e.g., by pyrrolizidine alkaloids and bifunctional furanocoumarins);

- reaction of the toxins with tubulin of the spindle apparatus, thus inhibiting the cell division (e.g., by colchicine, taxol, podophyllotoxin, vincristine, hemiasterlin, and dolastatins);
- inactivation of ribosomal components (e.g., by ricin, abrin, and ribosome-inactivating proteins like diphtheria toxin).

Depending on the toxin concentration, on the duration of exposure of the genetic apparatus to the toxin and on the type of cells affected, different effects may occur. Acute poisoning leads to the destruction of all cells with a high protein biosynthesis rate, e.g., of liver cells (cytotoxic effect) but also of tumor cells. Long-term but also only single intake of small amounts of toxins can trigger mutations. This may result in heritable malignant degeneration of cells, carcinogenic effect, or if germ cells are affected, teratogenic effect.

### 1.8.3.5 Influence on Enzymes

Many biologically active compounds act by altering enzyme activities. This results in different effects. For example, inhibition of the cytochrome oxidase of the respiratory chain by cyanide ions leads to interruption of the energy supply and to death of cells. Inhibition of cardiac muscle $Na^+/K^+$-ATPase by cardioactive steroid glycosides increases in therapeutic doses the contractility of the heart, but in toxic doses arrhythmias occur. 1-Amino-cyclopropanol, a cleavage product of coprin, prevents oxidation of the toxic acetaldehyde formed during ethanol degradation by reacting with acetaldehyde dehydrogenase. Phorbol esters exert their cocarcinogenic effect via stimulation of protein kinase C.

Some toxins, mainly of bacterial or animal origin, possess enzymatic activity themselves. The α-toxin from *Clostridium perfringens* is a phospholipase C, and the EF toxin from *Bacillus anthracis* is an adenylate cyclase. Diphtheria toxin and cholera toxin catalyze the transfer of adenosine ribosyl phosphate residues from nicotinamide adenine nucleotide to target proteins, thereby altering their functionality. By interfering with the blood coagulation cascade, the hemorrhagically active venoms of snakes lead to consumptive coagulopathies, resulting in blood incoagulability. The so-called auxiliary enzymes in animal venoms, e.g., hyaluronidase, destroy extracellular molecules of the tissue, e.g., hyaluronic acid, and promote the penetration of toxins into the tissue.

### 1.8.3.6 Triggering of Allergies

Many ingredients of microorganisms, plants, and animals can act as allergens (marked by ✳). These are normally harmless macromolecules, but if they have penetrated the human or animal body, they can act as antigens under special conditions and trigger the formation of specific antibodies. The first contact with the affected person, sometimes several contacts, leads to the formation of antibodies against the allergen, i.e., to the sensitization of the affected person. The initial contact or contacts do not cause any symptoms of disease. But upon later renewed contact with the allergen, an exaggerated, pathological immune reaction occurs, which is referred to as an allergy. Depending on the number of prior contacts with the allergen necessary for sensitization, the sensitization potential of an allergen is described as weak, medium, or strong.

The preconditions for sensitization vary between individuals. Insufficient demands on the immune system due to excessive hygiene or insufficient contacts with potential allergens in the phase of imprinting of the immune system in the childhood, changes in the microbiome of the body, for example, due to antibiotics or unsuitable diet, environmental pollution, or genetic factors are some conditions which are blamed for the increasing occurrence of allergies.

Symptoms of allergies may include respiratory symptoms, e.g., asthma, in the case of inhalation allergens; disorders of the digestive tract, e.g., vomiting, diarrhea, abdominal colic, and constipation, in the case of ingestion allergen-containing foods; and skin symptoms, e.g., urticaria, in the case of dermal contact with allergens. Irrespective of the type of contact with the allergen, systemic reactions may occur, e.g., anaphylactic shocks.

Most important allergens are proteins with a relative molecular mass of about 10 kDa or more, polysaccharides, nucleic acids, and carriers of such macromolecules like viruses, bacteria, fungal spores, animal dander, animal hairs, mite droppings, and pollen.

Small-molecule compounds ($M_R$ below 1 kDa) can trigger the formation of antibodies as so-called haptens if they have been bound by chemical reaction to a macromolecular carrier, usually a protein of the body. After sensitization, also the unbound hapten itself can react with the antibody formed.

**Allergenic plants** containing ingredients with strong sensitizing potency are, e.g., Peruvian lilies (Alstroemeria species), ragweed (*Ambrosia artemisiifolia*), celery (*Apium graveolens*), peanut (*Arachis hypogaea*), birch (Betula species), hazelnut (*Corylus avellana*), Coleus hybrids, Chrysanthemum species, marguerites (Leucanthemum species), elecampagne (*Inula helenium*), marsh-elder (*Iva xanthiifolia*), Phacelia species, German primrose (*Primula obconica*), sumac (Rhus species), tansy (Tanacetum species), and tulips (Tulipa species [**65, 156**]).

Sensitization by plants mostly occurs through their contact with the skin of humans. In most cases, haptens are responsible for plant allergies. Examples of haptens are sesquiterpene lactones, butane-4-olides, and alkylphenols. These highly reactive, low-molecular-weight, mostly lipophilic compounds penetrate the skin and react with endogenous proteins, mostly from the skin, to form allergens. Sensitization is also possible by the aerogenic route, i.e., by inhalation of allergens or allergen carriers from plants, e.g., pollen, spores, or other dusts. Food allergies can be triggered by consumption of allergenic plants.

The aerogenic allergen carriers such as pollen often carry on their surface the birch pollen allergen Bet v 1 (155 amino acid residues, more than 20 isoallergens known), or the allergen profilin (124–153 amino acid residues, many isoallergens are known). These antigens will be recognized by the immune system of the mucosa and lead not only to local but also to systemic allergic reactions. About 50–93% of individuals allergic to birch pollen develop allergic reactions when ingesting foods and spices containing such allergens.

Allergens from venoms are almost peptide toxins, proteotoxins, auxiliary enzymes, often hyaluronidases, and proteins whose function is still unknown. Because of the molecular size of these allergens, sensitization with venoms is caused usually by percutaneous application, e.g., by stings or bites. Sensitization to allergens from animals can also be elicited by aerogenic contact, e.g., when inhaling mite feces or dander from animals found in house dust. Allergies caused by animal hair are a separate category and named animal hair allergies.

**Type I allergies (immediate type, anaphylactic type)** are usually triggered by sensitization with proteins that have relatively small molecular masses (10–70 kDa). A well-known example is wasp venom allergy. Sensitization leads to the formation of IgE-type antibodies that bind to mast cells or basophilic granulocytes. Upon subsequent contact with the allergenic agent, these cells, thus equipped, release inflammatory mediators, e.g., histamine, leukotrienes, prostaglandins, and kallikrein. Contact with dusts containing allergens leads to conjunctivitis, rhinitis, and asthma attacks. Ingestion of allergens with food causes swelling of the lips, tongue, and throat but also symptoms in the gastrointestinal tract, the skin (urticaria), the respiratory tract (asthma), and the cardiovascular system. In severe cases, anaphylactic shock can be evoked, which can trigger circulatory collapse and lead to death within a few minutes.

**Type II allergies (cytotoxic type)** and **type III allergies (immune complex type, Arthus type)** hardly play a role when an organism is exposed to plant toxins.

**Type IV allergies (allergies of the delayed type)** occur mostly after sensitization with haptens. The hapten–protein complexes, equipped with specific antibodies, capable of recognizing the haptens, sensitize T lymphocytes. Upon renewed contact with a corresponding hapten, these T cells release cytokines that trigger skin reactions, mostly eczemas, via an effector chain. These reactions only become visible after a delay, i.e., after 12–96 h.

In opposite to allergic contact dermatitis, the **irritant contact dermatitis** is elicited directly, without the involvement of the immune system, by skin-irritating substances, e.g., by the phorbol esters of the lactic juices of the spurges. Irritant contact dermatitis can also be caused by phototoxic substances like furanocoumarins that sensitize the skin to UV radiation.

## 1.9 Clinical Toxicology

### 1.9.1 General

The symptoms of severe acute poisonings explained by us in special sections in the chapters of this book series are marked with ♨, in case of low risk of poisoning with (♨), in case of chronic poisonings with ⊗, case descriptions with ⚕, allergies with ✳, and intoxications of animals with 🐾. These sections are mainly intended to provide information about the course of poisoning, the dangers, and possible consequential damages. The passages marked with ☙ are not intended to be treatment concepts (!). They give only suggestions to the first aider and the physician about possible measures in treatment and about measures which should be avoided. Besides, they should warn the layman against careless self-help.

Poisoning by actively toxic organisms is usually noticed immediately. Symptoms after ingestion of passively toxic organisms, which include the most poisonous plants but also mushrooms, foodstuffs, pharmaceuticals, or recreational drugs, are often not immediately recognizable. The first signs of poisoning after ingestion of toxic plants are often burning sensation and redness in and around the mouth, nausea, vomiting, salivation, crampy abdominal pain, diarrhea, apathy, confusion, dizziness, shortness of breath, and temporary unconsciousness. Special care must be given to children 1–4 years of age. They often show sudden behavioral changes. Infants of drug-addicted mothers can develop withdrawal symptoms after birth and need intensive treatment. If sudden illness is observed in previously healthy people, especially if several people are affected at the same time, poisoning by poisonous mushrooms, other toxic constituents in food, or by misuse of drugs can be assumed.

### 1.9.2 Measures in Case of Acute Poisoning

#### 1.9.2.1 Contacting Poison Information Centers

If poisoning is suspected, an information center for poisoning cases of the respective country should be called immediately.

The addresses and telephone numbers of the poison information center (poison control) with 24-h service are given in the appendix of the book. The poison emergency numbers can of course also be used from abroad. These telephone numbers, fax numbers, addresses, and e-mail addresses can also be obtained from a doctor's office, a clinic, a pharmacy, on the internet, and via the emergency numbers of the countries. It is advisable, especially if you are frequently in the woods and meadows with children, to store the telephone number required for a poison emergency call in your cell phone.

The information center must be informed on (prepare for the call!):
- details about the person presumed poisoned: age, sex, approximate body weight, and body size;

- the type of poisoning source: in case of ingestion of plants it is helpful to have a branch with leaves, flowers, and fruits to the hand; in case of insect bite or snake bite, if possible, a description of the responsible animal;
- the mode of contact with the toxin: ingestion, skin contact, inhalation, sting, bite;
- the approximate ingested amount, for example, about five berries;
- the time passed since presumed poisoning;
- the location of the patient, transportability;
- the reason for poisoning: accidentally, overdosage, intentional intoxication, suicidal intent, abuse, etc.;
- symptoms of poisoning already occurred;
- preexisting medical conditions and medications; and
- first aid measures undertaken, if any.

If necessary, according to the poisoning information center, a doctor should be called or visited, even if the poisoned person seems to have recovered. If severe symptoms of poisoning occur, transport to a hospital must be arranged immediately.

Depending on the poisoning severity the physician will either recommend appropriate symptomatic measures or make a referral to an appropriate hospital. Numerous guidebooks are available to make the right decision on appropriate measures: [25, 37, 62, 63, 64, **51, 97, 161**].

### 1.9.2.2 First Aid (Immediate Actions)

In the case of a possible poisoning, first aid must be administered as soon as possible and already before the arrival of a doctor or ambulance. Eating and drinking are not allowed. Only some water can be administered. Milk accelerates the absorption of toxins and is forbidden (exception: ingestion of oxalic acid-containing plants). Important measures of first aid are:

#### – Reassurance of the affected person

The first aider should avoid hectic, keep calm, and reassure the victim, especially if it is a child. The poisoned person should be shown that something is done to help him or her.

#### – Prevention of further ingestion of poisons

Suspicious substances should be removed from the mouth (if the patient is responsive) or after skin contact from the skin, possibly after taking off contaminated clothing, by washing preferably with body–warm water containing a detergent (liquid soap, shampoo), with wine–red potassium permanganate solution, or with pure running water. Transdermal patches, which are often playfully applied by children, should be removed from the skin and the application site washed with water containing a detergent. The roof of the mouth where such patches can stick should also be checked. If the eyes are affected, they must be rinsed with body–warm water containing 9 g table salt/L (two teaspoons spread (!) full) or with clear running water for about 10–15 min. If neither remedy is available, a handkerchief soaked in water should be squeezed several times into the eyelid slit for rinsing. For information about measures after stings or bites of poisonous animals, see Vol. 4.

– **Ensuring the vital functions**
- Constant observation of breathing and heartbeat. In case of respiratory arrest, the airways must be cleaned and rescue breathing performed (12 times per minute), and in case of cardiac arrest, cardiac massage performed (80–100 times per minute).
- In case of unconscious persons with preserved spontaneous respiration, the affected person must be placed in the recovery position and protected from heat loss.
- In case of vomiting, the head of the lying patient should be positioned to the side to prevent suffocation.
- Poisoned persons with respiratory problems should be positioned in a sitting position.
- In case of imminent or existing anaphylactic shock, characterized by clouding of consciousness, pale skin, sweating, and accelerated heartbeat, shock positioning must be performed: Legs should be positioned slightly raised, maximum 45 degrees; head should be positioned low, maximum 20 degrees.
- In case of convulsions, a handkerchief or similar should be clamped between the teeth, especially to prevent biting the tongue.

### 1.9.2.3 Clinical Measures
In cases of peroral ingestion, the physician will perform **primary poison removal** in addition to symptomatic treatment.

Initially and only after strict consideration by a physician (!), activated charcoal (Carbo activatus, medicinal charcoal) can be given, only as a suspension in water (!). The single dose is about 0.5–1.0 g/kg BW. If possible, the suspension should be drunk, if not given via a nasogastric indwelling tube. It should never be applied via a gastric tube. Aspiration of activated charcoal suspension leads to severe lung damage! The administration of activated charcoal should be useful to interrupt the enterohepatic circulation of the toxin even for a longer time after toxin ingestion.

The combination of activated charcoal with a saline or osmotic laxative serves to rapidly excrete the charcoal loaded with the toxin and prevents the constipation triggered by the charcoal. Sodium sulfate (Glauber's salt, 0.25 g/kg BW, max. 20 g, in 100 mL water) or sorbitol (2–5 mL 70% sorbitol solution/kg BW) can be given as a laxative.

Endoscopic removal of plant residues is a suitable way to remove plant parts that are difficult to digest, e.g., yew needles.

In the case of p.o. ingestion of potentially life-threatening amounts of poison only a short time before (maximum 60 min), the physician, not the emergency physician

(!), can perform gastric lavage. Before gastric lavage, 1 mg of atropine is administered i.m. to prevent reflex laryngospasm.

The usefulness of inducing vomiting for primary toxin removal is discussed controversial. Only little use is made of this option today. Induced vomiting is useful only immediately and for up to 2 h after ingestion of the toxin, before complete absorption has occurred. Lay persons should induce vomiting only if the action has been suggested by a poison control center or a physician. Vomiting should never be induced in those patients who are clouded in consciousness, somnolent, or unconscious (risk of aspiration!), in circulatory depression, bradycardia, and convulsions.

Vomiting can be provoked by the administration of Sirupus ipecacuanhae, but not in children younger than 9 months. The recommended dosages are 10–15 mL until 2 years of age, 15–30 mL in children over 2 years of age, and 20–30 mL in adults. After that, the patient should drink plenty of fluids. Vomiting occurs after 10–20 min. If vomiting does not occur, the administration must not (!) be repeated. In Germany, *Sirupus ipecacuanhae* requires a prescription because of potential risks. In the USA, its use is prohibited. Vomiting can also be induced by mechanical irritation of the posterior pharyngeal wall, e.g., with a finger, a spoon handle, or a stomach tube. The use of a hypertonic saline solution as an emetic (two tablespoons to a glass of water) is risky and in children forbidden (fatal cases occurred!). If this application has had no effect after 10 min, the application must not be repeated under any circumstances (!). Plenty of water must be drunk afterward. The i.m. application of apomorphine to induce vomiting is rejected because of the risk of strong side effects.

The routine use (!) of gastric emptying measures, the administration of activated charcoal and of laxatives for the purpose of poison removal, is obsolete today and should only be carried out in individual cases. The use should be made only after examination of the anamnesis, the data on the ingested poison, and the time between ingestion and start of treatment and other marginal conditions.

Primary toxin removal can be completed by methods of **secondary toxin elimination** like forced diuresis, hemodialysis, hemoperfusion, hemofiltration, peritoneal dialysis, membrane plasma separation, plasma perfusion, and blood exchange transfusion available only in clinics. The effectiveness of these measures has not been proven in every case.

Symptomatic treatment may include artificial respiration in case of respiratory failure, application of plasma expander or plasma in case of circulatory failure, and administration of a vasoconstrictor in the case of dilatation of peripheral vessels or a vasodilator in the case of centralization of circulation. Convulsions can be treated with benzodiazepines or barbiturates, cardiac arrhythmias with extracorporal pacemaker, and circulatory failure with lidocaine or β-blockers.

A specific causal treatment of poisoning with **antidotes** or **antisera** is only possible for a few poisonings so far. Examples for antidotes against plant toxins are inhibitors of acetylcholinesterase, e.g., physostigmine salts or pyridostigmine, that antagonize hyoscyamine/atropine, and conversely, naloxone against opiate intoxications (morphine, codeine, and heroin) or vitamin K against intoxications with dicoumarol. Antisera are available against poisonings with some proteotoxins, e.g., botulinum toxins or venoms of spiders, snakes, scorpions, and other animals. Administration of specific antibody fragments is helpful in treatment of poisoning with cardioactive steroid glycosides.

Selected reviews on treatment of poisoning: [64, **82, 97, 107, 169**], treatments of small animals after poisoning by plants: [**61, 78, 83**].

## 1.9.3 Diagnostics of Poisonings

The diagnosis of poisonings is based on the
– anamnesis, i.e., knowledge of the previous history (source of poisoning, mode of uptake or contact, quantity of absorbed toxin, time of toxin uptake or contact);
– symptoms encountered (syndromic poisoning, toxidrome); and
– qualitative and quantitative toxicological-chemical analysis.

In most cases, all three factors will need to be analyzed to definitively conclude the cause of poisoning.

The easiest way to clarify the history is to interview the patient or, especially in the case of children, others who may know the history. Usually, only imprecise information is available about the appearance and location of ingested plants, the appearance of animals that attacked the patient, or any spoiled food. Remains of plant, fungal, or animal material, and cleaning waste, meals, or vomit should be analyzed and stored for more detailed toxicological-chemical analysis. Vernacular names given by laymen require verification. Even seemingly clear statements may not always be accurate. Folk names of plants, fungi, or animals given by the affected person or helpers should be critically questioned, as they do not always allow conclusions to be drawn about the real source of the poison.

To clarify the source of poisonings, it is necessary to identify the plant or the mushroom causing the poisoning by botanical or mycological keys. They may be particularly helpful if they are well illustrated. The books of Cullen [11], Altmann [**4**], and Frohne and Pfänder [**49**] should be mentioned for plants, and Gerhard [17], Bresinsky and Besl [7], Flammer [14], Marshall [36], Michael et al. [42], and Spoerke [59] for mushrooms. Easily available apps and photos taken by smartphone as well as the internet can quickly provide initial information. If only parts of plants or mushrooms are present, a microscopic examination can be helpful for those who are experienced in this [7, 51, 64, **49**].

However, one should not overestimate these possibilities. The inexperienced person in this field will hardly come to a sure and above all fast success. The safest way is to consult a botanist, a mushroom expert, a pharmacist, in case of ornamental plants a gardener or a florist, and in case of plants from forests or mushrooms a forester or a mycologist.

The plant or fungal material should be presented as completely as possible, i.e., in the case of smaller plants the whole plant, in the case of shrubs and trees, branches with leaves and flowers or fruits, not only the berries alone suspected as the cause of poisoning, in the case of mushrooms, the entire available remains, e.g., cleaning waste and remains of the meal.

Poisonings by animals in Central Europe, with the exception of those caused by the commonly known poisonous snakes like adder and horned viper, require only symptomatic treatment, which means identification is not necessary. Numerous books including the volume about animal toxins in this series (Vol. 4) provide information on dangers from animals in temperate, subtropical, and tropical countries [20, 39, 40, 67, 68]. Information about toxins produced by microscopic fungi, bacteria, and microalgae is given in volume 5 of this series and besides, e.g., in [7, 14, 21, 42, 43, 51].

In addition to the determination of the source of poisoning, an evaluation of the clinical picture of the poisoning (toxidrome) should be performed [64]. Certain leading symptoms provide clues to the cause of poisoning. For example, convulsions may be signs of cicutoxin or strychnine poisoning. Severely constricted pupils and unconsciousness suggest morphine poisoning, and dilated pupils suggest intoxication with tropane alkaloids. Cardiac arrhythmias may indicate intoxication with cardioactive steroid glycosides. However, the leading symptoms have only a probabilistic value, since a symptom is never specific for only one cause; never all symptoms of intoxication are completely and simultaneously present, and the intoxication picture may be uncharacteristic, e.g., due to overlapping with other diseases. Under certain circumstances, the elicitation of symptoms may also be very difficult (e.g., in the case of unconsciousness or young children).

Symptom-based diagnosis is most meaningful if a combination of symptoms that is relatively characteristic of a cause of poisoning can be evaluated (e.g., the four main symptoms of poisoning by atropine). General symptoms such as nausea and vomiting triggered by gastrointestinal irritation can hardly be used for diagnostic purposes, since they occur in most acute poisonings. Overviews of leading symptoms and their possible causes are given for example in the books of Moeschlin [107] and Ludewig and Regenthal [97].

The diagnosis of poisoning on the base of the medical story and symptoms is corroborated, and in most cases only possible by toxicological-chemical analysis. This concerns especially chronic poisonings.

A prerequisite for a flawless toxicological-chemical analysis is above all the proper preservation of the test material. This includes remains of plants or mushrooms not yet ingested, vomit, urine, stool, gastric lavage fluid, dialysis fluid after hemodialysis, and venous blood. These materials should be packed in an airtight container, without added preservatives, and should reach the hands of the toxicological chemist immediately.

Poison detection by toxicological-chemical analysis is carried out qualitatively and quantitatively, specifically (if a particular poison is suspected) or nonspecifically with the aid of methods similar to those used for the isolation and identification of natural substances (see Section 1.3). In doing so, not only poisons must be searched for but also biotransformation products must be recorded.

## 1.10  Precautions to Avoid Poisoning

### 1.10.1  General Precautions

– Knowledge about poisonous plants, mushrooms, and animals should be acquired already in childhood, and continuously extended and applied.
– Collection and ingestion of unknown plants and mushrooms must be avoided.
– Contact with skin-irritant plants, e.g., giant hogweed or Christ thorn, only with gloves, protective clothes, and protected eyes.
– Correct and clearly labeled storage of plant parts, e.g., bulbs, during wintertime to avoid confusion.
– Spoiled food must not be eaten.
– Toxic products must not be stored in containers intended for food.
– It is advisable to store the telephone number required for a poison emergency call in the cell phone.

### 1.10.2  To Prevent Poisonings in Children

– Children should be instructed early not to put plant parts in their mouths.
– Children should be instructed to avoid contact with unknown animals.
– Tobacco products and pharmaceuticals should not be in the reach of children.
– Poisonous plants, including poisonous houseplants, flowers, and other plants fallen on the ground, should be not reachable by children. Do not place flowerpots or vases with poisonous plants in the house at all.
– Cultivate only plants in the garden that you are sure are nontoxic. Ask a gardener, florist, or pharmacist.
– Find out which plants grow already in your garden are poisonous. Remove them or make them not accessible to children.
– Walk with your children to kindergarten or school and warn them about poisonous plants growing on roadsides, field margins, and open gardens.
– Possibly poisonous plants in the home and garden should be labeled with signs so that the poison information center can be informed of the correct name of the plant in case of an emergency.
– Per child, 5 g of powdered medicinal charcoal should be stocked in the household as an emergency remedy (use only as a suspension and only after consultation with doctor or a poison information center).

### 1.10.3 To Prevent Poisoning in Pets

–   Poisonous plants in house, garden, and outside must be out of reach of the pets.
–   Poisonous products like chocolate for dogs or pyrethroid-containing insecticides for cats must be out of reach.
–   Moldy food must not be fed.

## References

(for numbers in bold see cross-chapter literature)

[1]   Adetunji CO et al. (2021) in: **109**, p. 367
[2]   Ahmed S, Kovinich N (2021) Phytochem Rev 20: 483
[3]   Anywar G (2021) in: **109**, p. 1
[4]   Atanasov AG et al. (2015) Biotechnol Adv 33(8): 1582
[5]   Atanasov AG et al. (2021) Nat Rev Drug Discov 20(3): 200
[6]   Beressa TB et al. (2021) in: **109**, p. 189
[7]   Bresinsky A, Besl H (1985) Giftpilze – Ein Handbuch für Apotheker, Ärzte und Biologen. WVG, Stuttgart
[8]   Brooks CJW, Watson DG (1985) Nat Prod Rep 2: 427
[9]   Choudhuri S et al. (2018) Toxicol Sci 161(1): 5
[10]  Cortinovis C, Caloni F (2015) Toxins 7: 5301
[11]  Cullen J (2006) Practical Plant Identification. Including a Key to Native and Cultivated Flowering Plant in Northern Temperate Regions. Cambridge University Press, Cambridge
[12]  David R (2020) Mutagenesis 35: 153
[13]  ECHA (2016) Practical guide. How to use alternatives to animal testing to fulfil your information requirements for REACH registration. https://echa.europa.eu/documents/10162/13655/pra10162/ctical_guide_how_to_use_alternatives_en.pdf/148b30c7-c186-463c-a898-522a888a4404
[14]  Flammer R (2014) Giftpilze. AT Verlag, Aarau
[15]  Fraenkel GS (1959) Science 129: 1466
[16]  Friedrich C, Müller-Jahncke WD, eds. (2012) Gifte und Gegengifte in Vergangenheit und Gegenwart. WVG, Stuttgart
[17]  Gerhardt E (2000) Mushrooms, a Quick Identification System. BLV, München
[18]  Gibbs FW (2018) Poison, Medicine, and Diseases in Late Medieval and Early Modern Europe. Routledge Abingdon-on-Thames, Oxfordshire
[19]  Gmelin JF (1803) Allgemeine Geschichte der Pflanzengifte. Raspesche Buchhandlung, Nürnberg
[20]  Habermehl GG (1994) Gift-Tiere und Ihre Waffen. Springer, Berlin
[21]  Hardegree MC, Tu AT, eds. (1988) Handbook of Natural Toxins, Vol. 4: Bacterial Toxins. Marcel Dekker New York and CRC Press, Boca Raton
[22]  Hartmann T (1985) Plant Syst Evol 150: 15
[23]  Havlíček V, Spižek J, eds. (2014) Natural Products Analysis: Instrumentation, Methods, and Applications. Hoboken, New Jersey
[24]  Hegnauer R (1984). Bedeutung der Chemotaxonomie für die Pharmazeutische Biologie. In: Czygan FC (ed.) Biogene Arzneistoffe. Vieweg, Braunschweig, p. 157
[25]  Heinmeyer G, Fabian U eds. (1997) The Poisoning and Drug Emergency, 3rd ed. Ullstein Mosby, Berlin

[26]  Jaroszewski JW (2005) Planta Med 71(8): 691

[27]  Jeandet P (2015) Molecules 20(2): 2770

[28]  Jones EN (2007) Poison Arrows. University of Texas Press, Austin

[29]  Kleinjans J (2014) Toxicogenomics-Based Cellular Models: Alternatives to Animal Testing for Safety Assessment. Elsevier Sci, Burlington

[30]  Knudsen I (1986) Genetic Toxicology of the Diet. Alan R Liss Inc, New York

[31]  Laroche C et al. (2019) Regulatory Toxicol Pharmacol 108: 104470

[32]  Leeuwen van CJ, Vereire TG (2007) Risk Assessment of Chemicals: An Introduction. Springer, New York

[33]  Linington R et al., eds. (2019) Nat Prod Rep 36: 942

[34]  Macinnis P (2011) A Brief History of Poisons. From Hemlock to Botox to the Killer Bean of Calabar. Arcade Publishing, New York

[35]  Macko P et al. (2021) Toxicol in Vitro 56: 105206

[36]  Marshall NL (2010, reprint) The Mushroom Book. A Popular Guide to Identification and Study of Our Common Fungi. Kessinger Publ, Whitefish

[37]  Martens F (2000) Guide: Poisonings. Urban & Fischer, München

[38]  Martinetz D, Lohs KH (1986) Giftmagie und Realität, Nutzen und Verderben. Nikol Verlagsges, Hamburg

[39]  Mebs D (1989) Gifte im Riff, Toxikologie und Biochemie eines Lebensraumes. WVG, Stuttgart

[40]  Mebs D (2004) Venomous and Poisonous Animals. A Handbook for Biologists, Toxicologists, Toxinologists, Physicians and Pharmacists. WVG, Stuttgart

[41]  Mebs D (2014) Heilende Gifte. WVG, Stuttgart

[42]  Michael E, Hennig B, Kreisel H (1981–1988) Handbuch für Pilzfreunde, Vols I–VI. Fischer-Verlag, Jena

[43]  Moss J et al., eds. (1995) Handbook of Natural Toxins. Bacterial Toxins and Virulence Factors. Taylor and Francis, London

[44]  Nahrstedt A (1989) Planta Med 55(4): 333

[45]  Nepovimova E, Kuca K (2019) Arch Toxicol 93(1): 11

[46]  Newman D, Cragg M (2020) J Nat Prod 83(3): 770

[47]  Penrith ML et al. (2015) J South Afr Vet Assoc 86(1): a1200

[48]  Pitschmann V, Hon Z (2016) Molecules 21: 556

[49]  Porter R, Teich M (1995) Drugs and Narcotics in History. Cambridge University Press, Cambridge

[50]  Přichystal J et al. (2016) Anal Chem 88(21): 10338

[51]  Rumack BH, Salzman MD (1978) Mushroom Poisoning: Diagnosis and Treatment. CRC Press, Boca Raton

[52]  Sachana M, Hargreaves AJ (2012) Toxicological Testing. In Vivo and in Vitro Models. In: **59**

[53]  Salem MA et al. (2021) in: **109**, p. 323

[54]  Schäfer SG (2021) Die Dosis macht das Gift. Heilende Pflanzen im Spiegel der Geschichte. Quelle und Meyer, Wiebelsheim

[55]  Schmidt U et al. (2009) Supertoxic: A comprehensive database of toxic compounds. Nucleic Acids Res 37(Database issue): D295

[56]  Schweder T et al. (2005) Adv Biochem Engin/Biotechnol 96: 1

[57]  Seigler DS (1995) Plant Secondary Metabolism. Kluwer Acad Publ, Boston, MA

[58]  Silva AS et al., eds. (2020) Recent Advances in Natural Products Analysis. Elsevier, Amsterdam

[59]  Spoerke DG, Rumack BH (1994) Handbook of Mushroom Poisoning. Diagnosis and Treatment. CRC Press, Boca Raton

[60]  Swain T (1977) Ann Rev Physiol 28: 479

[61]  Teuscher E (1984). Zur möglichen Funktion von Sekundärstoffen in biologischen Systemen. In: Czygan FC (ed.) Biogene Arzneistoffe. Vieweg, Braunschweig, p. 61

[62]  Weilemann LS, Reinecke HJ (1996) Emergency Manual Poisonings. Thieme, Stuttgart

[63]  Weilemann S et al. (2006) Giftberatung Pflanzen. 3rd ed. Govi, Eschborn

[64]  Wendt S et al. (2022) Dtsch Ärztebl Int 119: 317

[65]  Wexler Ph (2014, 2019) History of Toxicology and Environmental Health. Toxicology in Antiquity I and II. Elsevier, Amsterdam

[66]  Wexler Ph, ed. (1996) Toxicology in the Middle Ages and Renaissance. Elsevier/Academic Press, Amsterdam

[67]  White J, Meier J (2017) Handbook of Clinical Toxicology of Animal Venoms and Poisons. CRC Press, Boca Raton

[68]  Williams J et al., eds. (1996) Venomous and Poisonous Marine Animals: A Medical and Biological Handbook. Blackwell, Hoboken, New Jersey

[69]  Wink M (2008) Nat Prod Commun 3(8): 1205

[70]  Wishart D et al. (2015) Nucl Acid Res 43(D1): D928–D934, www.t3db.org

[71]  Wittstock U, Gershenson J (2002) Curr Opin Plant Biol 5(4): 300

[72]  Xu C et al. (2005) Arch Pharm Res 28(3): 249

[73]  Yadesa TM et al. (2021) in: **109**, p. 79

# 2 Terpenes

## 2.1 Chemistry and Terminology

Terpenes are natural products whose carbon skeleton consists of units with the carbon skeleton of isoprene. Due to methyl group migration, elimination of methyl groups, ring opening, ring expansion, and/or ring constriction, the isoprene building blocks are only partially or no longer recognizable, especially in the case of polycyclic terpenes (Fig. 2.1). Terpenoids (also known as isoprenoids) are terpenes with an oxygen moiety and additional structural rearrangements. Nonetheless, the terms 'terpenes' and 'terpenoids' are used interchangeably. The estimated number of distinct compounds is about 80,000 [1, 3, 13].

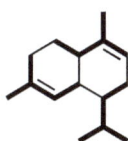

Monoterpene    Sesquiterpene
($\alpha$-Pinen)      ($\alpha$-Cadinen)

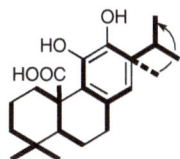

Diterpene
▬▬▬ original position of methyl group
(carnosolic acid)

Fig. 2.1: Building principles of terpenes.

The recognition of the building principle and the grouping of these compounds into a common class of substances, initially applied only to the terpenes built up from two isoprene units, goes back to Wallach (1887) and Ružička (1921) [19].

According to the number of isoprene units involved in the structure, one divides into:

- hemiterpenes (one isoprene unit, $C_5$ skeleton),
- monoterpenes (two isoprene units, $C_{10}$ skeleton),
- sesquiterpenes (three isoprene units, $C_{15}$ skeleton),
- diterpenes (four isoprene units, $C_{20}$ skeleton),
- sesterterpenes (five isoprene units, $C_{25}$ skeleton),
- triterpenes (six isoprene units, $C_{30}$ skeleton),
- tetraterpenes (eight isoprene units, $C_{40}$ skeleton), and
- polyterpenes (more than eight isoprene units).

https://doi.org/10.1515/9783110724738-002

The terpenes are present either as hydrocarbons or oxygenated compounds. Haloge-nated terpenes are also found, particularly in marine organisms. The incorporation of N atoms leads to terpene alkaloids. Steroids (see Chapter 8) are derived from the triter-penes (see Chapter 6). In addition, terpenoid residues (prenyl residues) also frequently react with non-terpenoid compounds to form substances of mixed biogenetic origin.

Reviews: chemistry of terpenes [1, 2, 12, 16], monoterpenes [10, 20], sesquiterpenes [8, 9], sesterterpenes [5], diterpenes [11, 23], and triterpenes [4, 6, 14, 17].

## 2.2 Biogenesis

The biogenesis of the building block of terpenes, the 'active isoprene', occurs in most organisms, in humans and animals exclusively, from three acetate residues (acetate-mevalonic acid pathway). First, acetoacetyl-CoA is formed from two molecules of acetyl-CoA, which combines with a third acetyl-CoA molecule to form 3-hydroxy-3-methyl-glutaryl-CoA. Reduction of a carboxyl group leads to mevalonic acid, which undergoes de-carboxylation via mevalonic acid diphosphate to isopent-3-ene-1-yl-diphoshate. The isopentenyl diphosphate is in equilibrium with the 3,3-dimethyl-allyl diphosphate, cata-lyzed by an isomerase. Both compounds constitute the 'active isoprene'.

In bacteria and some plants, biogenesis can also occur from pyruvic acid or its salts, pyruvates, and glyceraldehyde-3-phosphate (triose-pyruvate pathway, 1-deoxy-D-xylulose-5-phosphate (DXP)/2-methyl-D-erythrol-4-phosphate (MEP) pathway, non-mevalonic acid independent way). Thus, DXP is formed first, which passes via MEP into isopent-3-en-1-yl-diphosphate.

In higher plants, the acetate-mevalonic acid pathway operates mainly in the cytosol and mitochondria. The non-mevalonic acid pathway takes place mainly in plastids [22].

Catalyzed by terpene synthases, higher terpenes are built according to the follow-ing scheme:

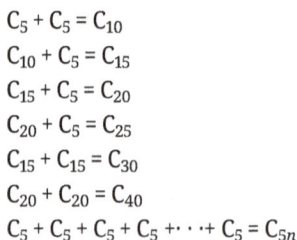

$$C_5 + C_5 = C_{10}$$
$$C_{10} + C_5 = C_{15}$$
$$C_{15} + C_5 = C_{20}$$
$$C_{20} + C_5 = C_{25}$$
$$C_{15} + C_{15} = C_{30}$$
$$C_{20} + C_{20} = C_{40}$$
$$C_5 + C_5 + C_5 + C_5 + \cdots + C_5 = C_{5n}$$

By terpene synthases and cytochrome P450 monooxygenases, the core scaffolds can be further changed.

Reviews: biogenesis of terpenes [1, 3, 7, 13, 15, 18, 22].

## 2.3 Distribution and Significance

Terpenes are widespread among living beings, including archaea, bacteria, fungi, plants, and animals. Highest diversity is found in plants. They are intermediates in the biogenesis of steroids and exhibit a broad spectrum of ecological functions, e.g., in defense against biotic and abiotic stresses, as plant hormones, or as signal molecules [3, 13, 22].

With the exception of polyterpenes, which are insoluble under physiological conditions, toxicologically significant representatives can be found in all other groups of terpenes (see Chapters 3–7).

The sesterterpenes, composed of five isoprene units, seem to be only sporadically distributed. They are pharmacologically little studied so far. Ophiobolin A (cochliophobin), the causative agent of spotted rice disease, was the first representative isolated in 1957 from *Helminthosporium oryzae*, the causative agent of rice flake disease and structurally elucidated in 1967 [24]. The ophiobolins with the basic ophiobolane structure have cytotoxic, antifungal, and antiviral effects [21]. Other representatives have been found sporadically in fungi, ferns, flowering plants, insects, and sponges. Pharmacological studies exist for the ester terpenes from sponges only. They are passed on through the food chain and possess, among other things, anti-inflammatory and antimicrobial properties.

## References

(for numbers in bold, see cross-chapter literature)

[1]   Barton D et al. (1999) Comprehensive Natural Products Chemistry. Volume 2: Isoprenoids Including Carotenoids and Steroids. Elsevier, Amsterdam
[2]   Bell EA, Chartwood BV, eds. (1980) Secondary Plant Products, Vol. 8. In: Encyclopedia of Plant Physiology. Springer Berlin, Heidelberg, New York
[3]   Boncan DAT et al. (2020) Int J Mol Sci 21: 7382
[4]   Connolly JD, Hill RA (2010) Triterpenoids. Nat Prod Rep 27: 79
[5]   Cordell GA (1974) Phytochemistry 13(11): 2343
[6]   Das MC, Mahato SB (1983) Phytochemistry 22(5): 1071
[7]   Dewick PM (2002) Nat Prod Rep 19: 181
[8]   Fischer NH et al. (1979) Prog Chem Org Naturst 38: 47
[9]   Fraga BM (2013) Nat Prod Rep 30: 1226
[10]  Grayson DH (2000) Monoterpenoids. Nat Prod Rep 17: 385
[11]  Hanson JR (2006), Diterpenoids. Nat Prod Rep 23: 875
[12]  Hill RA (1993) Terpenoids. In: Thomson RH (ed.) The Chemistry of Natural Products, 2nd ed., 106. Springer, Netherlands
[13]  Karunanithi PS, Zerbe P (2019) Front Plant Sci 10: 1166
[14]  Kulshreshtha MJ et al. (1972) Phytochemistry 11(8): 2369
[15]  Kuzuyama T, Seto H (2003) Nat Prod Rep 20: 171
[16]  Newman AA (1972) Chemistry of Terpenes and Terpenoids. Acad Press, London

[17]  Paut P, Rastogi RP (1979) Phytochemistry 18(7): 1095
[18]  Rohmer M (1999) Nat Prod Rep 16: 565
[19]  Ružička L (1963) Pure Appl Chem 6: 493
[20]  Schütte HR (1984) Prog Bot 46: 119
[21]  Shen X et al. (1999) J Nat Prod 62(6): 895
[22]  Singh B, Sharma RA (2015) 3 Biotech 5: 129
[23]  Sukh D et al. (1986–1989) CRC Handbook of Terpenoids I–IV. CRC Press, Boca Raton
[24]  Tsuda K et al. (1967) Tetrahedron Lett 8(35): 3369

# 3 Monoterpenes

## 3.1 General

The toxicologically important monoterpenes can be divided according to their basic structure into:
- aliphatic monoterpenes of the 2,6-dimethyloctane type, e.g., the halogenated aliphatic monoterpenes of red algae and the cycloethers of these compounds such as the aplysiapyranoids of some Mollusca;
- aliphatic monoterpenes of the 3,3,6-trimethylheptane type such as the chrysanthemum monocarboxylic acids found in the pyrethrins of Chrysanthemum species;
- monocyclic monoterpenes with a cyclohexane ring, e.g., of the *p*-menthane type such as 1,8-cineol, pulegone, or carvone in essential oils;
- bicyclic monoterpenes, e.g., of the thujan type such as thujone and sabinene, and of the bornan type such as camphor;
- monoterpenes with cyclopentane ring, e.g., the iridoids occurring in 50 plant families.

Particularly large is the number of volatile monoterpenes found in almost all essential oils of higher plants. The acute toxicity of the most representatives is low. Some of them, e.g., 1,8-cineol, camphor, or pulegone, can cause toxic effects in larger doses (see Section 3.2). Some components of essential oils have high toxicity. These include thujan derivatives (see Section 3.2.1), especially thujone. Oxidation products of monoterpenes contained in essential oils can act as allergens (see Section 3.2.6). In addition, many components of essential oils have an irritating effect on the skin and mucous membranes when applied locally. They represent a hazard, particularly for children [44]. If the in vitro and in *Caenorhabditis elegans*, a free living nematode, observed cytotoxic effects of the essential oils of rosemary, citrus, and eucalyptus [78] are relevant for human needs further investigations.

> ⛑ In case of ingestion of small amounts of most essential oils (up to 5 mL), no treatment measures are necessary (but observation!). For larger amounts, administration of activated charcoal, possibly laxatives, and symptomatic measures are recommended [**49**].

Other volatile monoterpenes are components of ants and termite venoms. Low-volatile monoterpene esters are pyrethrins, which are used as insecticides (see Section 3.3). The toxicological hazards from the iridoid valepotriates are likely to be rather low (see Section 3.4). Cantharidin, the defensive venom of blister beetles, is only formally a monoterpene chemically, biogenetically, it is a degradation product of a sesquiterpene [141]. Also noteworthy are the halogenated monoterpenes of red algae (see Section 3.5).

https://doi.org/10.1515/9783110724738-003

## 3.2 Monoterpenes as Toxins of Essential Oils

### 3.2.1 Thujane Derivatives

#### 3.2.1.1 General

The thujane derivatives thujone, umbellulone, sabinene, sabinol, and its esters (Fig. 3.1) are responsible for the toxicity of the essential oils and plants in which they are contained.

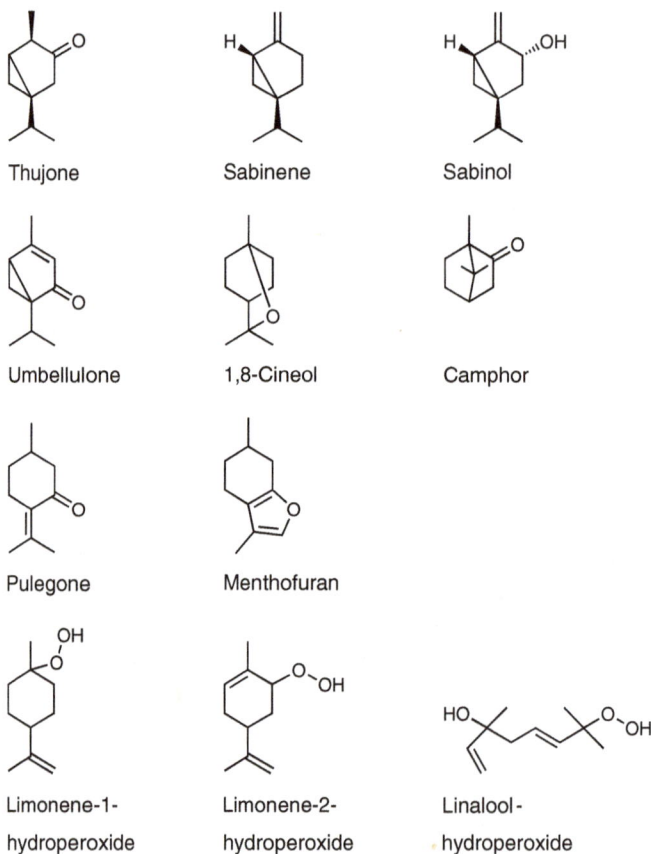

Thujone          Sabinene          Sabinol

Umbellulone      1,8-Cineol        Camphor

Pulegone         Menthofuran

Limonene-1-      Limonene-2-       Linalool-
hydroperoxide    hydroperoxide     hydroperoxide

Fig. 3.1: Toxicologically noteworthy monoterpenes and monoterpene hydroperoxides in essential oils.

**Thujone** (constitution elucidated in 1900 by Semmler [111]) is a mixture of the stereo-isomers (epimeric forms) of (–)-α-thujone and (+)-β-thujone. In the pharmacological literature, no distinction is usually made between the two isomers, although the isomers have considerable differences in toxicity.

Toxicologically notable plants in which (–)-α-thujone dominates in the mixture of isomers are Thuja species and some chemotypes of *Tanacetum vulgare* L., tansy, and of *Salvia officinalis* L., sage. Preferentially (+)-β-thujone is formed by Artemisia species, as well as by other chemotypes of *T. vulgare* and *S. officinalis*.

Thujone is rapidly absorbed through the mucous membranes and through intact skin because of its lipophilic nature and eliminated mainly through the kidneys and lungs. (–)-α-Thujone and its metabolite 7-hydroxy-α-thujone act as noncompetitive antagonists at GABA$_A$ receptors [27, 94, 99]. (+)-β-Thujone targets other neurotransmitter systems, e.g., the serotoninergic one, or distinct sites of the GABA$_A$ receptor [99]. Neurotoxicity is the principal toxic outcome in acute and chronic intoxications. Local irritant effects can occur [58].

The Committee on Veterinary Medicinal Products of the European Medicines Agency (EMA) reports the LD$_{50}$ for thujone as 87.5 mg/kg BW, s.c., mouse, and 240 mg/kg BW, i.p., rat [38]. Up to a dose of 75 mg/day for an adult thujone is considered safe for humans. An adult of 60 kg BW consuming 1 L of an alcoholic beverage containing 5 mg thujone/L (EU: maximum allowable content in beverages containing up to 25% alcohol) would take approximately 0.08 mg thujone/kg BW. This is approximately 100 times less than the no-observed-adverse-effect level (NOAEL) derived from a 14-week study in rats [41]. The Committee on Herbal Medicinal Products of the EMA has recommended an upper daily thujone intake of 6 mg derived from products used for medicinal purposes.

**(–)-Umbellulone** is the toxic constituent of the essential oil of *Umbellularia californica* (Hook. et Arn.) Nutt., headache tree (Lauraceae), which is native to North America [36]. Snuffing the powder of the dried leaves of the tree can cause sneezing and headache [**99**]. A chemotype of *Tanacetum vulgare* also contains umbellulone as a major constituent of the essential oil.

**Sabinene** (constitution elucidated by Semmler in 1902 [112]) occurs as (+)-sabinene (*R,R*-sabinene, α-sabinene) in larger amounts in the essential oil of Juniperus species. It is thought to be responsible for the strong irritant effect of the essential oil of *Juniperus sabina* L., along with (+)-sabinol (*trans*-sabinol, constitution elucidated in 1900 by Semmler) and its esters.

### 3.2.1.2 Thujone in Arborvitae Species (Thuja Species) and Platycladus Species

The genus Thuja, Arborvitae, Cupressaceae, includes five species. Two of them are native to North America and three to East Asia. *Thuja occidentalis* L., red cedar (Ph. 3.1), has its origins in eastern North America and is widely cultivated as an ornamental tree. The other North American species is *Th. plicata* Donn ex D. Don (*Th. gigantea* Nutt.), Western red cedar. Species native to East Asia are *Th. koraiensis* Nakai, Korean thuja, *Th. standishii* (Gordon) Carriére, Japanese thuja, and *Th. sutchuenensis* Franch., Szechuan thuja. The natural habitat of *Platycladus orientalis* (L.) Franco, formerly known as *Th. orientalis* L., Chinese arborvitae, is in the northeastern part of Asia. Many varieties exist.

Ph. 3.1: *Thuja occidentalis* (source: WALA Heilmittel GmbH).

The leaves of the above-mentioned species contain 0.1–1.0% essential oil in which thujone is present: in *Th. occidentalis* 49–65% (–)-α-thujone [150], in *Th. plicata* only traces of thujone and as main constituents α-pinene and car-3-ene [131], and in *P. orientalis* about 50% (–)-α-thujone [9]. *Th. standishii* contains mainly fenchone, fenchyl acetate, bornyl acetate, and β-eudesmol in its essential oil [90].

Other active compounds present in the wood and leaves of Thuja species are diterpenes, lignan derivatives, e.g., thujaplicatin, and tropolone derivatives, the thujaplicins [21, 122].

Poisonings by arborvitae trees occurred in case of overdosing of drugs or of its misuse as abortifacients. They have been observed in children who had eaten fresh leaves or shoots of the plants [161]. Accidental poisoning is also possible during horticultural handling of the plants or after topical use of preparations.

♟ Symptoms after ingestion of leaves of Thuja species can be bloody vomiting, abdominal pain, diarrhea, cramps, and severe irritation of the uterus, liver, and kidney. External contact may cause violent irritant effects, e.g., redness and inflammation of the skin [38, 91, **49**, **161**].

❡ A 7 month-old child was treated with homeopathic preparations of Thuja for the purpose of providing a calming effect around times of vaccination. The child developed generalized tonic-clonic seizures. They stopped after discontinuation of Thuja and brief treatment with pentobarbital [118].

✳ Inhalation of wood dust from *Th. plicata* or *Th. occidentalis* can, triggered by the thujaplicins, cause asthma attacks [**65**].

🐾 Intoxications of dogs and horses by *Th. occidentalis* are reported [108, **90**].

Preparations from *Th. occidentalis* serve, e.g., in combination with extracts of Echinacea species, as immunostimulants for colds [91]. Several Thuja species are used in the traditional medicine of many countries because of their antimicrobial properties [18]. Some lignans and diterpenes of Thuja showed in vitro cytotoxic effects against tumor cells [21].

### 3.2.1.3 Thujone in Wormwood Species (Artemisia Species)

The genus Artemisia, wormwood, Asteraceae, comprises about 500 species. They are distributed all over the world, especially in the temperate zones. For Europe, the occurrence of about 50 species is indicated [37]. Among them are the widespread *Artemisia vulgaris* L., mugwort, *A. absinthium* L., common wormwood, *A. campestris* L., field wormwood, *A. dracunculus* L., tarragon, estragon, and *A. abrotanum* L., southernwood. Most of them are used for culinary and medicinal purposes. *A. annua* L., sweet wormwood, qinghao, has been used in traditional Chinese medicine for the treatment of fever. Today, it is famous as origin of the antimalarial sesquiterpene artemisinin (Fig. 4.9). Further, Artemisia species and their traditional use are described in [1].

Artemisia absinthium (Ph. 3.2) is of particular toxicological interest, as its thujone-rich essential oil is used as an ingredient in the alcoholic beverage absinthe. The aqueous wormwood extracts used for production of wormwood wines and liqueurs contain only traces of thujone in addition to bitter sesquiterpene lactones. The small amounts added to traditional dishes such as Korean rice cake are toxicologically not relevant.

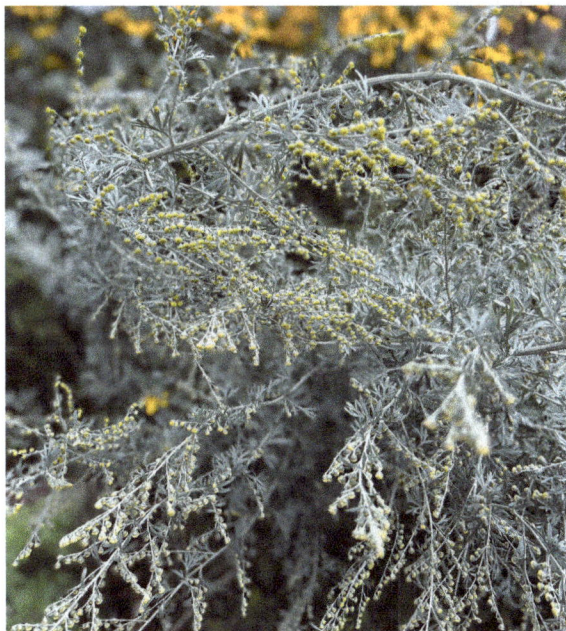

Ph. 3.2: *Artemisia absinthium* (source: bgwalker/iStock/Getty Images Plus).

The plant is a semishrub that grows up to 1.2 m tall. The three-pinnately lobed leaves with lanceolate tips have dense silvery-gray hairs on both sides. The yellow flower heads are 2–4 mm wide. It is found especially in dry ruderal places.

*A. absinthium* yields 0.3–1.3% essential oil. Its composition is very complex (about 60 compounds have been isolated) and strongly dependent on the chemotype. Main components are thujone (mainly (+)-β-thujone), 10–80% in total, thujyl alcohol and its esters, especially the acetate, and more rarely also sabinyl acetate (up to 28%) or epoxy-ocimene (16–57% [104, 123]).

Other noteworthy constituents of wormwood are sesquiterpene lactones of the guaianolide type (artabsin and absinthin, both are the main bitter substances of the herb), germacranolide type (artabin), and eudesmanolide type (arabsin). These sesquiterpene lactones lack an exocyclic methylene group on the lactone ring, and thus appear to possess no allergenicity [104]. Polyines, coumarins, and lignans are noteworthy as further constituents [120, 123].

Some chemotypes *of A. verlotiorum* Lamotte, Verlot's mugwort, *A. campestris*, and *A. abrotanum* also contain significant amounts of thujone. *A. pontica* L., Roman wormwood, which is also used in the preparation of wormwood wines, contains about 30% thujone in the essential oil [9, 92, 120].

About 2% of wormwood extract in the diet of rats had no adverse effects on the animals after 13 weeks [89]. Topical application of undiluted essential oil of *A. absinthium* did not cause irritant effects on the skin of healthy volunteers [84, 123].

Despite the common use, there are only few reports about human poisoning due to *A. absinthium*.

Fatal intoxications occurred with the use of wormwood herb as a vermifuge and abortifacient, which was common in folk medicine in the past **[169]**. Today, uncontrolled preparations traded over the Internet pose a problem.

♀♦ The symptoms of poisoning by wormwood essential oil are illustrated by the following case: a 31-year-old healthy man, who had ordered wormwood oil via the Internet and ingested about 10 mL undiluted, suffered from acute rhabdomyolysis with kidney failure and could only be saved by intensive medical treatment [143].

♦ A 10-month-old male infant progressed with severe diarrhea and persistent metabolic acidosis after ingesting home-prepared extract which was given for the treatment of common cold [66].

⊗ Another possibility of poisoning is the repeated consumption of absinthe. This alcoholic drink (50–72% by volume ethanol) is made from alcoholic extracts of wormwood herb, anise fruits, fennel fruits, hyssop herb, lemon balm leaves, more rarely angelica roots, marjoram herb, juniper berries, and nutmeg. Usually, the drink is diluted with cold water, which was poured into the absinthe over a piece of sugar cube in a special silver strainer (absinthe spoon) above the glass. This turns the initially green clear drink (chlorophyll, 'green fairy') into milky yellow (emulsion formation of the water-insoluble components of the essential oil). It is also popular to dip a piece of sugar with a spoon into absinthe and lights it over the glass. The alcohol burns and the melted sugar drips into the glass. When the flame goes out, the remaining sugar is stirred into the glass and the absinthe is quickly drunk.

The consumption of absinthe was especially popular in the second half of the nineteenth century in military and artistic circles in France and increasingly spread among the population. In 1913, 40 million liters were drunk in France alone. The psychotomimetic effect of the drink (increased well-being, increased mental activity and creativity, and hallucinations) led artists in particular to chronic abuse in the past. Among others, van Gogh, Gaugin, Toulouse-Lautrec, Oscar Wilde, Picasso, and Hemingway were inspired by absinthe. Absinthe had been banned in the USA and many European countries in the first decades of the twentieth century. It began to reappear in the 1990s. The content of thujone is restricted. In the EU, the limit is 5 mg/mL thujone in spirits up to 25% alcohol by volume, 10 mg/mL above that, and 35 mg/L in bitter spirits [74]. In the USA, foods and beverages must be thujone-free, which in practice means that they contain less than 10 mg/L thujone.

Chronic consumption of the absinthe in the past led to the appearance of absinthism, which was characterized by tremors, stomach irritation, emaciation, visual disturbances, headaches, depression with suicidal tendencies, convulsions, and finally personality deterioration, paralysis, and death. It is considered proven that some of van Gogh's hallucinations were due to the consumption of absinthe [56, 133, 138]. Analysis of the contents of historical absinthe bottles shows that the thujone content hardly exceeded today's limits even at that time and suggests that more probably alcohol was the cause of the most poisoning symptoms [75].

Pharmaceutically, wormwood herb extracts are mainly used as a bitter digestive [11, 123]. The suggested application of A. absinthium, e.g., in the treatment of Crohn's disease requires careful risk–benefit analysis [76]. The essential oil of A. absinthium is contraindicated in pregnant females, nursing mothers, and individuals with hyperacidity and peptic ulcer [11].

### 3.2.1.4 Thujone in Common Tansy (*Tanacetum vulgare*)

The genus Tanacetum, tansy, Asteraceae, includes about 160 species. The species are native to many regions of the northern hemisphere. Some authors assign the species to the genus Chrysanthemum. In Europe, six species occur. Of them, *T. vulgare* (*Chrysanthemum vulgare* (L.) Bernh.), common tansy (Ph. 3.3), is of toxicological interest. It grows widely in Europe and Asia and was introduced to North America in the eighteenth century.

Common tansy is a perennial, rhizome-forming plant, growing 0.6–1.2 m tall with leaves 15–25 cm long and 8–11 mm yellow flower heads that lack ray florets and are in umbrella panicles. It is widespread, especially along roadsides and ruderal places.

The herb contains about 0.2–0.8%, and the flower heads up to 1.5% essential oil. The composition of the essential oil varies widely [59, 126]. The following compounds may

Ph. 3.3: *Tanacetum vulgare* (source: Wolfgang Kiefer).

occur, among others, as major essential oil constituents: (–)-α-thujone, (+)-β-thujone, sabinene, umbellulone, camphor, bornyl acetate, α-pinene, and 1,8-cineol [47, 57, 119, 126, 139].

Other active ingredients of tansy are sesquiterpene lactones of the guaiacolide, germacranolide, and eudesmanolide type, e.g., tanacetin, reynosin, parthenolide, and vulgarolide, the latter with unusual basic body [20, 57, 119, 126, 139]. The herb of *T. parthenium* (L.) Sch. Bip. (*Ch. parthenium* (L.) Sch. Bip.), feverfew, is used for the prophylaxis of migraine because of the sesquiterpenes it contains. All sesquiterpene lactones have an α-methylene-γ-lactone ring and are therefore possible allergens (see Section 4.2.5). Some polyynes, partially containing sulfur, are present [117].

Poisoning of humans has been reported when the drug was used as an anthelminthic, an abortifacient, or for epilepsy treatment [6].

❢ Frohne and Pfänder [49] describe the case of poisoning of a patient to whom a naturopath prescribed 10 g of Oleum Tanaceti in 30 g of castor oil instead of 10 drops (= gtt.) as an anthelmintic. The patient suffered severe seizures.

The lethal dose of tansy essential oil for humans is 15–30 g [93]. Because of the varying thujone content of different chemotypes, poisoning may occur even at lower doses.

> ☠ Symptoms of poisoning by tansy essential oil described in older literature include vomiting, diarrhea, convulsions, arrhythmias, mydriasis, and pupillary rigidity, as well as signs of liver and kidney damage. Death may occur 1–4 h after ingestion due to circulatory and respiratory arrest or organ damage [93].

✱ The sesquiterpene lactones contained in the plants can cause contact allergies and probably aerogenic allergies [65].

🐾 The plants are largely avoided by animals [90].

Because of the thujone content, common tansy herb, flowers, and oil are considered substances of concern that must not be used in formulations. The use as a vermifuge should be avoided.

### 3.2.1.5 Thujone in Sage Species (Salvia Species)

The genus Salvia, sage, is with about 900 species the most extensive genus of the Lamiaceae. Ten species occur in Central Europe. Of these, however, only *Salvia officinalis* L., common sage, is of toxicological interest. The leaves of the species *S. divinorum* Epling and Játiva, sage of the diviners, native to Central America, are used as intoxicants because of their psychostimulant diterpenes (see Section 5.5).

*Salvia officinalis* (Ph. 3.4) is a semishrub to 70 cm tall with barky branches and dull green, densely hairy shoots. This Mediterranean plant, native to the Balkan Peninsula (Dalmatia), thrives in many countries as a cultivated plant; rarely it occurs feral.

Ph. 3.4: *Salvia officinalis* (source: Eberhard Teuscher).

The leaves of common sage contain 1.5–3.5% essential oil [96, **148**]. The content and composition of the essential oil are race-specific and depend on the developmental state of the plant. The essential oil of *S. officinalis* ssp. *minor* (Gmelin) Gams and ssp. *major* (Garsault) Gams, Dalmatian sage, contains 30–60% thujone in addition to camphor and 1,8-cineol as other major components. The oil of *S. lavandulifolia* Vahl (*S. officinalis* ssp. *lavandulifolia* Gams), Spanish sage, is almost free of thujone; major constituents are 1,8-cineol and camphor. The essential oil of *S. fruticosa* Mill. (*S. triloba* L. fil.), Greek sage, carries about 5% thujone in addition to the major component cineol. Essential oil of *S. sclarea* L., clary, native to Syria, may contain a maximum of 0.2% thujone [73, 96, **148**].

Furthermore, the antimicrobial diterpenes picrosalvin (carnosol), diterpene quinones, and salvin (carnosolic acid) are worth mentioning [17, 35, 69, **148**].

The $LD_{50}$ of sage oil is 45 mg/kg BW, mice [100].

Intoxications by sage oil have been reported after accidental ingestions, especially by children, overdosage [52], or after abuse of larger amounts for abortion purposes [**53**].

> ♀ ♦ A 5½-year-old girl ingested about 5 mL of sage oil and subsequently experienced a generalized tonic-clonic seizure lasting 10 min [52].

Chronic poisoning can be ruled out with the small amounts of thujone that enter the human body when sage leaves are used as a spice or medicinal drug, especially since use over a long period of time is unlikely. The use of sage as a home tea for continuous use, however, seems questionable. There are no objections to the widespread use of sage extracts for gargling in cases of inflammation of the oral and pharyngeal mucosa (antiseptic effect of the essential oil and the diterpenes, astringent effect of the tannins) and as an antihidrotic. At most, frequent gargling with sage extracts may cause stomatitis in individual cases [**53, 93**].

## 3.2.2 Sabinene, Sabinol, and Sabinol Esters in Juniper Species (Juniperus Species)

The genus Juniperus, juniper, Cupressaceae, includes about 60 species, all occurring in the northern hemisphere. *Juniperus communis* L., common juniper (Ph. 3.5), *J. communis* var. *saxatilis* Pall. (*J. sibirica* Burgsd.), alpine juniper, and *J. sabina* L., sade tree, are native to Central Europe. Some other Juniperus species are planted as ornamental shrubs, e.g., *J. virginiana* L., red juniper (North America), *J. chinensis* L., Chinese juniper, *J. oxycedrus* L., cade juniper, and *J. horizontalis* Moench, creeping juniper (North America).

While the therapeutically used berry cones of the common juniper and their essential oil are harmful to humans only in case of overdose, the dried branch tips of the sade tree and the essential oil distilled from them are said to have high toxicity.

The sade tree (Ph. 3.6) is a prostrate shrub or tree with an ascending trunk. The young plants develop only needle-like, protruding leaves. The leaves of the older specimens are scaly, cross-opposite, about 2 mm long, and unpleasantly odorous. The female flowers develop, usually a year after pollination, into a green berry cone formed by the fusion of the three upper scale leaves, which turns blue-black the following spring. The tree is found on sunny mountain slopes, in scrub and in dry pine forests of the Alps, the Caucasus, and other mountain regions till Central Asia. It is sometimes grown as an ornamental shrub.

Ph. 3.5: *Juniperus communis* (source: WALA Heilmittel GmbH).

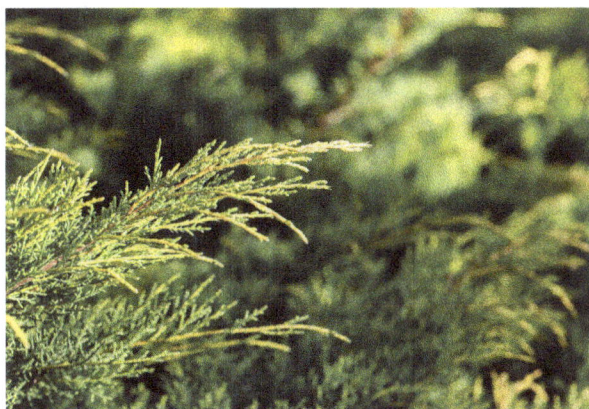

Ph. 3.6: *Juniperus sabina* (source: Anna Nelidovai/iStock/Getty Images Plus).

The main constituents are (+)-sabinene (10–55%), sabinol, sabinyl acetate, β-cadinene, terpineol, and terpinen-4-ol (formed during distillation from sabinene [46, 107]). From the berries, 2.5% essential oil was obtained [30].

Relatively high levels of sabinene may also be found in the essential oils of the scale- or needle-shaped leaves of other Juniperus species, e.g., in that of *J. communis* 0.2–50% (high values in the high mountains, [7, 136]), of *J. communis*, var. *depressa* Pursh, Canadian juniper, up to 48% [49], of *J. chinensis* L., of *J. pseudosabina* Fisch. and C.A. Mey, Turkestan juniper, up to 38% [46, 107], and of *J. horizontalis* up to 37% [26]. The therapeutically used essential oil of the mock berries of common juniper also contains sabinols (about 20% [136]), as does that of several nontoxic plants (e.g., *Piper nigrum*, black pepper, about 25% in the essential oil of the fruits [146]). Sabinol and sabinyl acetate are also found in the essential oils of *J. phoenicea* L., Phoenician juniper, and *J. thurifera* L., Spanish juniper [49], and in those of representatives of other genera, e.g., Artemisia and Salvia.

The role of sabinene, sabinol, and sabinyl acetate as active ingredients of the tissue-destroying essential oil of *Juniperus sabina* is very questionable in view of their distribution even in nontoxic plants. The nonvolatile lignans of the leaves (about 0.2%), podophyllotoxin and savinin [55] can also hardly be held responsible for the effect. Presumably, the effect is due to the rapid postmortem formation of peroxides from the unsaturated terpenes in the sade tree tips, possibly catalyzed by the chlorophyll of the leaves. In subjects allergic to terpenes, very severe skin reactions were observed even after contact with very small amounts of terpene peroxides. However, in nonallergic subjects, even 20% solutions did not elicit any effects [51]. According to our own observations, rubbing the fresh leaves of *J. sabina* on the skin does not cause any irritation.

The essential oil of the branch tips of the sade tree was formerly used externally to remove warts and condylomas. Extracts were used as abortifacients. Poisoning occasionally occurred during these applications. In a few cases, poisoning was observed in children who had put plant parts in their mouths or swallowed them. The lethal dose in humans is said to be about 20 g of the drug [97], and about six drops of the essential oil [107]. The $LD_{50}$ of the oil is with about 3,000 mg/kg BW, mice, i.p., much higher than the dose necessary for showing hepatoprotective effects (200 mg/kg BW) in rats [2].

🔆 Symptoms of poisoning observed with ingestion of sade tree leaves or cones, extracts from these organs, or sade tree essential oil were severe vomiting, bloody diarrhea, tenesmus, discharge of bloody urine, later oliguria, and anuria, and after resorption convulsions, coma, and finally central respiratory paralysis. Accidental poisoning in children by ingestion of sade tree leaves or sade tree cones led to strong salivary secretion, retching, abdominal pain, sometimes violent, even bloody vomiting, and occasionally to tachycardia [161]. External application of extracts from the (dried!) twig tips resulted in blistering and necrosis [49, 53, 107].

🔆 Treatment of poisoning with the sade tree leaves should be generous fluid intake, in case of external contact cleaning of the affected skin areas, in case of ingestion administration of charcoal, and in case of larger amounts gastric emptying. Further treatment is symptomatic [161].

*Juniperus communis* essential oil has a diuretic effect. In large quantities, it can cause kidney irritation. In rats, intake up to 1,000 mg/kg BW of juniper oil for 28 days was not toxic [110]. In humans, in cases of preexisting renal impairment its use should be cautious. Because of possible abortive effects, pregnant women should avoid consuming large amounts of juniper berries.

Tar (Pix Juniperi) obtained by dry distillation of the wood of *Juniperus oxycedrus* is still used in folk medicine in Asia under the name of cade oil (juniper tar) because of its anthelmintic effect.

✦ A healthy man developed fever, hypotension, renal failure, hepatotoxic symptoms, and severe facial skin damage after ingesting a homemade extract (one teaspoon) of *J. oxycedrus*. The man was saved by intensive treatment [68].

### 3.2.3 1,8-Cineol in Eucalyptus Species

1,8-Cineol (Fig. 3.1) is the main component of essential Eucalyptus oils obtained by steam distillation from fresh leaves and twigs of various Eucalyptus species, e.g., *Eucalyptus globulus* Labill., blue gum (Ph. 3.7), Myrtaceae. Their oils are often used as remedies for treatment of cold and locally as analgesics. Eucalyptus species are native to Australia, and they are cultivated in many tropical and subtropical countries.

Ph. 3.7: *Eucalyptus globulus* (source: WALA Heilmittel GmbH).

1,8-Cineol is an agonist at the ion channels TRPM8, a sensor for cold, and an antagonist at TRPA1, a sensor for pain [124].

Poisoning by Eucalyptus oils, caused by medicinal use [28] or by uncontrolled intake in children, has been reported worldwide, particularly in Australia [29, 142] and India [32, 61, 70, 82, 114]. Most cases may be attributed to ingestion. Even topical application or inhalation can cause adverse effects. Individual sensitivity appears to vary widely.

The $LD_{50}$ of 1,8-cineol, given intragastric, is 2,480 mg/kg BW, rat, and 3,849 mg/kg BW, mice [31, 63]. Pathological observations indicated that the target organs are the liver and kidney [147]. In humans, even small ingestions of pure essential oil (5 mL) can lead to severe symptoms [25].

> ☂ After ingestion of Eucalyptus oil, onset of symptoms is usually within 30 min but can be delayed up to 4 h after exposure. After ingestion, it comes to clouding of consciousness (in children ready to amounts of 2–3 mL essential oil), vomiting, ataxia, and respiratory distress. Seizures can occur. Application on the skin leads to skin irritation and contact dermatitis. The symptoms usually resolve within 24 h [25, 61, 82].

❢ Severe pulmonary and renal disturbances developed in one neonate to whom Eucalyptus oil was applied [102]. Death occurred in two cases after accidental ingestion of 3.5–5 mL of Eucalyptus oil; in other cases, much higher doses were survived [31].

Rubbing infants and young children up to 2 years of age with cough ointments containing essential Eucalyptus oil under the nose or in the chest area developed glottic edema (laryngeal edema with closure of the glottis, risk of suffocation!).

Bronchoconstriction may occur in asthmatics. In pregnant women, preparations containing essential Eucalyptus oil must not be used in the first trimester of pregnancy, otherwise for a maximum of 1 week. It should not be used in children under 6 years of age.

> ⚕ Treatment of intoxications after inhalation or ingestion of essential Eucalyptus oil should be symptomatic. Noninvasive ventilation or intubation can be necessary. Asymptomatic children should also be carefully observed [25].

### 3.2.4 D-Camphor as Possible Toxin in Essential Oils

D-Camphor (*R,R*-(+)-camphor, Fig. 3.1) is obtained from the essential oil of the wood of the camphor tree, *Cinnamomum camphora* (L.) J. Presl. (Ph. 3.8), Lauraceae. The camphor tree occurs naturally in East Asian countries and has been naturalized in other parts of the world. Camphor is also contained in the essential oils of *Salvia officinalis*, *Hyssopus officinalis* L., hyssop, *Thuja officinalis*, and *Juniperus* species.

D-Camphor is used externally for muscle and joint pain, and internally for cardiovascular problems. Besides, it is used because of its typical aroma as a fragrance, e.g., in cosmetics and household cleaners.

It is well absorbed after ingestion and after inhalation. Camphor activates thermosensitive TRPV channels, especially TRPV3, and antagonizes TRPA1 channels [85]. A noncompetitive inhibition of nicotinergic acetylcholine receptors has also been shown [93]. In rats, it triggers renal and testicular inflammation through activation of NF-κB and upregulation of inflammatory markers [116].

Poisoning can occur after ingestion or local application of essential camphor oil or camphor ointments for medicinal purposes or by accidental ingestion of camphor-containing household products and repellents [4, 71, 80]. The lethal dose for infants is about 1 g. In adults, severe poisoning occurs from 10 to 20 g, p.o. The lethal dose is reported to be 50–500 mg/kg BW, p.o., or 6 g/kg BW, s.c. Camphor inhalation may cause irritation of the mucous membranes above 2 ppm in the breathing air [22, 71, **34**, 97].

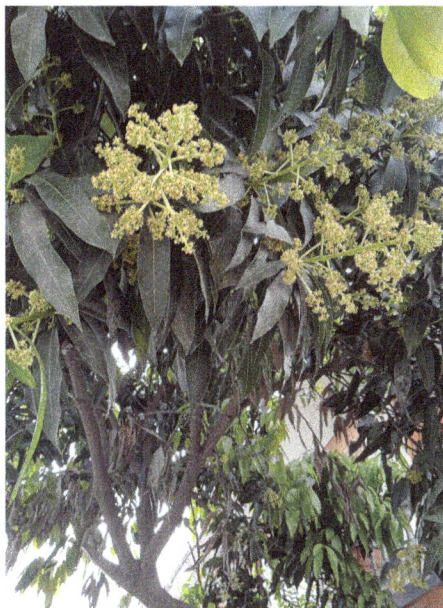

Ph. 3.8: *Cinnamomum camphora* (source: Meena Rajbhandari).

☠ After ingestion of camphor oil, the toxic effects begin within 5–20 min reaching a peak after about 90 min after ingestion. Symptoms of poisoning with camphor by ingestion are nausea, vomiting, colic, headache, dizziness, blurred vision, delirium, and as main features convulsions and shortness of breath. The breath has a characteristic aromatic odor. Camphor can cause skin and eye irritation on contact. Ingestion may lead to abortion [22, 71, **34**].

Camphor preparations must not be used at all in infants, and at most in low doses (<5% camphor in the preparations) in young children. According to FDA, camphor-containing products must not exceed a concentration of 11% camphor [116].

🚑 In cases of camphor poisoning, primary toxin removal and, in case of convulsions, administration of diazepam and possibly ventilation are necessary [**34**]. In case of ingestion of camphor by children, inpatient monitoring should be performed. For quantities of 0.5 g camphor/kg BW or more, gastric lavage should be performed if ingestion occurred no more than 1 h ago [**161**].

### 3.2.5 Pulegone and Menthofuran as Possible Toxins in Essential Oils

Pulegone and menthofuran (Fig. 3.1) are ingredients of the essential oil of the Lamiaceae *Mentha pulegium* L., pennyroyal, native to Europe, North Africa, and the Middle East (Ph. 3.9, up to 97% pulegone in the essential oil), of *Hedeoma pulegioides* (L.) Pers., American pennyroyal (up to 82% pulegone of the essential oil), native to North

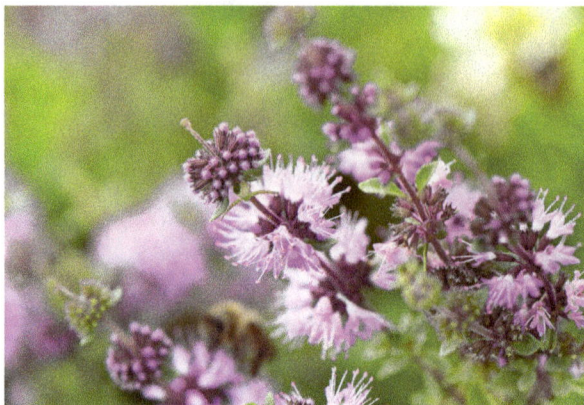

Ph. 3.9: *Mentha pulegium* (source: Drbouz/iStock/Getty Images Plus).

America, and of *Calamintha nepeta* (L.) Savi, common calamint, native to the Mediterranean region. Besides there, pulegone has been found in other Mentha species, among them *M. × piperita* L., *M. spicata* L., and *M. arvensis* L., further Hedeoma and Calamintha species (all Lamiaceae), plants of other families, e.g., Asteraceae, and some invertebrate animals (bryozoan *Conopeum seurati*) [14].

Pennyroyal is used in cooking and in traditional medicine for colds and dyspepsia. Its essential oil is widely used in dyes, perfumery, and aromatherapy, besides as an abortifacient. Limits in its use in food products have been issued, but there are currently no limits for medicinal products. Its concentration in cosmetics should not exceed 1% [14].

In humans, pulegone is metabolized under the influence of CYP enzymes to menthofuran and other hepatotoxic metabolites. Toxic effects on the lung, kidney, and brain have been demonstrated. Carcinogenic activity cannot be excluded. Large interindividual differences exist [5, 14, 16, 151].

The $LD_{50}$ values for pulegone are 150 mg/kg BW, i.p., rat, and 330 mg/kg, BW, i.v., dog [14].

The EMA determined the no-observed-adverse-effect level (NOAEL) of pulegone as 37.5 mg/kg BW/day, basing on the assessment of pulegone's liver and kidney toxicity. FDA sets NOAEL as 13.39 mg/kg BW/day, derived from carcinogenicity studies in rodents [40, 42, 62].

Cases of poisoning by pulegone with hepatotoxic and pulmotoxic symptoms, some even fatal, have been reported. Severe poisoning has been observed after ingestion of 5 or 10 mL of essential oil of *Mentha pulegium*; in other cases, 30 mL has been survived [5, 8, 14]. A further risk is the use of mint-flavored E-cigarettes [62]. The use as home tea is not recommended.

⚑ Two children were found to have severe liver damage and neurological injury after drinking the tea. One child died [8, 148].

⚱ Symptoms of poisoning by pulegone or pulegone-rich essential oils are hepatic and renal damages, seizures, and coma [14].

### 3.2.6 Oxidation Products of Monoterpenes as Allergens

✱ Oxidation products of unsaturated monoterpenes formed by oxidation, e.g., peroxides, epoxides, and endoperoxides, are held responsible for the sensitizing effects of some essential oils (see also *Juniperus sabina*, Section 3.2.2). Other components of essential oils, e.g., carvone, can also become allergens.

Contact allergies are particularly frequently triggered by tea tree oil, which is obtained by steam distillation from leaves and twig tips of Melaleuca species native to Australia, e.g., *Melaleuca alternifolia* (Maiden and Betche) Cheel (Ph. 3.10), tea tree, Myrtaceae. The essential oil contains terpinen-4-ol, γ-terpinene, α-terpinene, and 1,8-cineol as its main constituents and is used externally, e.g., for treatment of acne and mycoses. Severe poisoning is possible if the oil is ingested. The $LD_{50}$ of the oil is 1.9–2.6 mL/kg BW, p.o., rat [137].

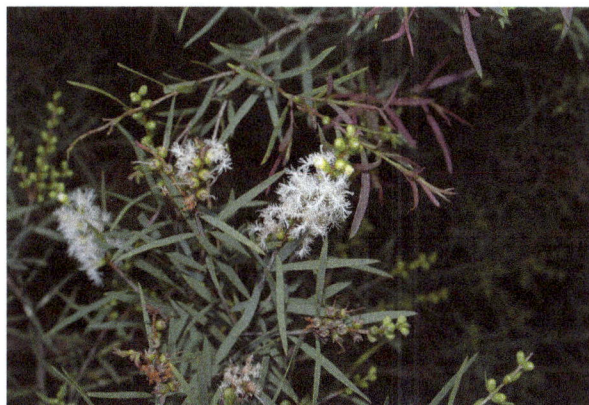

Ph. 3.10: *Melaleuca alternifolia* (source: Hans Wohlmuth).

🜄 A 4-year-old boy showed primarily neurotoxic symptoms within 30 min of ingesting a small amount of the oil but recovered quickly [87].

🕮 To avoid adverse effects, only properly stored and diluted tea tree oil should be applied topically, and ingestion should be avoided [54, 109].

🐾 Animal intoxications have been reported after local application in cats, dogs, and birds [13, 137]. Cats develop weakness, depression, ataxia, incoordination, and muscle tremors after external application of essential tea tree oil. A pet cockatiel (*Nymphicus hollandicus*) showed severe liver damage and a comatose state after application of three drops of tea tree oil directly on its wing [137]. The use of grooming products containing tea tree oil on cats should be avoided.

✱ D-**Limonene**, which is contained in the essential oil of Citrus species, is one of the fragrance substances most often used in cosmetic and consumer products. Upon air exposure, oxidation occurs. Oxidation products, mainly hydroperoxides (Lim-OOH, Fig. 3.1), are responsible for the growing number of contact allergies following contact with limonene-containing products like shampoos, soaps, or deodorants. Hydroperoxides of **linalool** (Fig. 3.1) and **geraniol** are also important contact allergens [12, 134]. Linalool is a component of the essential oils of Lavandula species and some other herbs, and geraniol is a component of the essential oil of Rosa species. These essential oils are often used as fragrances. In a study with 821 patients, 77 patients showed positive patch test reactions to limonene hydroperoxides and 96 to linalool hydroperoxide. Thirty-eight patients reacted to both. Most reactions were considered possibly or probably clinically relevant (66.3% and 68.8%, respectively [34]).

Hairdressers and therapists are the occupations with the highest incidence rates of contact allergies caused by fragrances [45].

📌 The most important prophylactic measure is the complete avoidance of contact with the corresponding preparations by the sensitized persons. Possible cross-allergies must be considered. Treatment of allergic manifestations is usually with glucocorticoids and antihistamines.

## 3.3 Pyrethrins

Pyrethrins are esters of the monoterpenic acids (+)-*trans*-chrysanthemic acid or (+)-*trans*-pyrethric acid with the alkyl cyclopentenolones (+)-pyrethrolone (pyrethrin I or pyrethrin II), (+)-cinerolone (cinerin I or cinerin II), or (+)-jasmolone (jasmolin I or jasmolin II) as alcohol moiety (Fig. 3.2). The enzyme producing the ester linkage was recently identified as a GDSL-type lipase-like protein [83, 148].

Pyrethrins are produced by several Asteraceae species, e.g., *Tanacetum cinerariifolium* (Trevir.) Sch. Bip. (*Chrysanthemum cinerariifolium* (Trevir.) Vis.), Dalmatian pyrethrum, native to the Adriatic coast, and *T. coccineum* (Willd.) Grierson (*Ch. coccineum* Willd.), painted daisy (Ph. 3.11), native to the Caucasus. Both species are cultivated in many countries of temperate and tropical climates. *T. coccineum* is grown in many varieties as an ornamental plant for production of cut flowers in gardens and nurseries.

The flower heads of these species contain 1.2–1.8% pyrethrins. The pyrethrins are localized in glandular hairs on the surface of the achenes, in their secretory ducts,

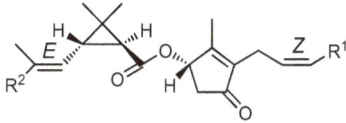

| Cinerin I or II | $R^1 = CH_3$ |
| Pyrethrin I or II | $R^1 = CH=CH_2$ |
| Jasmolin I or II | $R^1 = CH_2-CH_3$ |
| Chrysanthemum acid ester | $R^2 = CH_3$ |
| Pyrethrum acid ester | $R^2 = COO-CH_3$ |

Fig. 3.2: Pyrethrins: insecticides of Asteraceae.

Ph. 3.11: *Tanacetum coccineum* (source: Ulrike Lindequist).

and on the leaves. The dried flower heads of the plants were used as officinal drug Flores pyrethri, insect flores, to control intestinal worms. Today, they are used to obtain pyrethrins, which serve as environmentally friendly insecticides. Synthetic insecticides, the pyrethroids, were derived from them. Pyrethrins and pyrethroids are used in agriculture and households, as well as directly on children for head lice, on pets for ticks and fleas, and on furniture for bedbugs [97]. They supply about 25% of the world's insecticide needs [113].

Pyrethrins are well absorbed from the gastrointestinal and respiratory tracts, but hardly through the skin. In humans, they are hydrolyzed by the esterases of the intestinal mucosa, and their cleavage products are rapidly converted by hepatic mixed-function oxidases to nontoxic substances that are excreted within 24 h after binding to glucuronic acid. The detection of the metabolite 3-phenoxybenzoic acid in urine can be used to follow the exposure [97].

The primary targets of pyrethrins are the voltage-gated sodium channels of irritable membranes. Prolongation of the sodium influx results in prolongation of the depolarization phase, thus preventing cell responsiveness. Voltage-gated calcium and chloride channels have been implicated as secondary sites of action [115]. Genotoxic effects on mucosal cells of the nose have been observed in vitro [129].

Pyrethrins have a much higher toxicity to insects (2,250 times) than to humans and most mammals due to the divergent structure and sensitivity of sodium channels, rapid detoxification in humans, and insect's smaller size and lower body temperature [23]. The $LD_{50}$ is 2,370 and 1,030 mg/kg BW, p.o., for male and female rats and 273–796 mg/kg BW, p.o., for mice. The WHO classifies pyrethrin insecticides as moderately hazardous (class II) [88].

Although the acute toxicity of pyrethrins to humans is low, symptoms of poisoning may occur with peroral ingestion (food and water), indoor use of spray preparations containing the compounds, and eye and skin contact. Children are the most vulnerable group [23, 95].

🕱 Symptoms of poisoning with pyrethrins after p.o. ingestion include anesthesia of lips and tongue, disturbance of taste sensation, facial paresthesia, nausea, vomiting, faintness, and abdominal pain and headache. Inhalation may cause rhinitis, cough, and chest pain; eye contact may cause pain, increased lacrimation, conjunctival irritation, and eyelid edema [**161**]. Few fatal cases have been reported [15]. Severe symptoms of poisoning are more likely to be caused by synthetic pyrethroids, less so by natural pyrethrins.

Studies have shown that there is an association between pyrethroid exposure during pregnancy and early childhood and neurodevelopmental and behavioral outcomes in children, e.g., higher prevalence of attention-deficit hyperactivity disorder. Neonatal exposure of rats to the pyrethroid deltamethrin leads to disruption of dopaminergic pathways [97]. For some pyrethroids, adverse effects on immune system, kidney, and liver have been shown [23].

✱ In the past, allergies (pyrethrum dermatitis and bronchial asthma) frequently occurred after application of insect powder. Possible allergens of pyrethrum flowers are the sesquiterpenes, especially pyrethrosines. Today, the allergens of the raw material are removed during the production of pyrethrins [**65**].

🖅 Treatment of poisoning with pyrethrins is symptomatic [98]. Pyrethroids must be used with caution, especially because of widespread and repeated exposition.

🐾 Among higher animals, cats are the most vulnerable. Compared to dogs, the activity of pyrethrin-metabolizing enzyme glucuronyl transferase in cats is decreased. Poisoning often occurs when preparations intended for dogs are accidentally eaten by cats. Poisoning is usually lethal. The use of preparations containing pyrethroids in cats is therefore prohibited. Fish also react sensitively. Lethal effects of pyrethroids on flower-pollinating insects such as bees are problematic.

## 3.4 Iridoids

### 3.4.1 General

Iridoids (iridanes) are monoterpenes with a *cis*-linked, partially hydrogenated cyclopenta[c]pyran base body, in the fully hydrogenated state also known as iridane. The initially formed $C_{10}$ iridoids can be secondarily converted into $C_9$ or $C_8$ iridoids by elimination of one (at C-4 or C-8) or two C atoms (at C-4 and C-8) or into secoiridoids by splitting the cyclopentane ring. They mostly occur as 1-*O*-monoglucosides or 1-*O*-acyl derivatives. Further, acid residues may be attached to the basic body or to the glucose residue. Since they partially change into blue or black compounds of unknown structure when treated with acids, they have also been called pseudoindicones. More than 200 compounds of this group are known [128].

Iridoids were isolated from plants as early as the end of the nineteenth century, but it was not until 1958 that the structure of the basic body was recognized by Halpern and Schmid [53].

About 600 iridoid glycosides are known. They occur in 57 plant families, frequently in Apocynaceae, Gentianaceae, Lamiaceae, Loganiaceae, Menyanthaceae, Plantaginaceae, Rubiaceae, Scrophulariaceae, Valerianaceae, and Verbenaceae.

Of toxicological interest are the iridoids of valerian and those that serve as defense toxins of numerous insects.

Of general interest may be the fact that many volatile iridoids are attractants for animals of the family Felidae, e.g., cats, lions, jaguars, and leopards. The most potent attractant is nepetalactone (Fig. 3.3), found in the essential oil of *Nepeta cataria* L., catnip, Lamiaceae [105]. It affects the animals via olfactory receptors in the nasal mucosa and induces the secretion of β-endorphin to the blood. That leads to activation of μ-opioid receptors and therefore to behavioral changes of most of the affected animals. The compounds are able to repel insects. On the other hand, aphids produce nepetalactone as a sex pheromon [79]. A variety of iridoids found in insect venoms, e.g., iridomyrmecin, as well as the iridoid alkaloid actinidin, found in Valeriana species, are also attractive to cats [132].

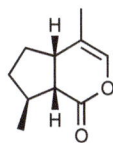

Nepetalactone    Fig. 3.3: A monoterpene of *Nepeta cataria* with attractant activity for cats.

### 3.4.2 Valepotriates as Possible Mutagens of Valerian Species (Valeriana Species) and Other Valerianoideae

The former family Valerianaceae is today classified as subfamily Valerianoideae in the family Caprifoliaceae. More than 400 species and hybrids are accepted. In Central Europe, the genera Valeriana, valerian, Centranthus, red valerian, and Valerianella, corn salad, are represented. Other members are native to Asia, North America, and South America (especially in the Andes). Valepotriates are found in all of the abovementioned genera. Because of its widespread use as a medicinal plant, valerian and its constituents deserve special attention.

The genus Valeriana includes till 200 species, with 12 species occurring in Central Europe. Of importance is especially *V. officinalis* L., common valerian (Ph. 3.12), a perennial plant growing up to 2 m tall with pale pinkish purple to white flowers. It is widespread in Europe, Asia, and North America, in moist locations. The aggregated species includes several diploid, tetraploid, and octoploid minor species.

Ph. 3.12: *Valeriana officinalis* (source: Wolfgang Kiefer).

The roots of *V. officinalis* used as a medicinal herb contain 0.2–2.0% valepotriates in addition to 0.2–2.0% essential oil. The roots of *V. wallichii* DC., Indian valerian, native to the Himalayas, and *V. edulis* Nutt. ex Torr. and Gray, edible valerian, native to western and central North America, have a higher content of valepotriates and were originally used to obtain the valepotriates for therapeutic purposes.

Valepotriates (Fig. 3.4, name is formed from valeriana epoxytriacylate) are 8,10-epoxy-iridoid triester with double bonds between C-3 and C-4 (monoene valepotriates) or between C-3 and C-4 and between C-5 and C-6 (diene valepotriates). Acetic acid, isovaleric acid, β-methylvaleric acid, α-isovaleric acid, α-acetoxyisovaleric acid, β-acetoxyisovaleric acid, β-acetoxy-β-methylvaleric acid, and β-hydroxyisovaleric acid occur as acid components.

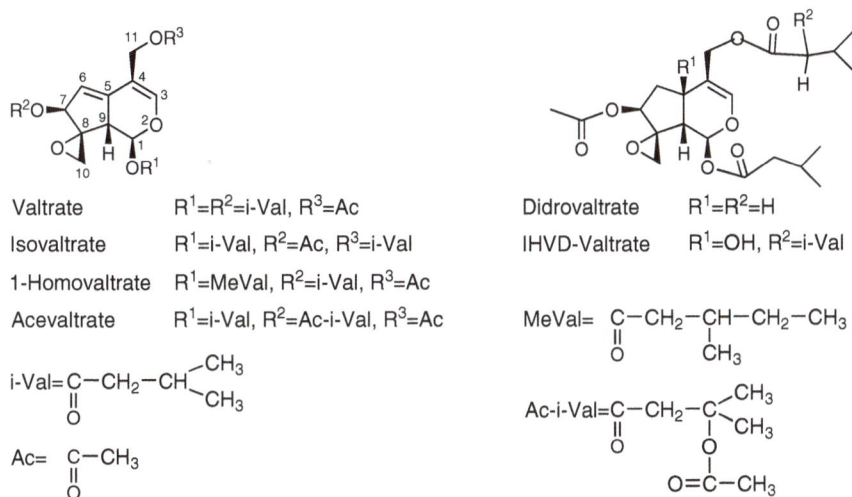

Valtrate          $R^1=R^2=$i-Val, $R^3=$Ac
Isovaltrate       $R^1=$i-Val, $R^2=$Ac, $R^3=$i-Val
1-Homovaltrate    $R^1=$MeVal, $R^2=$i-Val, $R^3=$Ac
Acevaltrate       $R^1=$i-Val, $R^2=$Ac-i-Val, $R^3=$Ac

Didrovaltrate     $R^1=R^2=$H
IHVD-Valtrate     $R^1=$OH, $R^2=$i-Val

Fig. 3.4: Valepotriates: monoterpenes of Valerianaceae.

The credit for the discovery of the valepotriates goes to the research groups of Thies [127] and of Poethke [81].

In the drug Valerianae radix, the diene valepotriates valtrate and isovaltrate constitute the major amount of valepotriates [130]. The valepotriates are very unstable and rapidly convert to degradation products without epoxide grouping, e.g., baldrinal and homobaldrinal. Accordingly, commercially available valerian tinctures do not contain valepotriates; the degradation products also disappear very quickly from the tinctures due to subsequent reactions. In tea infusions of valerian root, the valepotriates are present only in traces because of their low solubility in water. (–)-Borneyl isovalerate is the main odor carrier of the essential oil. Other components are sesquiterpenes, e.g., valerenic acid, monoterpenoid pyridine alkaloids, e.g., actinidine, lignans, e.g., 4'-O-β-glucosyl-9-O-(6"-deoxysaccharosyl)-olivil, caffeic acid derivatives, and GABA in relatively high concentrations.

*Centranthus ruber* (L.) DC. (Ph. 3.13) contains 2.5–3.0% valepotriates, predominantly valtrate and homovaltrate, in its underground organs [43].

The desired sedative effect of valerian is mainly attributed to lignans present in the hydrophilic extracts. They act agonistically at $A_1$-adenosine receptors. Valerenic acid acts on $GABA_A$ receptors [64] and is also a partial agonist at 5-HT5A receptors

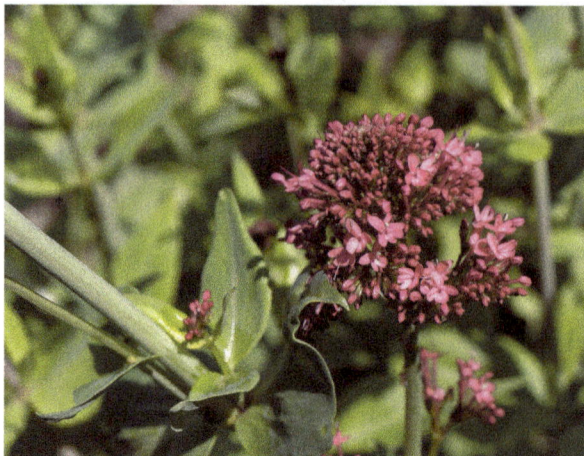

Ph. 3.13: *Centranthus ruber* (source: Wolfgang Kiefer).

[33]. In vitro studies show that isovaltrate has an inverse agonistic effect at these receptors, i.e., it partially cancels the effect of agonists [77].

The acute toxicity of valepotriates is low. In mice, the $LD_{50}$ is 4,600 mg/kg BW, p.o., and 64 mg/kg BW, i.p. [140]. However, the compounds prove to be alkylating, cytotoxic, and mutagenic in vitro. Baldrinal and homobaldrinal, the degradation products of valepotriates, also affect DNA and protein synthesis of various cell lines and cause mutations. Thus, in addition to the epoxide group present in valepotriates, functional groups first formed during biotransformation or spontaneous degradation, e.g., the α,β-unsaturated aldehyde group, can be also responsible for the mutagenic effect [135]. On the other hand, daily feeding of pregnant female rats with an ethanolic extract of *V. officinalis* on gestational days 1–8 or 8–15 showed that doses up to 65 times those used in humans had no adverse effects on fertility and fetal development of the animals [149].

A genotoxic risk to humans from prolonged use of finished medicinal products containing valerian is unlikely. Since valepotriates are only very poorly absorbed and are subject to a strong first-pass effect, the mucosa of the gastrointestinal tract or the liver would be the most likely target organs. In any case, only very small amounts of valepotriates and their degradation products are contained in finished drugs and tea extracts [60]. In the fruit fly *Drosophila melanogaster*, the preparations even show antigenotoxic activity [101].

Valerian preparations are used mainly due to their sedative activity. Symptoms of an approximately 20-fold overdose were overcome within 24 h [145]. However, preparations from the roots of *V. wallichii* and *V. edulis*, which are particularly rich in valepotriates, should not be used.

## 3.5 Monoterpenes of Seaweeds (Marine Macroalgae)

Seaweed is a term for macroscopic, multicellular benthic marine algae. They are one of the largest producers of biomass in the marine environment and constitute an important part of diet mainly in some parts of Asia [106].

For protection against microorganisms and herbivorous aquatic animals, the algae of the Chlorophyta (green algae), Phaeophyta (brown algae), and Rhodophyta (red algae) divisions have developed a variety of secondary metabolites, including compounds derived from fatty acids, some with acetylene grouping, polyketides, monoterpenes, sesquiterpenes (see Section 4.5), diterpenes (see Section 5.10), unusual amino acids, sulfur-containing compounds, and alkaloids. The monoterpene content of the seaweeds depends on the collection region.

In **green algae** and **brown algae**, the found monoterpenes were predominantly those also known from higher plants, e.g., geraniol, linalool, α-pinene, terpinolene, carvone, 1,8-cineol, and limonene (Fig. 3.1).

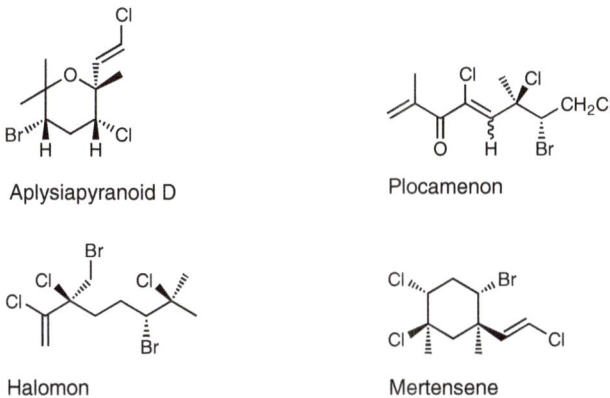

Aplysiapyranoid D

Plocamenon

Halomon

Mertensene

Fig. 3.5: Halogenated monoterpenes of Rhodophyceae.

**Red algae** produce, either via the acetate-mevalonate or the triose-pyruvate pathway, structurally and pharmacologically interesting substances in addition to the above-mentioned compounds [68]. Acyclic monoterpenes with bromine or chlorine atoms or both are likely the results of haloperoxidase action on either myrcene or ocimene [24]. Aliphatic 2,6-dimethyloctane derivatives [67, 86], cycloethers of these compounds, e.g., the aplysiapyranoids A to D [72], cyclic dimethyl-ethyl hexane derivatives, e.g., aplysiaterpenoid A [86], and cyclohexane derivatives [19] have been found in Plocamium species (Fig. 3.5). The aplysiapyranoids A to D [72] and aplysiaterpenoid A [86] were first isolated from the lumpfish *Aplysia kurodai*. These animals ingest the substances in their food and store them as defense toxins [144].

Red algae of other genera, e.g., Ochtodes, Portieria, Microcladia, Chondrococcus, Laurencia, Rhodomela, and Symphyocladia, are also capable of producing halogenated monoterpenes [10].

Monoterpenes from Plocamium and Portieria species are especially well investigated. The only species of Plocamium found in the North and Baltic Seas, *P. cartilagineum* (L.) Dixon (*P. coccineum* (Huds.) Lyngb.), common comb kelp, contains numerous halogenated monoterpenes with dimethyl-ethyl-cyclohexane parent bodies in addition to aliphatic monoterpenes [3]. Some strains of *Portieria hornemannii*, found in the Caribbean, form the aliphatic monoterpene halomon (Fig. 3.5 [48]).

Many of the halogenated monoterpenes have antimicrobial, cytotoxic, and insecticidal effects. Plocamenon (Fig. 3.5), a halogenated monoterpene ketone from Plocamium species, acts in vitro as a strong mutagen [121].

Halomon is the best-studied halogenated monoterpene. Structure–activity investigations between related halogenated monoterpenes showed that the halogen at position 6 was essential for cytotoxic activity and that a higher number of halogen atoms resulted in higher activity against breast cancer cells [24]. Halomon is under preclinical investigation as a potential antitumor agent with promising effects in a mouse brain cancer model. It acts as a DNA methyl transferase-1 inhibitor [103]. The lethal dose of halomon for mice is 200 mg/kg BW, i.v. [50].

Another candidate for drug development is the polyhalogenated monoterpene mertensene (Fig. 3.5), isolated from the red alga *Pterocladiella capillacea*. It inhibits the proliferation of human cancer cell lines through the modulation of ERK-1/-2, AKT, and NF-κB signaling pathways [125].

## References

(for numbers in bold, see cross-chapter literature)

[1]     Abad MJ et al. (2012) Molecules 17: 2542
[2]     Abdel-Kader MS et al. (2019) Saudi Pharm J 27: 945
[3]     Abreu PM, Galindro JM (1996) J Nat Prod 59: 1159
[4]     Alsaad AMS et al. (2015) BMB Compl Alt Med 15: 40
[5]     Anderson IB (1996) Ann Int Med 124: 726
[6]     Antonow N, Jotob G (1984) Nevrol Psichiat Neurochirurg 23: 281
[7]     Baerheim-Svendsen A et al. (1985) Sci Pharm 53: 159
[8]     Bakering JA et al. (1996) Pediatrics 98(5): 944
[9]     Banthorpe DV et al. (1973) Planta Med 23(1): 64
[10]    Barahona LF, Rorrer GL (2003) J Nat Prod 66(6): 743
[11]    Batiha GE et al. (2020) Antibiotics 9: 353
[12]    Bennike NH et al. (2019) Contact Dermat 80(4): 208
[13]    Bischoff K, Guale F (1998) J Vet Diagn Investig 10: 208
[14]    Božović M, Ragno R (2017) Molecules 22: 290
[15]    Bradberry SM et al. (2005) Toxicol Rev 24: 93

[16]  Brewer CT, Chen T (2017) Int J Mol Sci 18: 2353
[17]  Brieskorn CH et al. (1964) J Org Chem 29: 2293
[18]  Caruntu S et al. (2020) Molecules 25: 5416
[19]  Castedo L et al. (1984) J Nat Prod 47(4): 724
[20]  Chandra A et al. (1987) Phytochemistry 26(5): 1463
[21]  Chang LC et al. (2000) J Nat Prod 63(9): 1235
[22]  Chen W et al. (2013) Molecules 18: 5434
[23]  Chrustek A et al. (2018) Medicina 54: 61
[24]  Cikoš AM et al. (2019) Marine Drugs 17: 537
[25]  Clinical Practice Guidelines: Eucalyptus Oil Poisoning. https://www.rch.org.au/clinicalguide/guide
      line_index/Eucalyptus_Oil_Poisoning
[26]  Coucham FM, Rudloff FV (1965) Can J Chem 43: 1017
[27]  Czyzewska MM, Mozrzymas JW (2013) Eur J Pharmacol 702: 38
[28]  Darben T et al. (1998) Australas J Dermatol 39(4): 265
[29]  LM D et al (1997) Austr J Publ Health 1: 297
[30]  De Pasqual TJ et al. (1978) Am Quim 74: 1093
[31]  DeVincenzi M et al. (2002) Fitoterapia 73(3): 269
[32]  Dhakad AK et al. (2018) J Sci Food Agric 98: 833
[33]  Dietz BM et al. (2005) Brain Res Mol Brain Res 138: 191
[34]  Dittmar D, Schuttelaar MLA (2019) Contact Dermat 80(2): 101
[35]  Dobrynin VN et al. (1976) Khim Prir Soedin 5: 686
[36]  Drake ME, Stuhr ET (1935) J Am Pharm Assoc 24: 196
[37]  Ekiert H et al. (2020) Molecules 25: 4415
[38]  EMEA-The European Agency for the evaluation of medicinal products. Committee for Veterinary
      Medicinal Products-Thuja occidentalis. Summary Report, 1999
[39]  EMA/HMPC (2011) Public statement on the use of herbal medicinal products containing thujone
      http://www.ema.europa.eu/docs/en_GB/document_library/Public_statement/2011/02/
      WC500102294.pdf
[40]  EMA/HMPC. Public statement on the use of herbal medicinal products containing pulegone and
      menthofuran. In: Ageny EM (ed.) EMA/HPMC/138386/2005 Rev. 12016
[41]  European Commission. Opinion of the Scientific Committee on Food on Thujone. 2002.
      http://europa.eu.int/comm/food/fs/sc/scf/out162_en.pdf
[42]  FDA. Food Additive Regulations; Synthetic Flavoring Agents and Adjuvants. In: FDA, ed. Vol. 83 FR
      50490 Federal Register 208: 50490–50503
[43]  Fink C (1982) Dissertation, Universität Marburg, Germany
[44]  Flaman Z et al. (2001) Paediatrics Child Health 6: 80
[45]  Foti C et al. (2007) Contact Dermat 56(2): 109
[46]  Fretz TA et al. (1976) Sci Hortic 5: 85
[47]  Gallino M (1988) Planta Med 54(2): 182
[48]  Ginsburg DW, Paul VJ (2001) Mar Ecol Prog Ser 215: 261
[49]  Gorgaev MI et al. (1963) Iswesst Akad Nauk Kaz SSR Ser Chim 2: 103
[50]  Graziano MJ (1988) Mol Pharmacol 33: 706
[51]  Grimm W, Gries H (1970) Berufsdermatosen 18: 165
[52]  Halicioglu O et al. (2011) Pediatr Neurol 45(4): 259
[53]  Halpern O, Schmid H (1958) Helv Chim Acta 41: 1109
[54]  Hammer KA (2006) Food Chem Toxicol 44: 616
[55]  Harada M, Yamashita A (1969) Yakugaku Zasshi 83: 1205
[56]  Hein J, Arnold WN (1992) Vincent van Gogh: Chemicals, Crisis and Creativity. Birkhäuser, Boston.
[57]  Héthelyi E et al. (1981) Phytochemistry 20(8): 1847

[58] Höld KM et al. (2000) Proc Natl Acad Sci USA 97: 3826

[59] Holopainen M et al. (1987) Planta Med 53(3): 284

[60] Houghton RJ (1988) J Ethnopharmacol 22: 121

[61] Ittyachen AM et al. (2019) J Med Case Rep 13: 326

[62] Jabba SV, Jordt SE (2020) JAMA Intern Med 180(3): 468

[63] Khamidulina KK et al. (2006) Toksikol Vestnik 1: 42

[64] Khom S et al. (2007) Neuropharmacol 53(1): 178

[65] Kimura M et al. (1984) Jap J Pharmacol 36: 275

[66] Kocaoglu C, Ozel A (2014) J Pak Med Assoc 64: 1081

[67] König GM et al. (1999) J Nat Prod 62(2): 383

[68] Koruk ST et al. (2005) Clin Toxicol 43: 47

[69] Krivut BA, Tolstykh LP (1980) Chim Farm Shur 14: 66

[70] Kumar KJ et al. (2015) Toxicol Int 22(1): 170

[71] Kumar S (2019) Indian J Crit Care Med 23(Suppl4): S278

[72] Kusumi T et al. (1987) J Org Chem 52: 4597

[73] Kuštrak D et al. (1984) J Nat Prod 47(3): 520

[74] Lachenmeier DW et al. (2004) Dtsch Lebensm Rundsch 100: 117

[75] Lachenmeyer DW et al. (2008) J Agric Food Chem 56: 3073

[76] Lachenmeyer DW (2010) J Ethnopharm 131: 224

[77] Lacher SK et al. (2007) Biochem Pharmacol 73: 248

[78] Lanzerstorfer P et al. (2021) Arch Toxicol 95: 673

[79] Lichman BR et al. (2020) Sci Adv 6(20): eaba0721

[80] MacKinney TG et al. (2015) BMJ Case Rep. 10.1136/bcr-2014-209101

[81] Mannetstätter H et al. (1967) Pharm Weekbl 3: 284

[82] Mathew T et al. (2017) Epilepsia Open 2(3): 350

[83] Matsui R et al. (2020) Scient Rep 10: 6366

[84] Mihajilov-Krstev T et al. (2014) Planta Med 80(18): 1698

[85] Mihara S, Shibamoto T (2015) Allergy Asthma Clin Immunol 11: 11

[86] Miyamoto T et al. (1988) Liebigs Ann Chem 1191

[87] Morris MC et al. (2003) Pediatr Emerg Care 19: 169

[88] Mossa ATH et al. (2018) Bio Med Res Internat 2018: 4308054

[89] Muto T et al. (2003) J Toxicol Sci 28: 471

[90] Nakatsuka I, Hirose Y (1955) J Jap Forest Soc 37. 496, zit. C.A. 50. 7404

[91] Naser B et al. (2005) eCAM 2: 69

[92] Pandey AK, Singh P (2017) Medicines 4: 68

[93] Park T et al. (2001) Biochem Pharmacol 61: 787

[94] Pelkonen O et al. (2013) Regul Toxicol Pharmacol 65(1): 100

[95] Perkins A et al. (2016) Int J Environ Res Public Health 13: 829

[96] Pitarević I et al. (1984) J Nat Prod 47(3): 409

[97] Pitzer EM et al. (2019) Toxicol Sci 169(2): 511

[98] Proudfoot AT (2005) Toxicol Rev 24: 107

[99] Rivera E et al. (2014) Brain Res 1555: 28

[100] Rodrigues MR et al. (2012) J Ethnopharmacol 139(2): 519

[101] Romero-Jimenez M et al. (2005) Mutat Res 585: 147

[102] Römhild W et al. (1999) Monatsschr Kinderheilk 147: 35

[103] Rosa GP et al. (2019) Marine Drugs 18: 8

[104] Sacco T, Chtalva F (1988) Planta Med 54(1): 93

[105] Sakurai K et al. (1988) Agric Biol Chem 52: 2369

[106] Salehi B et al. (2019) Molecules 24: 4182

[107]   Satar S (1984) Pharmazie 39: 66
[108]   Schediwy M et al. (2015) SAT/ASMV 157(3): 147
[109]   Schempp CM et al. (2002) Hautarzt 53: 93
[110]   Schilcher H, Leuschner F (1997) Arzneim Forsch 47: 855
[111]   Semmler FW (1900) Ber Dtsch Chem Ges 33: 275
[112]   Semmler FW (1902) Chem Ber 35: 2945
[113]   Shafer TJ et al. (2005) Environ Health Perspect 113: 123
[114]   Sitaraman R, Rao G (2019) Cureus 11(9): e5734
[115]   Soderlund DM (2012) Arch Toxicol 86(2): 165
[116]   Somade OT (2019) Toxicol Rep 6: 759
[117]   Sörensen NA (1961) Pure Appl Chem 2: 569
[118]   Stafstrom CE (2007) Pediatr Neurol 37(6): 446
[119]   Stahl E, Scheu D (1967) Arch Pharm 300: 456
[120]   Stangl R, Greger H (1980) Plant Syst Evol 136: 125
[121]   Stierle DB, Sims JJ (1984) Tetrahedron Lett 25: 153
[122]   Swan EP et al. (1969) Phytochemistry 8(2): 345
[123]   Szopa A et al. (2020) Plants 9: 1063
[124]   Takaishi M et al. (2012) Mol Pain 8: 86
[125]   Tarhouni-Jabberi S et al. (2017) Mar Drugs 15: 221
[126]   Tetenyi R et al. (1975) Phytochemistry 14(7): 1539
[127]   Thies PW, Funke S (1966) Tetrahedron Lett 11: 1155
[128]   Tietze LF (1983) Angew Chem 95: 840
[129]   Tisch M et al. (2005) Am J Rhinol 19: 141
[130]   Titz W et al. (1982) Sci Pharm 50: 309
[131]   Traud J, Musche H (1983) Fresenius Z Anal Chem 315: 221
[132]   Tucker AO, Tucker SS (1988) Econ Bot 42: 214
[133]   Turkheimer FE et al (2020) Front Psychiat 11: 685
[134]   Uter W et al. (2020) Int J Environ Res Public Health 17: 2404
[135]   Vanderhude W et al. (1985) Arch Toxicol 56: 267
[136]   Vernin G et al. (1988) Phytochemistry 27(4): 1061
[137]   Vetere A et al. (2020) BMC Vet Res 16: 29
[138]   Vogt DD, Montagne M (1982) Int J Addict 17: 1015
[139]   von Eickstedt KW, Rahmann S (1969) Arzneim Forsch 19: 316
[140]   van Schantz M et al. (1966) Planta Med 14(4): 421
[141]   Wang GS (1989) J Ethnopharmacol 26: 147
[142]   Webb NJ, Pitt WR (1994) J Paediatr Child Health 30: 190
[143]   Weisbord SD et al. (1997) N Eng J Med 337: 825
[144]   Wessels M et al. (2000) J Nat Prod 63(7): 920
[145]   Willey LB et al. (1995) Vet Hum Toxicol 37: 364
[146]   Wrolstad RE, Jennings WGJ (1965) J Food Sci 30: 274
[147]   Xu J et al. (2014) Int J Clin Exp Pathol 7(4): 1495
[148]   Xu H et al. (2019) New Phytol 223: 751
[149]   Yao M et al. (2007) J Ethnopharmacol 113: 204
[150]   Yatagai M et al. (1985) Biochem Syst Ecol 13: 377
[151]   Zárybnický T et al. (2018) Int J Mol Sci 19: 1805

# 4 Sesquiterpenes

## 4.1 General

The sesquiterpenes ($C_{15}$ compounds), which are formally composed of three isoprene units, form a large group of terpenes with more than 11,000 representatives. So far, more than 100 basic compounds of this group are known. Widespread are compounds of the

– bisabolane-type, e.g., the sesquiterpenes of Matricaria species,
– germacrane-type, e.g., the germacranolide costunolide,
– guaiane-type, e.g., the guaianolide cynaropicrin,
– pseudoguajane-type, e.g., the pseudoguajanolide spathulin, and
– eudesmane-type, e.g., the eudesmanolide frullanolide [190].

In addition, representatives of the aromadendrane, cadinane, eremophilane, farnesane, and xanthane types, among others, occur in higher plants. In microscopic fungi, eremophilane and trichothecane derivatives, in basidiomycetes compounds of the drimane-, guajane-, illudane-, protoilludane-, lactarane-, secolactarane-, marasmane-, and sterpurane-type [5], and in red algae such of the chamigrane- and laurane-type were found.

A variety of volatile sesquiterpene are components of essential oils. For example, the essential oil of Myoporum species (Scrophulariaceae) derives its toxicity from the presence of toxic furanosesquiterpenes, e.g., ngaione and dehydrongaione (Fig. 4.1).

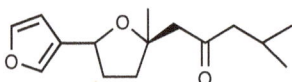

Ngaione

**Fig. 4.1:** A furanosesquiterpene of Myoporum species.

**Ph. 4.1:** Myoporum species (source: Horst Pilgrim).

https://doi.org/10.1515/9783110724738-004

🐏 Myoporum species, e.g., *Myoporum insulare* R. Br., common boobialla, *M. tetrandrum* (Labill.) Domin, boobialla, *M. laetum* G. Forst, ngaio, and *M. montanum* R. Br., water bush (Ph. 4.1), are shrubs native to Australia, New Zealand, and Pacific Islands. They are the cause of animal poisoning characterized by hepatogenic photosensitization with periportal or centrilobular hepatic degeneration and necrosis. Kidney and gastrointestinal tract are also targeted [162]. Experimental feeding of the leaves of *M. laetum* to cows and sheep resulted in anorexia, digestive disturbances, and photodermatitis. Histopathological examination revealed liver necrosis. Five of nine sheep that received 40 g of the fresh plant/kg BW per day for 10 days died [138].

The Maoris rub their skin with the leaves of *M. laetum* to protect themselves from mosquitoes and sandflies [171].

Of particular interest are toxic sesquiterpene lactones (see Section 4.2), norsesquiterpenes from ferns (see Section 4.3), sesquiterpenes in Ledum species (see Section 4.4) and in macroalgae (see Section 4.5). A dimeric sesquiterpene of the cadinane type is gossypol from the cotton plant. Since it can formally be regarded as a triterpene, gossypol is discussed in the chapter on triterpenes (see Section 6.6). Sesquiterpenes also play a role as phytoalexins, e.g., in Solanaceae and as allelochemicals (see Chapter 1).

Noteworthy is also the occurrence of sesquiterpenes in fungi and in marine animals, e.g., in water lilies (Actinaria), leather corals, sponges, sea hares, and in the defense toxins of ants, swift-swimming beetles, butterflies, and of termites.

## 4.2 Poisonous Sesquiterpene Lactones

### 4.2.1 Chemistry, Distribution, and Pharmacology

The sesquiterpene lactones, of which more than 5,000 representatives are known, possess predominantly the basic bodies eremophilane, eudesmane germacrane, guaiane, pseudoguaiane, and xanthane, whose isopropyl side chain is integrated into a fused lactone ring. The biologically active sesquiterpene lactones are almost always α-methylene-γ-lactones, that is, they have an exocyclic methylene group adjacent to the carbonyl group on the lactone ring. In addition to hydroxy groups, some of which may be esterified or glycosidically linked to sugar; epoxy groups, carboxy groups, halogen, or sulfur atoms occur in the molecule.

Sesquiterpene lactones are relatively stable, nonvolatile, lipophilic, and characterized by bitter taste. Some of them, the so-called proazulenes, react on heating in solution with opening of the lactone ring, decarboxylation, and dehydration or dehydrogenation and pass into the water vapor volatile azulenes.

The first sesquiterpene lactone to be isolated was santonin in 1830 by Kahler [79] from the flower heads of *Artemisia cina* Berg ex Poljakov, Levant wormseed, which were used as Flores Cinae, wormseed, to control intestinal worms. The structure was elucidated in 1929/30 by Clemo, Haworth, and Walton [31].

Sesquiterpene lactones are ubiquitously distributed in Asteraceae (except for the Tribus Tageteae) [43]. They occur in approximately 15 further plant families, among them Apiaceae, Amaranthaceae, Aristolochiaceae, Lauraceae, Lamiaceae, Magnoliaceae, and Menispermaceae. They are also present in numerous species in the liverworts, Frullania species. Sesquiterpene lactones play an important role in communication between plants and interaction with insects, microorganisms, and animals. They act as attractants, deterrents, and antifeedants [71].

In higher plants, the sesquiterpene lactones are formed in glandular hairs and coat the surface of the plant as exudates so that they are easily transferred to the skin or to the mucosa of the mouth of animals or humans.

The presence of α,β-unsaturated carbonyl structures, e.g., an α-methylene-γ-lactone grouping and/or an β-unsubstituted cyclopentenone ring, is primarily responsible for the diverse biological effects of sesquiterpene lactones. Depending on whether one or two of these functional groups are present, a distinction is made between mono- and bifunctional sesquiterpene lactones. The α,β-unsaturated carbonyl groups react selectively with nucleophilic groups, preferably with exposed SH groups of proteins, and alkylate them in the manner of a Michael addition (Fig. 4.2). This chemical reaction induces steric and chemical changes in enzymes, receptors, and transcriptional factors leading to diverse biological responses. Antimicrobial, cytotoxic, antineoplastic, antiphlogistic, and antihyperlipidemic effects as well as allergic reactions result. Bifunctional compounds, e.g., helenalin, are usually more potent than monofunctional ones, e.g., dihydrohelenalin.

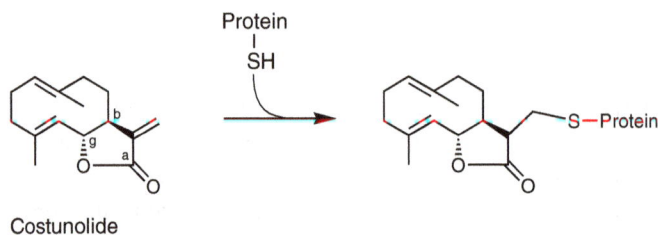

Costunolide

Fig. 4.2: Reaction of a sesquiterpene lactone with the SH group of a protein.

A well-investigated target of sesquiterpene lactones is the transcription factor NF-κB. It is a key factor of several biological processes including proliferation, differentiation, apoptosis, autophagy, and inflammation. Sesquiterpene lactones alkylate the cysteine residues 38 and 120 in the p65 subunit of the factor and prevent its activation [28, 39, 116, 155, 156, 140]. For the specificity of action of individual sesquiterpene lactones and the large differences in their bioavailability and biological activity, the conformation of the molecule, its lipophilicity, the presence of other electrophilic centers in the molecule, and other factors are decisive in addition to the reactive groups mentioned above.

Reactions with amino acids lacking SH groups, e.g., lysine, and the formation of noncovalent bonds influence the plasma protein binding of the individual compounds [180].

The cytotoxic effect of many sesquiterpene lactones is mainly based on the inhibition of enzymes catalyzing energy-producing and signal-transducing processes and on the blockage of nucleic acid and protein synthesis [121]. Helenalin as one of the best investigated sesquiterpene lactones inhibits human telomerase reverse transcriptase (hTERT) and telomerase in cancer cells and induces apoptosis through the mitochondrial apoptosis pathway. By downregulating NF-κB p65 it induces autophagic cell death [95].

The anti-inflammatory activity of sesquiterpene lactones is based on attack at multiple targets. The best studied is the inhibition of activation of NF-κB and other transcription factors, e.g., NF-AT. Sesquiterpene lactones prevent the release of IκB, an inhibitory subunit of NF-κB. The transfer of the activated factor into the nucleus and the binding to DNA cease. Cell signaling pathways are interrupted and synthesis of inflammatory cytokines cannot occur [28, 164]. Sesquiterpene lactones from *Tanacetum parthenium* inhibit platelet aggregation and prostaglandin biosynthesis in addition to preventing NF-κB activation. This explains the use of the plant in arthritis and migraine. Dehydroleucodin is the sesquiterpene lactone responsible for the gastroprotective effect of *Artemisia douglasiana* Besser ex Besser, California mugwort, a plant used in South America to treat gastric ulcers [50].

The bishemiacetal moiety of hymenoxone can convert nonenzymatically to the dialdehyde form. This reacts with amino groups of macromolecules to form a Schiff base. Hymenoxone thus not only alkylates the SH groups of enzymes but also leads to cross-links of DNA and consequently has a mutagenic effect [169].

✱ Sesquiterpene lactones are incomplete antigens, so-called haptens. By conjugation with sulfhydryl groups of proteins complete antigens are formed that may cause sensitization in individuals. On renewed contact with the allergen, allergic reactions of type IV (contact dermatitis with delayed reaction) are triggered ([72], see Section 4.2.5).

In addition to the allergizing effect, the effect of some sesquiterpene lactones on the cardiovascular system may be toxicologically significant. It has been well studied with helenalin and its esters and is initially manifested by an increase in the contractility of the heart, with arrhythmias and atrial arrest occurring at higher concentrations. The mechanism of action is presumably membrane stabilization and the resulting intervention in the regulation of intracellular calcium ion concentration. The $LD_{50}$ of helenalin in mice is 10 mg/kg BW, i.p., and 150 mg/kg BW, p.o. [72].

Some sesquiterpene lactones possess specific pharmacological effects (see Section 4.2.6).

Sesquiterpene-containing plants and isolated sesquiterpenes and their derivatives are broadly used in medicine and cosmetics, especially because of their anti-inflammatory effects. They obtain growing interest because of their promising anticancer, antihyperlipidemic, hypoglycemic, and further beneficial activities [39, 107, 116, 150].

## 4.2.2 Sesquiterpene Lactones as Toxins of Arnica Species

The genus Arnica, Asteraceae, is subdivided into about 30 species. They are distributed in the temperate regions of Europe, North America, and Asia. In Central Europe, only *Arnica montana* L., mountain arnica (Ph. 4.2), is indigenous. *Arnica chamissonis* Less, Chamisso arnica, is native to North America.

Ph. 4.2: *Arnica montana* (source: WALA Heilmittel GmbH).

The main anti-inflammatory active ingredients of *A. montana* are the helenanolide-type pseudoguajanolides present in all parts of the plant (Fig. 4.3), especially the esters of helenalin and 11α,13-dihydrohelenalin [91]. The concentration of sesquiterpene lactones in flower basket was determined to be 0.3–1%. While helenalines dominate in central European plants, dihydrohelenalines were predominantly found in plants of Spanish origin [184, 185]. Concomitant substances include polyols, thymol esters and ethers, hydroxycoumarins, phenylacrylic acids, their esters with quinic acid, lignans, the iridoid loliolid, and immunostimulatory polysaccharides [132, 143, 158]. Main components in the essential oil of rhizomes and roots are 2,5-dimethoxy-*p*-cymene, 2,6-diisopropylanisole, and thymol methyl ether [167].

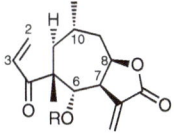

Helenalines

| R=H | (Helenalin) |

R=CO–CH₃ (Acetyl-)

$$R=CO-CH\overset{CH_3}{\underset{CH_3}{<}}$$ (Isobutyryl-)

$$R=CO-CH-CH_2-CH_3$$
$$\qquad\quad |$$
$$\qquad\quad CH_3$$ (2-Methyl-butyryl-)

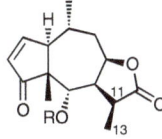

11α,13-Dihydrohelenalines

R=H (11α,13-Dihydrohelenalin)

$$R=CO-CH_2-CH\overset{CH_3}{\underset{CH_3}{<}}$$ (Isovaleryl-)

$$R=CO-C=CH_2$$
$$\qquad\quad |$$
$$\qquad\quad CH_3$$ (α-Methacryl-)

$$R=CO-C=CH-CH_3$$
$$\qquad\quad |$$
$$\qquad\quad CH_3$$ (Tigloyl-)

Fig. 4.3: Helenanolides of *Arnica montana*.

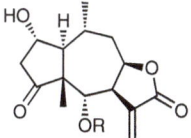

Arnifolines

$$R=CO-C\overset{E}{=}CH-CH_3$$
$$\qquad\quad |$$
$$\qquad\quad CH_3$$
(Tigloyl-)

$$R=CO-C\overset{Z}{=}CH-CH_3$$
$$\qquad\quad |$$
$$\qquad\quad CH_3$$
(Angelicoyl-)

$$R=CO-CH=C\overset{CH_3}{\underset{CH_3}{<}}$$
(Senecioyl-)

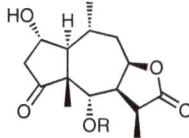

11α,13-Dihydroarnifolines

$$R=CO-C\overset{E}{=}CH-CH_3$$
$$\qquad\quad |$$
$$\qquad\quad CH_3$$
(Tigloyl-)

$$R=CO-C\overset{Z}{=}CH-CH_3$$
$$\qquad\quad |$$
$$\qquad\quad CH_3$$
(Angelicoyl-)

$$R=CO-CH=C\overset{CH_3}{\underset{CH_3}{<}}$$
(Senecioyl-)

$$R=CO-CH_2-CH\overset{CH_3}{\underset{CH_3}{<}}$$
(Isovaleryl-)

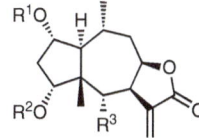

Chamissonolides

R¹, R²= H or
$$\qquad CO-CH_3$$

R³= H, OH or
$$\qquad OCO-CH_3$$

Fig. 4.4: Helenanolides of *Arnica chamissonis*.

In *A. chamissonis* ssp. *foliosa*, 0.07–1.4% sesquiterpene lactones were detected in the flower basket. The constituents show greater variability than those of *A. montana*. Additional main components are arnifolines, dihydroarnifolines, and chamissonolides. The angelic acid esters of helenalin and dihydrohelenalin and the propionic acid ester of dihydrohelenalin are present only in *A. chamissonis* (Figs. 4.3, 4.4 [92, 186, 187, **140**]).

Both the allergic skin reactions after external application of Arnica tincture or after touching the plants (see Section 4.2.5) and former poisonings during internal use as analeptic [7, 152] or misuse as abortifacient [29, 106] can be attributed to the esters of helenalin and dihydrohelenalin. The uncontrolled ingestion of tea containing Arnica flowers is also not harmless because of the narrow therapeutic range of the drug [160].

> ☠ External application of undiluted (!) Arnica tincture may cause severe skin irritation with blistering. Due to the resulting sensitization, skin rashes and other allergic symptoms may develop when Arnica-containing preparations or cosmetics containing extracts of other Asteraceae (cross-allergies!) are used several times [**65**]. When an Arnica preparation is used internally, vomiting, diarrhea, dizziness, arrhythmias, and respiratory disturbances, even death, may occur after ingestion of large amounts [160].

⫯ In a 25-year-old woman with myeloproliferative disease, a cream containing Arnica promoted the onset of Sweet's syndrome, an acute febrile neutrophilic dermatosis with necrotic changes on the face [37].

> 🚜 First measure is toxin removal and after that symptomatic treatment [**127**].

In the form of the diluted (!) tincture, Arnica is used externally for the treatment of sprains, bruises, rheumatism, and wounds. Arnica extracts are also ingredients of various cosmetics. External use of Arnica preparations must be done with caution because of the risk of existing sensitization. It should not be applied to broken skin where absorption can occur. Internal use is not recommended or should only take place under medical supervision.

### 4.2.3 Sesquiterpene Lactones as Toxins of Sneezeweed Species (Helenium Species), Rubberweed Species (Hymenoxys Species), and Geigeria Species

Helenalin-type sesquiterpene lactones have also been detected in the genus Helenium, sneezeweed, Asteraceae. Approximately 40 species are known.

**Helenium** species are native to North and South America. They are annual or perennial herbs with numerous flower heads arranged in umbel clusters. *H. autumnale* L., common sneezeweed, *H. bigelovii* A. Gray, Bigelow's sneezeweed, and some Helenium hybrids (Ph. 4.3) are popular garden flowers. *H. microcephalum* DC., smallhead sneezeweed, occurring as a weed in the southern United States, is blamed for frequently observed poisonings of grazing animals.

**Ph. 4.3:** Helenium hybrid (source: Ulrike Lindequist).

More than 30 different sesquiterpene lactones of the eudesmanolide, germacranolide, and guaianolide types, as well as their 4,5-*seco* derivatives and those of the pseu-
doguaianolide type, have been isolated from the various chemotypes of *H. autumnale*,
including a large number with an exocyclic methylene group on the lactone ring (Fig. 4.5).
Bis-sesquiterpene lactones and sulferalin, a methyl sulfonate of a pseudoguajanolide and

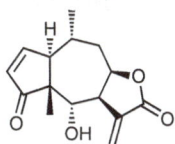

Helenalin

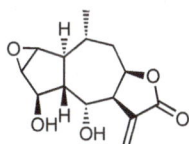

Autumnolide

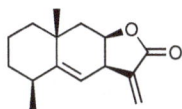

Alantolactone

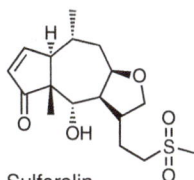

Sulferalin

**Fig. 4.5:** Sesquiterpenes of *Helenium autumnale*.

its esters, were also detected [23, 70, 88, 103, 104, 142, 161]. *H. microcephalum* contains as sesquiterpene lactones, e.g., helenalin and hymenoxone [39, 68].

The genus **Hymenoxys**, rubberweed, Asteraceae, contains about 30 species, native to North and South America. *H. odorata* DC., bitter rubberweed, *H. richardsonii* (Hook.) Cockerell, Colorado rubberweed, and *H. subintegra* Cockerell, Arizona rubberweed, thrive in the southern United States and Mexico. The plants are poisonous to livestock, especially sheep.

*Geigeria ornativa* var. *filifolia* (Mattf.) S. Ortiz and Rodr. Oubiña (*Geigeria filifolia* Mattf.) Asteraceae, is a member of the genus **Geigeria** (about 30 species) and occurs in South Africa. Like Hymenoxys species, it is responsible for significant losses to grazing animals.

Several guaianolides, e.g., florilenalin, pseudoguaianolides, e.g., hymenograndine, modified pseudoguaianolides (pentane ring extended to pyran ring by hemiacetal formation), e.g., hymenoxone, also named hymenovin (Fig. 4.6), have been isolated from *Hymenoxys odorata* [72, 73, 85], *H. richardsonii*, and *H. subintegra* [2].

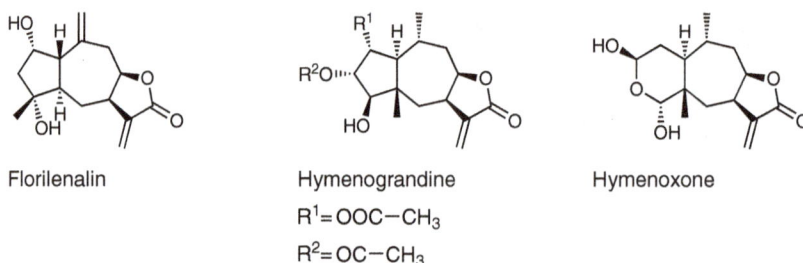

Florilenalin

Hymenograndine
$R^1 = OOC-CH_3$
$R^2 = OC-CH_3$

Hymenoxone

Fig. 4.6: Sesquiterpenes of *Hymenoxys odorata*.

In *Geigeria ornativa*, the xanthanolides gafrinin, griesenin, and dihydrogriesenin, among others, have been detected [161].

Helenalin and hymenoxone contribute significantly to the toxicity of Helenium and Hymenoxys species. Helenalin (25 mg/kg BW) or hymenoxone (30 mg/kg BW) administered to immature male ICR mice were lethally toxic to more than 60% of the animals within 6 days [108]. The $LD_{50}$ of hymenoxone is 75 mg/kg BW, p.o., for goats or sheep [73, 85]. Mutagenic activity has been detected [102].

In humans, we are not aware of any poisoning by Helenium and Hymenoxys species.

🐾 Poisonings by Hymenoxys species occur in sheep, in the United States especially frequently after dry years between December and May. They cause up to several million dollars in losses annually. The $LD_{50}$ of dried bitterweed for sheep ranges from 2.5 to 8.0 g/kg BW, p.o [145, 169]. Sheep avoid the plant, but they eat it in the absence of other forage [174].

Signs of poisoning usually develop within 1–2 weeks when sheep are grazing rubberweed-infested pasture. The poisoning is characterized by feed refusal, foaming around the nose as well as the mouth, vomiting, drooling, weakness, edema, and disturbances of blood coagulation. Inflammation of stomach, renal necrosis, and toxic hepatitis can occur. Death results from circulatory paralysis with secondary respiratory failure [174, 189].

*Geigeria ornativa*, when found en masse in South Africa, causes the poisoning known as 'vermeersiekte' in sheep and less commonly in goats and cattle. In 1954, about 50,000 animals died. Another wave of poisoning was observed in 1981 [77]. 'Vermeersiekte' can also be caused by other Geigeria species [25, 176]. Symptoms of poisoning include vomiting, bloating, stiffness, and paralysis. Histopathological examination reveals, among other things, degeneration of myofibrils.

There is no specific treatment for poisoning by the named plant species [174].

## 4.2.4 Sesquiterpene Lactones as Toxins of Lettuce Species (Lactuca Species)

The genus Lactuca, lettuce, Asteraceae, includes approximately 100 species. They are nearly worldwide distributed. About 17 species are represented in Europe and 10 in North America. *Lactuca sativa* L., lettuce, is the most often cultivated species. Varieties of this species are widely used for salads.

Of toxicological interest is *Lactuca virosa* L., bitter lettuce. It is widespread across central and southern Europe and is also found in Asia, North America, and Australia. Formerly it was cultivated as a medicinal plant to obtain Lactucarium (lettuce opium), the dried, secreted milky fluid, which was used as a cough suppressant and sleep aid.

Bitter lettuce contains up to 3.5% of the guaianolides lactucin and lactupicrin (lactucopicrin, intybin, Fig. 4.7 [13, 66, 112]) in its milky sap.

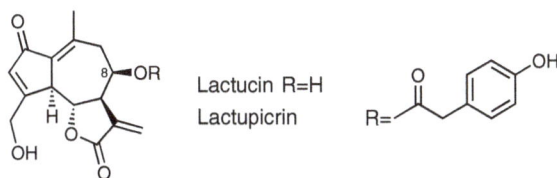

Lactucin R=H
Lactupicrin

Fig. 4.7: Sesquiterpene lactones of Lactuca species.

Lactucin and lactupicrin act via a GABAergic mechanism sedative [84]. Their toxicity is low. The lethal dose in mice is 0.5–0.6 g/kg BW, s.c [90]. Poisoning has been reported to occur with overdose of lactucarium or consumption of the leaves as vegetable [17]. The i.v. application of an extract of the herb for hallucinogenic purposes is reported [117].

♣ Symptoms of poisoning by lactucarium or leaves of *L. virosa* are sweating, ringing in the ears, dizziness, anxiety, headache, mydriasis, urinary retention, increased need for sleep, and acceleration of respiration and pulse. Hallucinations and euphoria can occur [17, **127**].

🛡 Treatment must be done symptomatically.

Lactucin and lactupicrin are also found in lower concentrations in other Lactuca species, e.g., *L. serriola* L., prickly lettuce, and *L. sativa* [83]. Furthermore, they have been detected in other species of Tribus Cichorieae (Lactucae), e.g., in *L. muralis* (L.) Gaertn. (*Mycelis muralis* (L.) Dumort.), wall lettuce, in *Cichorium intybus* L., chicory (Ph. 4.4) (0.02% in *C. intybus* L. var. *foliosum* Hegi, [82]), *C. endivia* L., endive, and in *Sonchus palustris* L., marsh sow thistle [154]. In the vegetables mentioned, the sesquiterpenes are noticeable by their bitter taste.

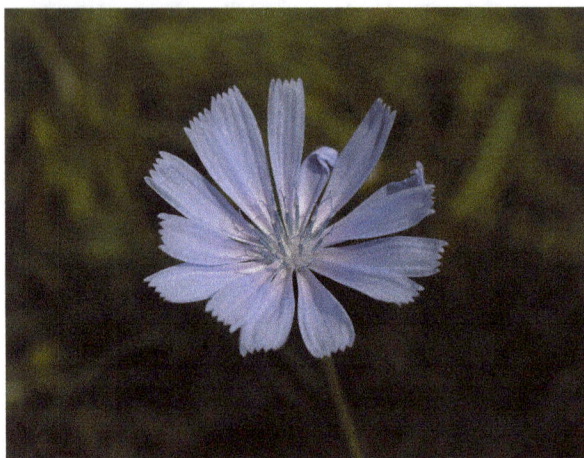

Ph. 4.4: *Cichorium intybus* (source: Wolfgang Kiefer).

Lactucin and lactupicrin have antimalarial activity [19]. Lactupicrin inhibits growth of glioma and other cancer cell types by induction of autophagy, oxidative stress, and apoptosis [144].

## 4.2.5 Sesquiterpene Lactones as Contact Allergens

Allergic diseases after repeated contact of humans with plants containing sesquiterpene lactones with an exocyclic methylene group on the γ-lactone ring have been described worldwide. In addition, those sesquiterpenes that possess other reactive structures, e.g., epoxy groups or β-unsubstituted cyclopentenone rings, also function as haptens [61]. Major allergenic plants that may be assumed to be active due to sesquiterpene lactones are summarized in Table 4.1 (for chemical structure see Figs. 4.3–4.8). They belong mainly to the Asteraceae. In Europe, Asteraceae-related allergy is among the top-10 contact sensitivities [38, 128, 153, **65**].

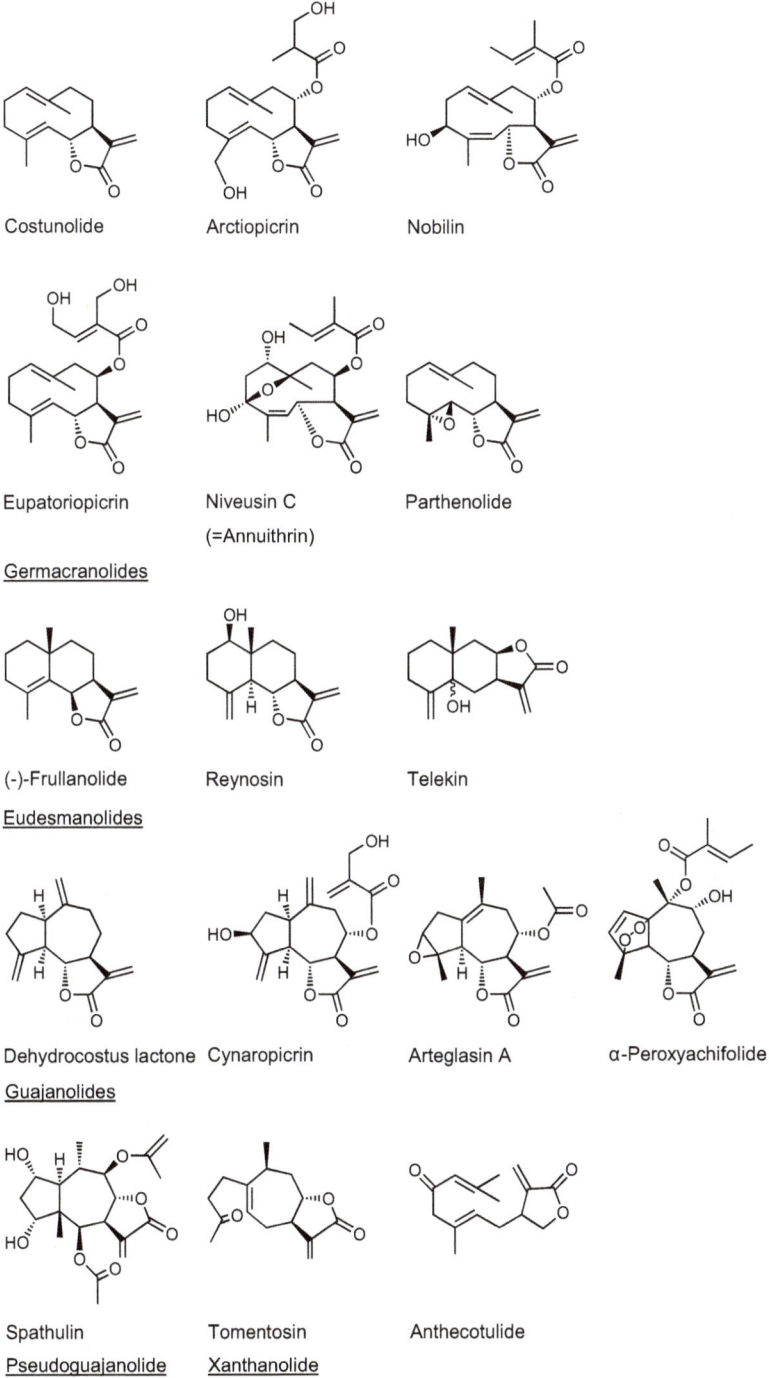

Costunolide

Arctiopicrin

Nobilin

Eupatoriopicrin

Niveusin C
(=Annuithrin)

Parthenolide

Germacranolides

(-)-Frullanolide

Reynosin

Telekin

Eudesmanolides

Dehydrocostus lactone  Cynaropicrin

Arteglasin A

α-Peroxyachifolide

Guajanolides

Spathulin
Pseudoguajanolide

Tomentosin
Xanthanolide

Anthecotulide

Fig. 4.8: Sesquiterpene lactones with sensitizing capacity (see also Figs. 4.3–4.7).

Tab. 4.1: Allergenic plants with sensitizing sesquiterpene lactones.

| Latin name | English name | Selected allergens | References |
|---|---|---|---|
| **Frullaniaceae** | | | |
| Frullania species | Liverwort | (+)-Frullanolide, costunolide | [9] |
| **Lauraceae** | | | |
| *Laurus nobilis* L. | Bay laurel | Dehydrocostus lactone, costunolide | [41, 193] |
| **Asteraceae** | | | |
| *Achillea millefolium* L. | Yarrow | α-Peroxyachifolide | [59] |
| *Ambrosia artemisiifolia* L. | Common ragweed | Artemisiifolin, ambrosin, psilostachyins | [86, 133] |
| *Anthemis cotula* L. | Stinking chamomile | Anthecotulide | [14, 109] |
| *Arctium lappa* L. | Great burdock | Arctiopicrin | [161] |
| *Arnica montana* | Mountain arnica | Helenalines, helenalinester | [184] |
| *Arnica chamissonis* | Chamisso arnica | Helenalines, helenalinester | [186] |
| *Calendula officinalis* L. | Pot marigold | Thiophene (?) | [140] |
| *Chamaemelum nobile* (L.) All. | Camomile | Nobilin | [161] |
| Chrysanthemum species | Chrysanthemum | Arteglasin A | [114] |
| *Cichorium intybus* L. ssp. *foliosum* Hegi | Chicory radiccio | Lactucin, lactupicrin | [135] |
| *C. endivia* L. | Endive | Lactucin, lactupicrin | |
| *Cynara cardunculus* L. var. *sylvestris* | Cardoon | Cynaropicrin, grossheimin | [16] |
| *Cynara cardunculus* L. var. *scolymus* | Globe artichoke | | |
| Dahlia hybrids | Dahlia | Unknown | |
| *Eupatorium cannabinum* L. | Hemp-agrimony | Eupatoriopicrin | [195] |
| Gaillardia species | Blanketflower | Spathulin, gaillardin | [161, 194] |
| *Galinsoga parviflora* Cav. | Gallant soldier | Unknown | |
| Helenium species | Sneezeweed | Florilenalin, autumnolide | [63, 105, 161] |
| *Helianthus annuus* L. | Common sun flower | Niveusin C, heliangin | [161] |
| *Inula helenium* L. | Elecampane | Alantolactone, costunolide | [22, 131, 161] |

Tab. 4.1 (continued)

| Latin name | English name | Selected allergens | References |
|---|---|---|---|
| *Iva xanthifolia* Nutt. | Marsh elder | Ambrosin, coronophilin, Ivoxanthin | **[65]** |
| *Lactuca sativa* L. | Lettuce | Lactucin, lactupikrin | [65, 101] |
| *Matricaria recutita* L. | German chamomile | Anthecotulide | [14, 59] |
| Rudbeckia species | Coneflowers | Costunolide, Reynosin | [75, 178, 179] |
| *Tanacetum parthenium* (L.) Sch. Bip. | Feverfew | Parthenolide, Reynosin | [15, 21, 54] |
| *Tanacetum vulgare* L. | Tansy | Reynosin, Parthenolide | [8, 20] |
| *Taraxacum officinale* L. | Dandelion | Taraxin acid glucoside | [139] |
| *Telekia speciosa* (Schreb.) Baumg. | Large yellow oxeye | Alantolactone, telekin | [24, 161] |
| *Xanthium strumarium* L. | California burr | Tomentosin, xanthanol | [161, 188] |
| *Zinnia elegans* Jaqu. | Youth- and old-age | Zinniolide | **[65]** |

✳ The allergies triggered by the allergens after repeated skin contact begin on the affected surfaces, hands, and forearms and later spread to the neck and face. They are characterized by erythema formation, swelling, inflammation, and rashes. Allergic rhinitis and conjunctivitis occur after repeated aerogenic contact with plant pollen. The sensitization potential of allergenic plants is influenced by their geographic origin and numerous environmental factors. The duration and frequency of contact with the allergen as well as the age and allergy susceptibility of the patient are also important for sensitization. In predamaged skin (neurodermatitis, exogenous influences, florid skin diseases), penetration of the allergens through the horny layer is facilitated and the risk of sensitization is increased [129].

Chrysanthemum and Leucanthemum species (Ph. 4.5), *Inula helenium* (Ph. 4.6), Argyranthemum species, *Tanacetum parthenium,* and *T. vulgare* (Ph. 3.3) have a high sensitization potential. Sensitization potency is medium for *Achillea millefolium, Cynara cardunculus* (Ph. 4.7), Helenium species (Ph. 4.3), *Helianthus annuus,* and *Laurus nobilis.* In the case of Frullania species (liverworts), *Calendula officinalis* (Ph. 4.8), *Chamaemelum nobile* (Ph. 4.9), *Cichorium intybus* (Ph. 4.4), *Eupatorium cannabinum* (Ph. 4.10), *Gaillardia aristata* (Ph. 4.11), *Galinsoga parviflora, Lactuca sativa, Matricaria recutita* L., *Rudbeckia hirta* (Ph. 4.12), and *Taraxacum officinale* (Ph. 4.13), the sensitizing potential is probably only weak [129, **65**].

The allergenic potential of *Arnica montana* depends on the chemical race. It is greater in Central European representatives than in Spanish ones, which have a higher proportion of dihydrohelenalins; overall, however, it is weaker than has long

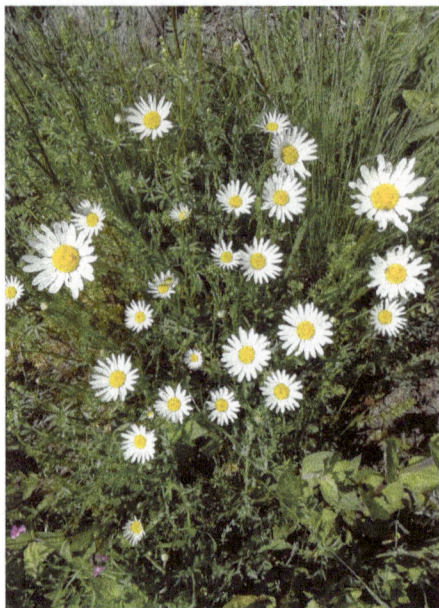

Ph. 4.5: *Leucanthemum vulgare* (source: Ulrike Lindequist).

Ph. 4.6: *Inula helenium* (source: Wolfgang Kiefer).

been assumed. Reactions to arnica may be partly due to other allergens than sesquiterpene lactones, e.g., polyacetylenes. Despite the frequent use of Arnica preparations, there are very few case reports in practice. On the other hand, in a study of 443 patients with known allergies to composite plants, 2% of the patients examined were allergic to *Calendula officinalis* [140].

Ph. 4.7: *Cynara cardunculus* (source: Wolfgang Kiefer).

Ph. 4.8: *Calendula officinalis* (source: WALA Heilmittel GmbH).

Ph. 4.9: *Chamaemelum nobile* (source: Wolfgang Kiefer).

Ph. 4.10: *Eupatorium cannabinum* (source: Wolfgang Kiefer).

*Anthemis cotula* contains about 0.7% anthecotulide and has a high allergy potential. Allergic reactions to *Matricaria recutita* which lacks anthecotulide, have been observed very rarely despite its frequent use in humans. Possibly the few cases are caused by contamination with *A. cotula* [59, 109, 146].

Quantification of allergens in Compositae mix, used in patch tests, revealed that feverfew extract contains 19%, tansy 11%, yarrow 2%, and arnica 3% of allergenic sesquiterpene lactones. There was a good correlation between estimated amounts and clinical results in patch tests [149].

Ph. 4.11: *Gaillardia aristata* (source: Eberhard Teuscher).

Ph. 4.12: *Rudbeckia hirta* (source: Ulrike Lindequist).

The highly allergenic potential of pollens of the rapidly spreading *Ambrosia artemisiifolia* is caused by peptides with IgE-binding capacity. Sesquiterpene lactones of the herb of this plant have been made responsible for nephrotoxicity in rats and on the other hand for beneficial effects on liver and on triglyceride level [86].

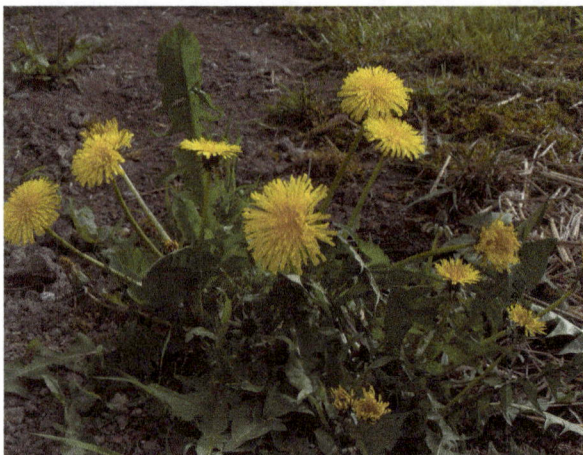

Ph. 4.13: *Taraxacum officinale* (source: Wolfgang Kiefer).

Allergic contact dermatitis develops particularly frequently during horticultural handling of the plants and during their harvesting and processing. Up to 30% of horticulturists involved in the cultivation and sale of chrysanthemums show allergies to these plants (Occupational Dermatosis [21, 38, 52, 62]). In Denmark, 10% of 250 gardeners examined with patch test showed sensitization to Asteraceae, especially chrysanthemums, daisies, and lettuce [130]. In France, artichoke allergy is recognized as an occupational disease in artichoke pickers [59]. Processing of Cynara or Cichorium species into food products may also cause sensitization. The Asteraceae *Iva xanthiifolia*, native to North America, is being studied in Europe as a potential raw material plant for obtaining biological insulation materials. Contact allergies have been observed during its processing, but rarely so far [65]. Allergies to Frullania species occur particularly in forest workers who have contact with the liverworts when felling and debarking trees (aerogenic contact dermatitis, 'woodcutter's disease' [136]).

The therapeutic use of preparations of allergen-containing plants may also be the cause of contact dermatitis. Examples are Arnica species [56, 166], bay laurel [58, 126], feverfew [60, 131], or stinking chamomile (confusion with chamomile, [64]). Further possibilities of contact with allergy plants are offered by cosmetics, skin care products, bath additives, shampoos, etc., which contain extracts from the previously mentioned Asteraceae [38].

Cross-reactions may occur between different Asteraceae with allergenic sesquiterpene lactones [21, 58, 99, 129, 152]. For example, application of Arnica-containing body oil resulted in exacerbation of a patient's contact dermatitis to a Gaillardia species (cause was spathulin [57]). Cross-reactivity between Asteraceae allergens and food allergens can occur [38].

📧 The most important prophylactic measure is the complete avoidance of contact with the corresponding allergy plants or preparations made from them by the sensitized persons. Possible cross-allergies must be considered. Treatment of allergic manifestations is usually with glucocorticoids and antihistamines [65]. Preparations containing potential allergens must never be used on previously damaged skin. A composite mix, consisting of feverfew, tansy, arnica, yarrow, and German chamomile, and a sesquiterpene lactone mix, consisting of alantolactone, costunolide, and dehydrocostuslactone, are suitable screening tools in several national patch test baseline series to prove potential allergies [149].

## 4.2.6 Sesquiterpene Lactones with Specific Pharmacological Effects

**Picrotoxinin** (Fig. 4.9 [32]) is a dilactone found together with the pharmacologically inactive picrotin (the 1:1 mixture is called picrotoxin) in the climbing shrub *Anamirta cocculus* (L.) Wight and Arn. (*A. paniculata* Colebr.), Levant berry (Menispermaceae, Ph. 4.14) native to South and Southeast Asia. Picrotoxinin is an antagonist at GABA receptors. It binds to proteins of the chloride channel coupled to the receptor and blocks it. This impairs presynaptic inhibitory mechanisms. The substance acts as an analeptic in small doses and as a convulsant in larger doses [120]. It is used, rarely today, as a respiratory analeptic in barbiturate poisoning and for therapy of peripherally induced forms of vertigo. The dried fruits of *A. cocculus*, known as Fructus Cocculi (German: Kokkelskörner), were used as a fish poison for fishing. The lethal dose to humans is about 20 mg of picrotoxinin or 2–3 g of the drug [97, 107]. Poisoning is at most medicinal.

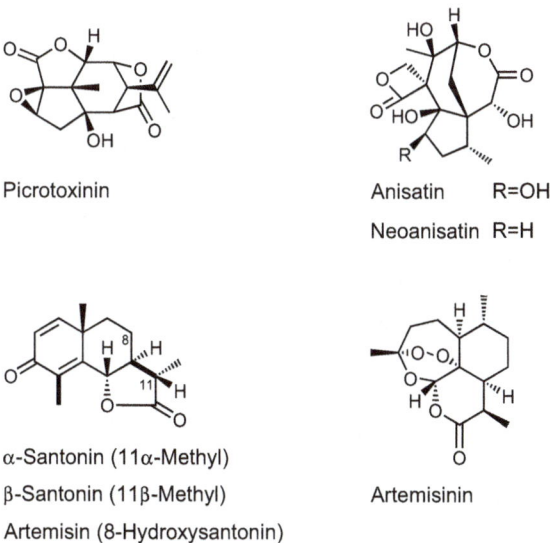

Picrotoxinin

Anisatin        R=OH

Neoanisatin  R=H

α-Santonin (11α-Methyl)

β-Santonin (11β-Methyl)

Artemisin (8-Hydroxysantonin)

Artemisinin

Fig. 4.9: Sesquiterpenes with specific pharmacological activities.

Ph. 4.14: *Anamirta cocculus* (source: Nadin Schultze).

☠ Poisonings are characterized by nausea, vomiting, miosis, bradycardia, convulsions, cyanosis, dyspnea, and finally respiratory paralysis [107].

**Anisatin** (shikimitoxin), neoanisatin (Fig. 4.9), and other diterpene lactones possessing a 8,9-*seco*-prezizaane skeleton are derived from the fruits of Japanese star anis, *Illicium anisatum* Sieb. and Zucc., in Japan also known as shikimi, Ph. 4.15), a tree native to Japan and Korea [92, 93, 119] and Florida star anise, *I. floridanum* J. Ellis, found in northern Florida and Georgia (Schisandraceae, formerly Illiciaceae [157]). The fruits of *I. lanceolatum* A.C. Smith, Guandong star anise, occurring in China, also contain highly oxidized sesquiterpenes from the same type, e.g., anisatinic acid [97]. The yield of anisatin from dried fruits of *I. anisatum* was 0.021%.

Structurally related veranisatins A–C occur in very low concentrations (<0.0002%) in the fruits of Chinese star anise, *I. verum* Hook. f., Bajiao, that are widely used for medicinal and culinary purposes [119, 181].

Anisatin and neoanisatin block the chloride ion channel of $GABA_A$ receptors and cause hyperactivity of CNS [80]. The $LD_{50}$ of anisatin ranges from 0.76 to 1 mg/kg BW, p.o. and i.p., mice. In mice, the $t\frac{1}{2}$ after oral application was 5.1 h, and the bioavailability was 22.6% [11]. The $LD_{50}$ of veranisatin A is 3 mg/kg BW, p.o., mouse [119].

Poisoning by anisatin and neoanisatin occurs when the fruits of the pharmaceutically or as spice used Chinese star anise, *I. verum*, are confused or adulterated with those of *I. anisatum* [18, 74, 76, 100, 113]. Because of the very low content of the veranisatins *I. verum* is not a hazard to humans when the drug is used in recommended doses. However, oral administration of high dose (500 mg/kg BW) of its ethyl acetate extract produced convulsions and death in mice [119, 181].

Ph. 4.15: *Illicium anisatum* (source: igaguri_1/iStock/Getty Images Plus).

♟ Symptoms of poisoning by Japanese star anise include vomiting, diarrhea, clonic–tonic convulsions, partial or complete urinary failure, and respiratory failure [113, 183].

Because of the risk of confusion with Japanese star anis, products with Chinese star anise must be carefully controlled. They are banned in some countries, e.g., in France and Spain. The FDA warns against the consumption of Chinese star anise tea. Because of safety reasons the administration of products from *I. verum* to children should be avoided [**148**].

2α-Hydroxyneoanisatin and some other sesquiterpenes from Illicium species possess neuroprotective and neurotrophic properties [49, 97].

**Santonin** (Fig. 4.9) is an eudesmanolide that exists in two forms: α-santonin and β-santonin. It occurs together with artemisin in some Artemisia and Seriphidium species, including *Seriphium maritimum* (L.) Poljakov (*Artemisia maritima* L.), sea wormwood (Ph. 4.16), Asteraceae, which thrives on the seashores of Europe. Its herb contains about 1.5% santonin. The main source of santonin, which was used as a treatment for intestinal nematode infections like ascaridosis up until the 1970s, is *Seriphidium cinum* (O. Berg) Poljakov (*Artemisia cina* O. Berg), Levante wormseed, santonica, a semishrub common in steppe areas of Central Asia.

Poisoning, including fatalities, has been observed with therapeutic use of santonin. The lethal dose for children is 60–300 mg of santonin [97].

♟ A typical symptom of poisoning by santonin is xanthopsia, i.e., bright white spots are first seen in violet, later in yellow. In addition, in severe cases, hematuria, albuminuria, temperature drop, micturition difficulties, convulsions, and unconsciousness occur [**127**].

⚕ Treatment of poisoning by the mentioned sesquiterpene lactones is by primary toxin removal and symptomatic.

Ph. 4.16: *Seriphium maritimum* (*Artemisia maritima*, source: Mantonature/iStock/Getty Images Plus).

Today, santonin and santonin derivatives are of interest because of their insecticidal, anti-inflammatory, and immunosuppressive activities [173].

**Artemisinin** is a sesquiterpene of great therapeutic interest. This secocadinane-type sesquiterpene endoperoxide is present in concentrations of 0.1–0.9% in the sweet wormwood, *Artemisia annua* L. The plant is distributed from temperate Asia to southeastern Europe, but adventitiously occurs sporadically in moderately dry, nitrogen-rich ruderal sites. *A. annua* has been used as a medicinal drug (qinghaosu) for thousands of years in China [1, 98].

After ingestion artemisinin is accumulated from the bloodstream in erythrocytes and kills the erythrocytic forms of *Plasmodium falciparum* (a protozoan unicellular parasite that causes malaria). Radicals formed during the cleavage of the endoperoxide bridge, catalyzed by the intraparasitic heme-iron, are responsible for this effect.

Toxicity studies in mice, rats, rabbits, dogs, and monkeys revealed neurotoxic, embryotoxic, genotoxic, hemotoxic, immunotoxic, cardiotoxic, and nephrotoxic effects of artemisinin and related derivatives, mostly after i.p. injection. Allergic reactions also occurred. However, neurotoxic effects in humans were observed only after long-term application of high doses. One reason for the difference may be the rapid elimination after oral intake practiced in humans. It is assumed that long-term availability rather than short-term peak concentrations of artemisinin or its derivatives cause toxicity [40, 94].

Artemisinin and some of its semisynthetic derivatives are important drugs in malaria treatment and get growing interest as potential anticancer drugs and for treatment of schistosomiasis.

## 4.3 Sesquiterpenes as Toxins of Bracken (*Pteridium aquilinum*)

*Pteridium aquilinum* (L.) Kuhn (*Pteris aquilina* L.), bracken, eagle fern, Dennstaedtiaceae (Ph. 4.17) is a forest weed widespread on all continents and absent only in the polar regions, deserts, steppes, and in mountains above the tree line. Bracken often occurs in huge stands and is currently spreading due to forest clearing caused by environmental damage.

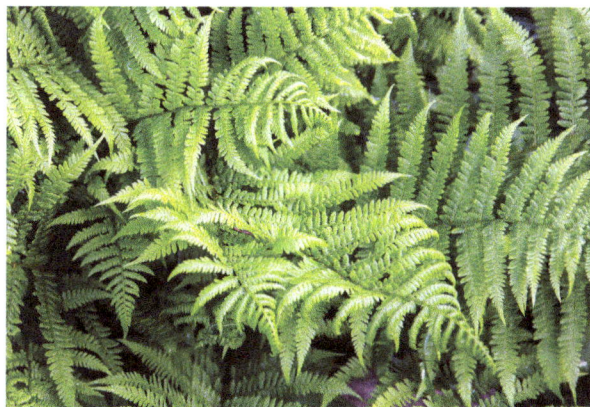

Ph. 4.17: *Pteridium aquilinum* (source: troika/iStock/Getty Images Plus).

Bracken is a perennial plant with a starch-containing rhizome that can grow to 2 m, rarely to 4 m tall. In South America, *Pteridium arachnoideum* (Kaulf.) Maxon, sometimes classified as *P. aquilinum* var. *arachnoideum* (Kaulf.) Brade, is of great toxicological importance.

The young, not up rolled shoots and the rhizomes of bracken are a widely eaten vegetable in Japan (warabi), Korea (gosari), Russian Far East, and in parts of China. Cooks detoxify the plant before eating by soaking in sodium bicarbonate solution overnight and boiling. The rhizome can be used for starch extraction. The fronds are also used as cattle feed, bedding, packing material, and for roofing [44].

Of toxicological interest are the contained carcinogenic norsesquiterpenes and the enzyme thiaminase.

In 1983, Van der Hoeven et al. [175] and Yamada and coworkers [122] independently found a very unstable glycoside of a norsesquiterpene with an illudane parent body, called aquilide A or ptaquiloside (Fig. 4.10). The yield is about 0.1–0.5% [124]. Sprouts of *P. arachnoideum* contain 1.3–1.9% ptaquiloside and 0.4–1% pterosin B. The levels in the mature green fronds ranged from 0.25% to 0.28% ptaquiloside and 0.07–0.09% pterosin B [34].

Ptaquiloside converts in alkaline environments to a conjugated dienone, which has strong alkylating properties toward a variety of nucleophilic compounds, e.g., amines, purine, and pyrimidine bases and hence nucleic acids. The final product of this

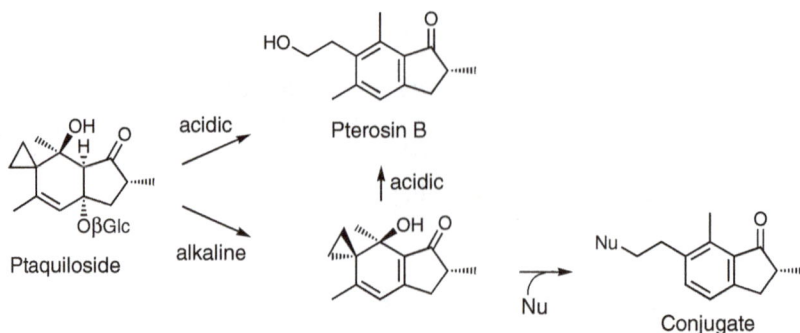

**Fig. 4.10:** Ptaquiloside of *Pteridium aquilinum* and its reaction products (Nu = nucleophilic substance).

transformation is pterosin B, which was known previously but had been shown to be nontoxic. In an acidic environment, pterosin B is formed immediately from ptaquiloside [125].

Ptaquiloside was also detected in *Pteris cretica* L., Cretan brake, Pteridaceae, native to Europe, Asia, and Africa and widely cultivated as houseplant, and *Histiopteris incisa* (Thunb.) J. Smith, bat's wing fern, Dennstaedtiaceae, widely distributed and abundant across most of the southern hemisphere. The related compounds hypacron and hypolosides A to C were found in the fern species *Hypolepis punctata* (Thunb.) Mett. and hypolosides B and C, all mutagenic, in *Dennstaedtia hirsuta* (Sw.) Mett. ex Miq. (Fig. 4.11, both Dennstaedtiaceae [148]). Ptaquiloside is also found in the ferns *Onychium contiguum* (Wall.) Hope (0.68 mg/kg DW) and *Cheilanthes farinosa* (Forssk.) Kaulf. (0.20 mg/kg DW), both Adiantaceae [137].

Accompanying these substances are noncarcinogenic pterosins (Fig. 4.11). The glycosides of these pterosins are called pterosides. More than 100 representatives of these 1-indanone derivatives with illudalan-(*seco*-illudan) basic bodies are known [115]. They occur in many other ferns, e.g., in the genera Dennstaedtia, Histiopteris, Monachosorum, Pityrogramma, Plagiogyria, and Pteris. Onitin ((2 *R*)-onitisin-14-*O*-α-D-glucoside) was also found in the common horsetail, *Equisetum arvense* L.

Caudatoside, a sesquiterpene glucoside with an illudane backbone, has been isolated from *Pteridium aquilinum* ssp. *caudatum* [27] and from *Adiantopsis chlorophylla* (Sw.) F., Pteridaceae, both native to Central and South America [159].

Other notable constituents include glycosides of *p*-hydroxy-styrene, the ptelatosides, α-ecdysone (hepatocarcinogenic in toads), tannins, and the cyanogenic glycoside prunasin [118].

The dienone formed from ptaquiloside in alkaline environments can react with DNA to form *N*-3-adenine and *N*-7-guanine adducts. Subsequent spontaneous cleavage of the *N*-glycoside bond occurs preferentially at the modified adenine residues, resulting in base-free positions, where the deoxyribose backbone is cleaved by β-elimination. The carcinogenic effect is thought to be due to mutation of the H-ras protooncogene,

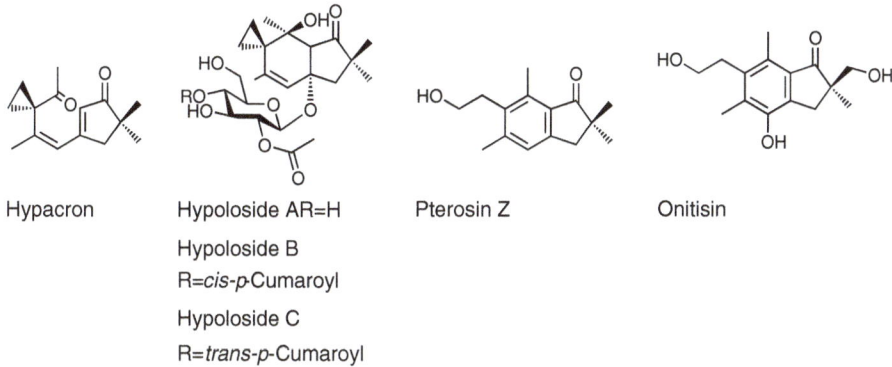

Hypacron    Hypoloside AR=H    Pterosin Z    Onitisin

Hypoloside B
R=*cis-p*-Cumaroyl

Hypoloside C
R=*trans-p*-Cumaroyl

**Fig. 4.11:** Indanone derivatives of ferns.

whose adenine residue in codon 61 is alkylated, followed by depurination [4, 134, 191]. The activation of the ATR-Chk1 DNA damage signaling pathway contributes to carcinogenicity. Pterosin B impairs the oxidation–reduction cycle of coenzyme Q in mitochondrial oxidative phosphorylation and by this way the hepatic gluconeogenic programs. Similar to pterosin A it has hypoglycemic effects [69].

A relation between ingestion of ptaquiloside and development of tumors, especially bladder cancer and upper alimentary squamous cell carcinoma, is considered accepted [34, 46]. Of 98 mice fed 200 mg of spores suspended in water by gavage, 53 developed malignancies [45]. A 7-day exposure to ptaquiloside in a feeding trial with bracken (2 mg ptaquiloside/kg BW) was sufficient to induce liver lesions and pathological precarcinogenic changes in the urinary bladder mucosa of cattle [141]. Long-term feeding of guinea pigs with food containing *Onychium contiguum* also resulted in tumor development [36].

Like some ferns and horsetails, bracken also contains thiaminase I (EC.2.5.1.2.). This enzyme transfers the pyrimidine portion of the thiamine (vitamin $B_1$) of the diet to alkaline substances, e.g., L-proline or nicotinic acid, when leaves or rhizomes of the plant are ingested and causes vitamin $B_1$ avitaminoses (Fig. 4.12). Ruminants whose rumen flora produces enough amounts of vitamin $B_1$ are not damaged [6].

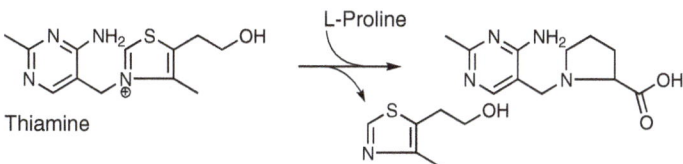

Thiamine

**Fig. 4.12:** Cleavage of thiamine catalyzed by thiaminase.

Poisoning of humans is caused by consumption of bracken as food, inhalation of the spores, drinking of contaminated water, and ingestion of sesquiterpene-containing products like milk or meat from animals that had eaten bracken [45, 48]. In Ireland, ptaquiloside was found in concentrations as great as 0.67 µg/L in drinking water from bracken stands surrounding drinking water abstraction sites [123]. Studies in Venezuela showed that when bracken was fed selectively to cows, about 8% of the ingested ptaquiloside was excreted in milk in the form of pterosin B [3]. Since bracken is not a foodstuff or a preferred animal feed in Central Europe, and since there is a high dilution of the contaminants in dairies due to the mixing of milk of different origins, the risk of poisoning is very low in Europe. At risk are owners of cows or goats in wooded areas who enjoy their animals' milk uncooked. Cooking in an alkaline environment destroys thiaminase and ptaquilosides [44].

⊗ While acute poisoning of humans by bracken is not known, epidemiological studies suggest a correlation between tumorigenesis and bracken ingestion. In the United Kingdom, ingestion of bracken ingredients during childhood has been found to increase the risk of developing stomach cancer later in life. In Costa Rica, the incidence of stomach cancer is three times greater in mountainous regions where bracken grows abundantly than in fern-free lowlands. The high incidence of esophageal cancer in Japan has also been attributed to consumption of the young fronds [4, 6, 44, 45, 48, 51, 192].

🐎 Acute poisonings by bracken are reported in animals. They occur when pastures are overstocked with bracken (>20%) or when the dried bracken is used as litter [**60**]. Free-range farming increases the risk of exposition to poisonous plants like bracken. Some of the poisoning symptoms (anorexia, neuromuscular disorders, developmental disorders) are observed only in nonruminants, e.g., horses and pigs, and can be remedied by the supply of thiamine. They are mainly explained by the thiaminase content of the fern [182]. Symptoms of poisoning caused by the ptaquilosides are also noticeable in ruminants. Cattle and sheep that have eaten bracken for several weeks develop hemorrhages, fever, leukopenia, thrombocytopenia, and blindness. Animals are more susceptible the younger they are [44, 48, 141]. *Adiantopsis chlorophylla* was a cause of hemorrhagic diathesis of cattle in Uruguay [159].

During the sporulation period of the ferns, forest workers in regions where bracken is abundant should wear protective masks [48]. If there is a hazard from drinking water from wells in the relevant areas remains unclear.

## 4.4 Aromadendrane Derivatives as Toxins of Labrador Tea (Ledum Species)

Although the genus Ledum, Labrador tea, was assigned as subsection to the genus Rhododendron in the early 1990s following the results of morphological cladistic analysis [55], it is to be treated here, according to World Flora Online [**166**], as a separate genus in the family Ericaceae.

The genus Ledum includes six species native to cool temperature and subarctic regions of the northern hemisphere. In Central Europe occur *Ledum palustre* L. (*Rhododendron tomentosum* Harmaja), wild rosemary (Ph. 4.18), and its ssp. *groenlandicum* (Oeder) Hultén, in North America also *L. glandulosum* Nutt. The mentioned species are shrubs up to 1.5 m tall with evergreen leaves. They are found on bogs, less frequently on rocky slopes.

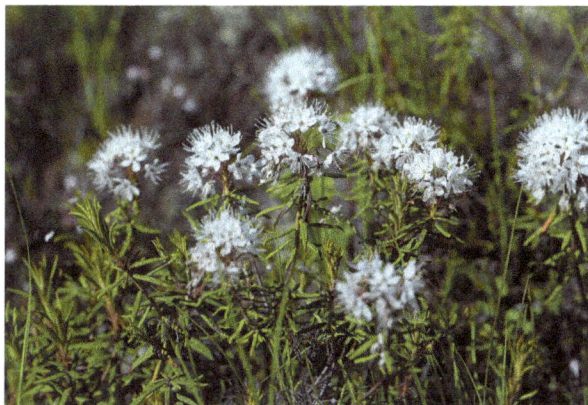

Ph. 4.18: *Ledum palustre* (source: Grigorii_Pisotckii/iStock/Getty Images Plus).

The plants contain essential oil localized in glandular hairs in all aerial parts (0.9–2.6% was obtained from leaves [53, 87, 111]). The composition is very complex and dependent on the variety or chemotype studied. About 80 compounds have been isolated. Major constituents are ledol and palustrol [78, 87, 90, 110, 177]. Grayanotoxins can be present.

Ledol (= ledum camphor) and palustrol (Fig. 4.13) are considered the active ingredients of the Labrador tea [170]. They lead initially to mild central excitation, later to paralysis [49].

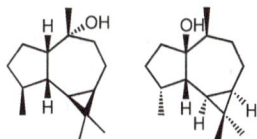

Ledol          Palustrol          Fig. 4.13: Sesquiterpenes of *Ledum palustre*.

Leaves of Ledum species are used as tea (Labrador tea) and as spice, originally by indigenous people in Northern area, now it is offered everywhere. Intoxications are caused by excessive consumption of the tea [33].

In the past (till the eighteenth century) the leaves of *L. palustre* (in German Sumpf-Porst) were added to beer ('Porstbier') to give the beer spice, to make it more intoxicating, and to get a state of agitation in drinking people [33, **49**]. Besides *L. palustre*, other plants, e.g., *Myrica gale* (Myricaceae), were also added to the beer. Such beers were named 'Grut beer'. Today, mostly *Humulus lupulus* spices the beer. Nevertheless, for several years, brewing grut beer gets increasing interest.

⚱ Symptoms after excessive consumption of ledol-containing tea or beer are vomiting, dizziness, drowsiness, convulsions, paralysis, and, in rare cases, death [33].

🐂 Lethal poisonings have been documented in livestock [33].

*L. palustre* has been used in folk medicine for healing infections, female disorders, arthrosis, and as abortifacient [78].

## 4.5 Sesquiterpenes of Seaweeds (Marine Macroalgae)

In the **green algae**, Chlorophyta, sesquiterpenes were found in 95% of the species of the order Bryopsidales (Caulerpales) in the families Caulerpaceae, Codiaceae, and Udotaceae. They have aliphatic or monocyclic, more rarely bicyclic structure and often a terminal 1,4-diacetoxybutadiene group, e.g., dihydrorhipocephalin from various Udotaceae, or a furan ring, e.g., furocaulerpin from *Caulerpa prolifera*. The 1,4-diacetoxybutadiene group and the furan ring are a masked dialdehyde constellation, which is responsible for the high biological activity of the substances (Fig. 4.14 [26, 47, 127, 165]). Similar structures can be found in diterpenes from seaweeds (see Section 5.10). It is anticipated that the 1,4-diacetoxybutadiene moiety may be regarded as a chemical defense [151].

In the **brown algae**, Phaeophyta, terpenophenols, composed of sesquiterpenes moieties and phenols, occur in addition to aliphatic, mono- or bicyclic sesquiterpenes, e.g., of the farnesane type. Examples are the farnesyl acetone epoxide from *Cystophora moniliformis*, zonarol, and the alkyl phenol farnesyl hydroquinone from *Dictyopteris undulata* (Fig. 4.14 [30, 81]).

In the **red algae**, Rhodophyta dominate substances isolated from the genus Laurencia. Unusual basic bodies, such as laurane or chamigrane, occur very frequently. In the laurane derivatives, the six-membered ring often has aromatic structure, e.g., isolaurinterol from *Laurencia filiformis* or aplysin from *L. decidua*, first isolated from *Aplysia kurodai*, a sea hare. Chamigrane derivatives can be highly reactive because of the presence of α,β-unsaturated ketone structures, e.g., in the norchamigrane derivative majusculone from *Laurencia majuscula*, or of oxirane rings, e.g., in a chamigran epoxide from *L. glomerata* [42, 168]. Sesquiterpenes from *Laurencia obtusa*, e.g., brasilenol, exhibit the unusual brasilane backbone [67].

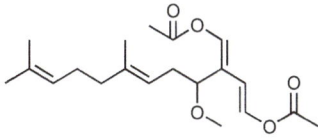

Dihydrorhipocephalin

<u>Chlorophyta</u>

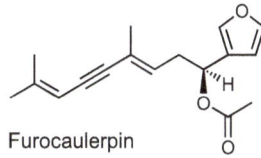

Furocaulerpin

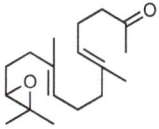

Farnesyl acetone epoxide

<u>Phaeophyta</u>

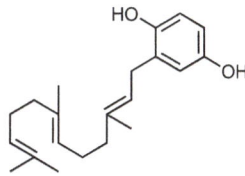

Farnesyl hydroquinone

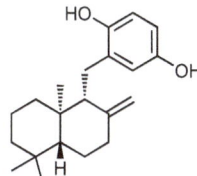

Zonarol

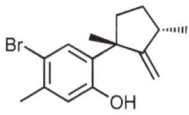

Isolaurinterol

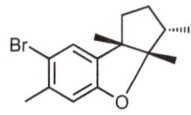

Aplysin

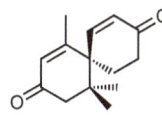

Majusculone

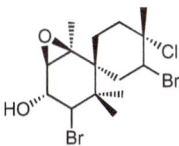

Chamigran epoxide

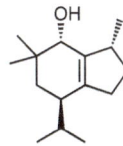

Brasilenol

<u>Rhodophyta</u>

Fig. 4.14: Sesquiterpenes of macroalgae.

Some of the above-mentioned compounds possess potent antimicrobial, cytotoxic, anthelmintic, and ichthyotoxic properties [10, 12, 35, 89, 147, 172]. Zonarol shows anti-inflammatory and neuroprotective properties [163]. Aplysin stabilizes intestinal barriers, regulates gut microbial composition, retards pancreatic necrosis [96], and suppresses tumor growth in animal assays [196].

We know nothing about the toxicity of sesquiterpenes from macroalgae to humans.

# References

(for numbers in bold see cross chapter literature)

[1]    Acton N, Klayman DL (1985) Planta Med 51(5): 441
[2]    Ahmed AA (1995) Phytochemistry 39(5): 1127
[3]    Alonso-Amelot ME et al. (1996) Nature 382: 587
[4]    Alonso-Amelot ME, Avendano M (2002) Curr Med Chem 9: 675
[5]    Andina D et al. (1980) Phytochemistry 19(1): 93
[6]    Anonym (1986) IARC Monographs on the Evaluation of the Carcinogenic Risk of Chemical to Humans 40. Lyon
[7]    Anonym (2001) Int J Toxicol 20(Suppl 2): 1
[8]    Appendino G, Gariboldi JR (1982) Phytochemistry 2l(5): 1099
[9]    Asakawa Y (1982) Prog Chem Org Nat Prod 42: 1
[10]   Bansemir A et al. (2004) Chem Biodivers 1: 463
[11]   Bao X et al. (2020) BioMed Res Int 2020: 8835447
[12]   Barnekow DE et al. (1989) J Am Chem Soc 111: 3511
[13]   Barton DHR, Narayanan CR (1958) J Chem Soc 1958: 963
[14]   Baruah RN et al. (1985) Planta Med 51(6): 531
[15]   Begley MJ et al. (1989) Phytochemistry 28(3): 940
[16]   Bernhard H (1982) Pharm Acta Helv 57: 179
[17]   Besharat S et al. (2009) BMJ Case Rep Bcr06 2008: 0134
[18]   Biessels GJ et al. (2002) Nederlands Tijdschrift Voor Geneeskunde 146(17): 808
[19]   Bischoff TA et al. (2004) J Ethnopharmacol 95: 455
[20]   Blaszyk E, Drozdz B (1978) Acta Soc Bot Polon 47: 3
[21]   Bleumink E et al. (1976) Contact Dermat 2(2): 81
[22]   Bohlmann F, Zdero C (1977) Phytochemistry 16(8): 1243
[23]   Bohlmann F et al. (1978) Phytochemistry 17(7): 1165
[24]   Bohlmann F et al. (1981) Phytochemistry 20(8): 1891
[25]   Botha CJ et al. (1997) J S Afr Vet Assoc 68: 97
[26]   Capon RJ et al. (1981) Aust J Chem 34: 1775
[27]   Castillo UF et al. (2003) J Agric Food Chem 51: 2559
[28]   Chadwick M et al. (2013) Int J Mol Sci 14: 12780
[29]   Ciganda C, Laborde A (2003) J Toxicol Clin Toxicol 41: 235
[30]   Cimino G (1975) Experientia 31: 1250
[31]   Clemo GR et al. (1930) J Chem Soc 1110
[32]   Conroy H (1957) J Am Chem Soc 79: 5550
[33]   Dampc A, Luczkiewicz M (2014) J Sci Food Agric 95(8): 1577
[34]   Da Silva Freitas R et al. (2020) Toxins 12: 288
[35]   Davyt D et al. (2001) J Nat Prod 64(12): 1552
[36]   Dawra RK et al. (2001) Vet Res Commun 25(5): 413
[37]   Delmonte S et al. (1998) Dermatology 197: 195
[38]   Denisow-Pietrzyk M et al. (2019) Environm Sci Pollution Res 26: 6290
[39]   Drogosz J, Janecka A (2019) Curr Drug Targets 20(4): 444
[40]   Efferth T, Kaina B (2010) Crit Rev Toxicol 49(5): 405
[41]   El-Feraly FS, Benigni DA (1980) J Nat Prod 43(3): 527
[42]   Elsworth JF, Thomson RH (1989) J Nat Prod 52(4): 893
[43]   Emerenciano V de P et al. (1985) Biochem Syst Ecol 13(2): 145
[44]   Evans IA (1987) Rev Environ Health 7: 161

[45]    Evans IA, Galpin OP (1990) Lancet 335: 231
[46]    Faccin TC et al. (2018) PLoS ONE 13(9): e0204656.
[47]    Fenical W, Paul VJ (1984) Hydrobiologia 116: 135
[48]    Fenwick GR (1989) J Sci Food Agric 46: 147
[49]    Fukuyama Y et al. (2020) J Nat Med 74: 648
[50]    Giordano OS et al. (1990) J Nat Prod 53(4): 803
[51]    Gomes J et al. (2012) Toxicol Sci 126(1): 60
[52]    Gonçalo S et al. (1996) Contact Dermat 35(5): 310
[53]    Greve P (1938) Dissertation, Universität Hamburg, Germany
[54]    Groenewegen WA et al. (1986) J Pharm Pharmacol 38: 709
[55]    Harmaja H (1991) Ann Bot Fenn 28: 171
[56]    Hausen BM (1980) Hautarzt 31: 10
[57]    Hausen BM (1985) Dermatosen 33: 62
[58]    Hausen BM (1985) Dtsch Med Wochenschr 110: 634
[59]    Hausen BM (1991) Dtsch Apoth Ztg 131: 987
[60]    Hausen BM, Osmund PE (1983) Acta Derm Venerol 63: 308
[61]    Hausen BM, Schmalle HW (1985) Contact Dermat 13(5): 329
[62]    Hausen BM, Schulz KH (1973) Berufsdermatosen 21: 119
[63]    Hausen BM, Spring O (1989) Contact Dermat 20(5): 326
[64]    Hausen BM et al. (1984) Planta Med 50(3): 229
[65]    Hausen BM et al. (1986) Contact Dermat 15(4): 246
[66]    Hendrickson JB (1963) Tetrahedron 19: 1387
[67]    Iliopoulou D et al. (2002) Org Lett 4: 3263
[68]    Imakura Y et al. (1980) J Pharm Sci 69(9): 1044
[69]    Itoh Y et al. (2016) Biochem Biophys Res Comm 473(2): 415
[70]    Itoigawa M et al. (1981) Yakugaku Zasshi 101: 605
[71]    Ivanescu B et al. (2015) J Anal Meth Chem 2015: 247685
[72]    Ivie GW, Witzel DA (1983) in: **77**, p. 543
[73]    Ivie GW et al. (1975) J Agric Food Chem 23(5): 841
[74]    Ize-Ludlow D et al. (2004) J Am Med Assoc 291: 562
[75]    Jakupovic J et al. (1986) Lieb Ann Chem 8: 1474
[76]    Johanns ES et al. (2002) Ned Tijdeschr Geneesk 146: 813
[77]    Joubert JPJ (1983) J S Afr Vet Assoc 54: 255
[78]    Judzentiene A et al. (2020) Molecules 25: 1676
[79]    Kahler (1830) Arch Pharm 34: 318
[80]    Kakemoto E et al. (1999) Biochem Pharmacol 58(4): 617
[81]    Kazlauskas R et al. (1978) Experientia 34: 156
[82]    Kilger F (1967) Dissertation. München, Germany
[83]    Kim H et al. (2017) Food Sci Biotechnol 26(3): 807
[84]    Kim H et al. (2019) Biol Pharm Bull 42: 1726
[85]    Kim KL et al. (1975) Res Commun Chem Pathol Pharmacol PMID 1179033
[86]    Kiss T et al. (2017) PLoS ONE 12(5): e0176818
[87]    Klokova MV et al. (1981) Khim Prir Soedin 802
[88]    Kondo Y et al. (1977) Tetrahedron Lett 18: 2155
[89]    König GM, Wright AD (1997) J Nat Prod 60: 967
[90]    Konovalova OA, Raybalko KS (1987) Rastit Resur 23: 295
[91]    Kos O et al. (2005) Planta Med 71(11): 1044
[92]    Kouno I et al. (1988) Chem Pharm Bull (Tokyo) 36: 2990
[93]    Kouno I et al. (1991) Phytochemistry 30(1): 351

[94] Li Q, Hickman M (2011). Toxicology 279(1-3): 1–9
[95] Lim CB et al. (2012) BMC Compl Altern Med 12: 93
[96] Liu Y et al. (2020) Mediators Inflammation 2020: 1280130.
[97] Liu YL et al. (2020) Phytochemistry 172: 112281
[98] Luo XD, Shen CC (1987) Med Res Rev 7: 29
[99] Machet L et al. (1993) Contact Dermat 28(3): 184
[100] Madden GR et al. (2012) Pediatric Emerg Care 28(3): 284
[101] Mahmoud ZE et al. (1986) Phytochemistry 25(3): 747
[102] Manners GD (1978) Toxicol Appl Pharmacol 45(2): 629
[103] Matusch R, Haeberlein H (1987) Liebigs Ann Chem 5: 455
[104] McPhail AT et al. (1975) Tetrahedron Lett 1229
[105] Melek FR et al. (1985) Phytochemistry 24(7): 1537
[106] Merdinger O (1938) Münch Med Wochenschr 85: 1469
[107] Merfort I (2011) Curr Drug Targets 12(11): 1560
[108] Merrill JC et al. (1988) J Toxicol Environ Health 23(2): 159
[109] Meyer A et al. (2005) J Nat Prod 68(3): 432
[110] Michailova NS et al. (1979) Khim Prir Soedin (3): 322
[111] Michailova NS et al. (1980) Rastit Resur 16: 393
[112] Michl H, Högenauer G (1960) Monatsh Chem 91: 500
[113] Minodier P et al. (2003) Arch Pediatr 10: 619
[114] Mladenova K et al. (1985) Planta Med 51(3): 284
[115] Mohammad RH et al. (2016) Phytochemistry 128: 82
[116] Moujir L et al. (2020) Appl Sci 10: 3001.
[117] Mullins ME, Horowitz BZ (1998) Vet Hum Toxicol 40: 290
[118] Murakami T, Tanaka N (1988), Prog Chem Organic Nat Prod 54: 1,51,104,176,302
[119] Nakamura T et al. (1996) Chem Pharm Bull (Tokyo) 44: 1908
[120] Napias C et al. (1980) Nature 283: 298
[121] Narasimhan TR et al. (1989) Gen Pharmacol 20: 681
[122] Niwa H et al. (1983) Tetrahedron Lett: 24: 5371
[123] O'Driscoll C et al. (2016) Molecules 21: 543
[124] Ojika M et al. (1985) J Nat Prod 48(4): 634
[125] Ojika M et al. (1987) Tetrahedron 43(22): 5261
[126] Özden MG et al. (2001) Contact Dermat 45(3): 178
[127] Paul VJ, Fenical W (1987) Biorg Mar Chem 1: 1
[128] Paulsen E (2002) Contact Dermat 47(4): 189
[129] Paulsen E et al. (1998) Contact Dermat 38(3): 140
[130] Paulsen E et al. (2001) Contact Dermat 45(4): 197
[131] Pazzaglia M et al. (1995) Contact Dermat 33(4): 267
[132] Pinchon TM, Pinkas M (1988) Plant Med Phytother 22: 124
[133] Porter TH et al. (1970) Phytochemistry 9(1): 199
[134] Prakash AS et al. (1996) Nat Toxins 4(5): 221
[135] Pyrek JS (1985) Phytochemistry 24(1): 186
[136] Quirce S et al. (1994) Contact Dermat 30(2): 73
[137] Raja Sekaran D et al. (2004) Indian J Anim Sci 74: 11
[138] Raposo JB et al. (1998) Vet Hum Toxicol 40: 132
[139] Rauwald HW, Huang JT (1985) Phytochemistry 24(7): 1557
[140] Reider N et al. (2001) Contact Dermat 45(5): 269
[141] Reis Aranha Dos PC et al. (2019) PLoS ONE 14(6): e0218628.
[142] Romo J, Romo de Vivar A (1967) Fortschr Chem Org Naturst 25: 90

[143]  Rosetti V et al. (1984) Int J Crude Drug Res 22: 53

[144]  Rotondo R et al. (2020 Molecules 25: 5843

[145]  Rowe LD et al. (1973) Southwest Vet 26: 287

[146]  Rudzki E, Jalońska S (2000) J Dermatol Treatment 11: 161

[147]  Ryu G et al. (2002) Nat Prod Sci 8: 103

[148]  Saito K et al. (1990) Phytochemistry 29(5): 1475

[149]  Salapovic H et al. (2013) Scientia Pharm 81: 807

[150]  Salazar-Gómez A et al. (2020) South Afr J Bot 135: 240

[151]  Salehi B et al. (2019) Molecules 24: 4182

[152]  Schelenz C (1938) Münch Med Wochenschr 85: 1791

[153]  Schempp CM et al. (2002) Hautarzt 53: 93

[154]  Schmidt A (1940) Botan Arch 40: 516

[155]  Schmidt TJ (1999) Curr Org Chem 3: 577

[156]  Schmidt TJ (2018) In: Sesquiterpene Lactones: Advances in Their Chemistry and Biological Aspects. Sülsen V, Martino V (eds). Springer Int Publ Cham Switzerland. P. 349

[157]  Schmidt TJ et al. (1998) J Nat Prod 61(2): 230

[158]  Schmidt TJ et al. (2006) Nat Prod Res 20: 443

[159]  Schneider de Olivera LG et al. (2020) Toxicon X5: 100024

[160]  Schoenemann H (1938) Münch Med Wochenschr 85: 787

[161]  Seaman FC (1982) Bot Rev 48(2): 121

[162]  Seawright AA et al. (1977) in: **63**, p. 241

[163]  Shimizu H et al. (2015) Biochem Biophys Res Commun 457(4): 718

[164]  Siedle B et al. (2004) J Med Chem 47: 6042

[165]  Smirniotopoulos V et al. (2003) J Nat Prod 66(1): 21

[166]  Spettoli E et al. (1998) Am J Contact Dermat 9: 49

[167]  Sugier P et al. (2020) Molecules 25: 1284

[168]  Suzuki M et al. (1987) Bull Chem Soc Jpn 60: 3795

[169]  Sylvia VL et al. (1987) Cell Biol Toxicol 3: 39

[170]  Tattje DHE, Bos R (1981) Planta Med 41(3): 303

[171]  The Encyclopedia of New Zealand. https://teara.govt.nz/en/sandflies-and-mosquitoes/page-3 (16.05.2021)

[172]  Topcu G et al. (2003) J Nat Prod 66(11): 1505

[173]  Trendafilova A et al. (2021) Foods 10: 65

[174]  USDA (U.S. Department of Agriculture): www.usda.gov

[175]  Van der Hoeven JCM et al. (1983) Carcinogenesis 4: 1587

[176]  Van der Lugt JJ, van Heerden J (1993) J S Afr Vet Assoc 64: 82

[177]  Van Schantz, Hiltunen R (1971) Sci Pharm 39: 137

[178]  Vasquez M et al. (1988) Phytochemistry 27(7): 2195

[179]  Vasquez M et al. (1990) Phytochemistry 29(2): 561

[180]  Wagner S et al. (2004) Planta Med 70(3): 227

[181]  Wang GW et al. (2011) J Ethnopharmacol 136: 10

[182]  Waret-Szkuta A et al. (2021) Porcine Health Management 7: 2

[183]  Wie DD et al. (2014) Mol BioSystems 10(11): 2923

[184]  Willuhn G et al. (1983) Planta Med 49(12): 226

[185]  Willuhn G et al. (1984) Planta Med 50(1): 35

[186]  Willuhn G et al. (1985) Planta Med 51(5): 398

[187]  Willuhn G et al. (1990) Planta Med 56(1): 111

[188]  Winters TE et al. (1969) J Org Chem 34: 153

[189]  Witzel DA et al. (1977) Vet Pathol 14: 73

[190]  Wu QW et al. (2006) Nat Prod Rep 23: 693
[191]  Yamada K et al. (1998) Angew Chem 110: 1918
[192]  Yamada K et al. (2007) Nat Prod Rep 24: 798
[193]  Yoshikawa M et al. (2000) Bioorg Med Chem 8: 2071
[194]  Yu SG et al. (1988) Phytochemistry 27(9): 2887
[195]  Zdero C, Bohlmann F (1987) Planta Med 53(2): 169
[196]  Zhang X et al. (2017 Oncotarget 8(38): 63923

# 5 Diterpenes

## 5.1 General

Diterpenes possess a formal chemical structure which consists of four isoprene units ($C_{20}$ compounds). Besides the aliphatic diterpenes of the 2,6,10,14-tetramethyl-hexadecane type, cyclic representatives are found preferentially. Abietane, kaurane, pimarane, and labdan frequently occur as hypothetical basic compounds in higher plants. Many toxicologically significant diterpenes of plants are derivatives of andromedane, daphnane, ingenane, taxane, and tigliane. In animals, derivatives of cembrane, kempane, rippertane, trinervitane, and other unusual ring systems are found. In many diterpenes, the methylation pattern of the parent compounds is obliterated by methyl group migration, ring closure, and ring cleavage [109].

The number of volatile diterpenes is small. Consequently, they are of minor importance as constituents of essential oils. Nonvolatile diterpenes occur as components of the resins of conifers, e.g., abietic acid. Other representatives are plant hormones, e.g., gibberellins. Abietane and labdan-type diterpene lactones are the bitter compounds of the Lamiaceae. Vitamin A and retinin may be considered formally chemically as diterpenes.

Of toxicological interest are the derivatives of andromedane as toxins of the Ericaceae (see Section 5.2), tigliane, ingenane, and daphnane derivatives as toxins of the Euphorbiaceae and Thymelaeaceae (see Section 5.3), the taxane derivatives as toxins of Taxus species (see Section 5.4), the opioid compounds of *Salvia divinorum* (see Section 5.5), forskolin as toxin of Coleus and Plectranthus species (see Section 5.6), hepatotoxic diterpenes in Teucrium and further plant species (see Section 5.7), diterpene glycosides as toxins of Carlina, Xanthium, Callilepis, and Weidelia species (see Section 5.8), ryanodine as toxin of Ryania species (see Section 5.9), and a variety of other diterpenes considered as possible toxins of macroalgae (see Section 5.10). Diterpenes have also been found in basidiomycetes, sponges, corals, or termites.

## 5.2 Andromedane Derivatives as Toxins of Heather Family (Ericaceae)

### 5.2.1 Distribution

The family Ericaceae, heather family, includes about 100 genera with about 2,500 species. Because of their content of andromedane derivatives, the genera Rhododendron (about 1,000 species and many hybrids, very complex taxonomy), Andromeda (3 species), Chamaedaphne (1 species), Kalmia (10 species), Pieris (7 species), Lyonia (about

https://doi.org/10.1515/9783110724738-005

35 species), Leucothoe (about 50 species), and Craiobiodendron (5 species) are of toxicological interest.

Ericaceae plants are dwarf shrubs, bushes, or trees with undivided, often evergreen leaves.

**Rhododendron** species are not only distributed in Asia but also in Europe, North America, and Australia. The two red-flowered species *Rhododendron hirsutum* L., hairy alpenrose, and *Rh. ferrugineum* L., rusty-leaved alpenrose, are native to Central Europe. They are mainly found in the Alps. While *Rh. hirsutum* has green leaves on both sides, those of *Rh. ferrugineum* are rust-colored underside. *Rh. ponticum* L., pontian (Ph. 5.1), native to South Europe and West Asia, has become invasive in Ireland and the United Kingdom.

A high number of Rhododendron species and their hybrids originate from Asia. They are grown as ornamental shrubs in many parts of the world. Among them are *Rh. luteum* Sweet, common yellow azalea, deciduous, golden-yellow flowering, *Rh.* × *obtusum* (Lindl.) Planch, deciduous to evergreen, white, blue, or red flowering, *Rh. kiusianum* Makino, Kyushu azalea, evergreen, pink, or purple flowering. *Rh. catawbiense* Michx., catawba rhododendron (Ph. 5.2), is native to the eastern United States. *Rh. simsii* Planch., Formosa azalea, native to China, is the stem form of so-called azaleas serving as houseplants. Rhododendron species are used in traditional medicine of several countries, e.g., to treat bronchitis, rheumatism, and other disturbances. Tartars used a tee of the flowers of *Rh. aureum* Georgi, Sibirian rhododendron, for ritual purposes [219]. Rhododendron is the national flower of Nepal and the symbolic flower of some states or provinces in the USA, India, and China.

Ph. 5.1: *Rhododendron ponticum* (source: Olena Lialina/iStock/Getty Images Plus).

Ph. 5.2: *Rhododendron catawbiense* (source: ADELART/iStock/Getty Images Plus).

Ph. 5.3: *Andromeda polifolia* (source: Grigorii_Pisotckii/iStock/Getty Images Plus).

*Andromeda polifolia* L., common bog rosemary (Ph. 5.3), is native to northern parts of the northern hemisphere. It is often found on Sphagnum bogs. The reddish flowers, standing in umbel clusters, are radial and globular-ovate. The plant grows to a maximum height of 30 cm.

*Chamaedaphne calyculata* (L.) Moench, leatherleaf, is the only representative of the genus Chamaedaphne. It is very similar to Andromeda but differs in having solitary flowers in the leaf axils and rust-colored leaves underneath. It is distributed throughout the cool temperature and subarctic regions of the northern hemisphere. It is commonly seen in cold, acidic bogs and grows in the form of large colonies.

**Kalmia** species are native to North America. Especially *Kalmia angustifolia* L., sheep laurel, and *K. latifolia* L., mountain laurel (Ph. 5.4), with red flowers 2–4 cm across, are cultivated as ornamental shrubs.

Ph. 5.4: *Kalmia latifolia* (source: seven/iStock/Getty Images Plus).

Ph. 5.5: *Pieris japonica* (source: Gert-Jan van Vliet/iStock/Getty Images Plus).

Plants of the genus **Pieris** are native to mountain regions of eastern and southern Asia and to eastern North America. The white-flowering shrubs *P. japonica* (Thunb.) D. Don ex G. Don, Japanese pieris (Ph. 5.5), *P. floribunda* (Pursh ex Sims) Benth. and Hook., fetterbush, and a hybrid between the American and Japanese species called 'Brouwer's Beauty' are cultivated as ornamental plants.

**Lyonia** species are shrubs and trees native to Asia and North America.

**Leucothoe** species are distributed in North and South America, Asia, and on Madagaskar. *Leucothoe fontanesiana* (Steud.) Sleumer, drooping laurel, is also found in parks in Europe.

The distribution of **Craiobiodendron** species is limited to southeast Asia.

## 5.2.2 Chemistry and Nomenclature

The toxic diterpenes of the Ericaceae, of which more than 30 are known, have an andromedane basic body (also known as a grayanan). It may carry up to eight hydroxy groups, some of which may be esterified with acetic acid, less commonly with propionic acid or lactic acid. Glucosides also occur. The oxygen function always present at C-3 can also be present as a C-2,3-epoxy group (Fig. 5.1).

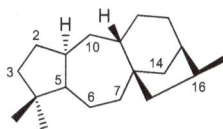

| Name | \[Structure Substituents\] 2β | 3β | 5β | 6β | 7α | 10β | 10α | 14β | 16 | \[LD$_{50}$ mg/kg BW\] a | b |
|---|---|---|---|---|---|---|---|---|---|---|---|
| Grayanotoxin I | H | OH | OH | OH | H | CH$_3$ | OH | OAc | OH | 1.3 | 1.3 |
| Grayanotoxin II (Andromedenol) | H | OH | OH | OH | H | =CH$_2$ | | OH | | | 26.1 |
| Grayanotoxin III (Andromedol) | H | OH | OH | OH | H | CH$_3$ | OH | OH | OH | 0.4 | 0.8 |
| Grayanotoxin IV | H | OH | OH | OH | H | =CH$_2$ | | OAc | OH | 3.1 | |
| Grayanotoxin XVIII | H | OH | OH | OH | H | =CH$_2$ | | H | OH | | |
| Rhodojaponin I | –O– | | OH | OAc | H | CH$_3$ | OH | OAc | OH | 3.1 | |
| Rhodojaponin III | –O– | | OH | OH | H | CH$_3$ | OH | OH | OH | 0.1 | 0.4 |
| Lyonol A (Lyoniatoxin) | –O– | | OH | OAc | OH | CH$_3$ | OH | H | OH | 0.4 | 3.0 |
| Kalmiatoxin I | H | OH | OH | OH | OH | CH$_3$ | OH | OH | OH | | |
| Kalmiatoxin VI | αOH | OH | OH | OH | OAc | CH$_3$ | OH | OH | OH | | |
| Grayanoside B | H | OGlc | OH | OH | H | =CH$_2$ | | H | OH | | |

OAc = OOC−CH$_3$, OGlc = O-β-Glucopyranoside, [a]LD$_{50}$ guinea pig, i. v., [b]LD$_{50}$ mouse, i. p.

Fig. 5.1: Andromedane derivatives of Ericaceae.

The nomenclature of the andromedane derivatives is confusing. The names are mostly derived from the generic names. Thus, they are called andromedol derivatives, andromedotoxins, asebotoxins, grayanotoxins, kalmiatoxins, lyoniatoxins, lyonols, pieristoxins, rhodojaponins, rhodomolins, rhodomosides, or rhodotoxins. Pieroside B is a leucothane derivative, and kalmanol is a B-homo-C-nor andromedane derivative (Fig. 5.2).

The most common agent of the toxic Ericaceae is grayanotoxin I (andromedotoxin, asebotoxin, acetylandromedol, and rhodotoxin; Fig. 5.1). It was first obtained simultaneously from *Pieris japonica* in 1882 by Plugge [217] and Eijkman [67]. The final structural elucidation was made in 1961 by Karisawa [138].

Subsequently, grayanotoxin I was isolated from a variety of different Ericaceae, such as Rhododendron species, e.g., *Rh. catawbiense, Rh. simsii, Rh. ponticum, Rh. luteum,* and *Rh. molle* G. Don (Ph. 5.6), Kalmia, Leucothoe, Lyonia, and Pieris species.

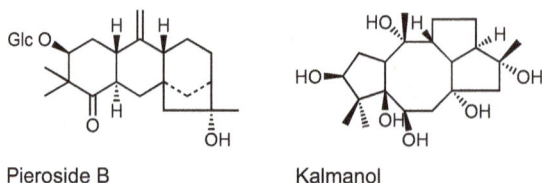

Pieroside B            Kalmanol

Fig. 5.2: Pieroside B, a leucothane derivative, and kalmanol, a B-homo-C-nor andromedane derivative.

Quantitative data on content are rarely reported. The content of grayanotoxin I in the leaves of *Rh. luteum* is 0.015%, of *Rh. ponticum* 0.0003–0.019%, of *Rh. catawbiense* 0.047% [275], and of *Kalmia angustifolia* 0.06–0.09% [257]. While Plugge [217] describes the isolation of grayanotoxin I from *Andromeda polifolia*, it has not been detected by Chung et al. [46]. It is possible that different chemotypes were studied. For example, Schindler [233] was able to detect grayanotoxin I only in *Kalmia latifolia* of North American provenance not in plants of European provenance.

In addition to grayanotoxin I, a number of other grayanotoxins and related diterpenes have been obtained from Rhododendron species. Grayanotoxins II–XXI [275, 294], rhodojaponins I–VII [118], grayanoside A and pieroside A [229], rhodomollins, rhodomolleins, and rhodomosides [17, 44, 163, 296, 297] should be mentioned. Possibly some of them are artifacts [275]. Main toxins of *Rh. molle* are the rhodojaponins [63]. Other Rhododendron species appear to be toxin-free. These probably include the two species native to Central Europe: *Rh. hirsutum* and *Rh. ferrugineum*. The cytotoxic activity of *Rh. ferrugineum* is attributed to a mixture of compounds including ferruginenes (Fig. 5.3). The compounds share structural similarities to cannabidiol (CBD). For ferruginenes B and C (C = methyl pentanol derivative of CBD) a weak interaction with cannabinoid type 2 ($CB_2$) and TRPV1 receptors could be shown [188, 219, 238].

Toxins have been detected in *Chamaedaphne calyculata* in older studies, but to our knowledge there are no findings about their structures.

Grayanotoxins II, III, and XVIII, lyonol A, leucothol A, and kalmiatoxins I–VI have been obtained from Kalmia species in addition to grayanotoxin I [33, 69]. Additionally, cytotoxic acylphloroglucides (2′,6′-dihydroxy-4′-methoxyacetophenone and phloretin) have been identified in *K. latifolia* [175].

Isolates from *Pieris japonica* included grayanotoxins VII and XVIII, grayanoside B, pieroside C, asebotoxin X, and the leucothane derivatives pieroside A and B (Fig. 5.2, [137, 230]), such from *P. formosa* (Wall.) D. Don the pierisformosides A–I, pierisformotoxin C, and the pierisoids A–E [161, 278].

The main active ingredient of *Leucothoe grayana* is grayanatoxin III, besides isograyanotoxin II is worth mentioning [259].

From Craiobiodendron species the craiobiotoxins and craiobiosides have been isolated [295].

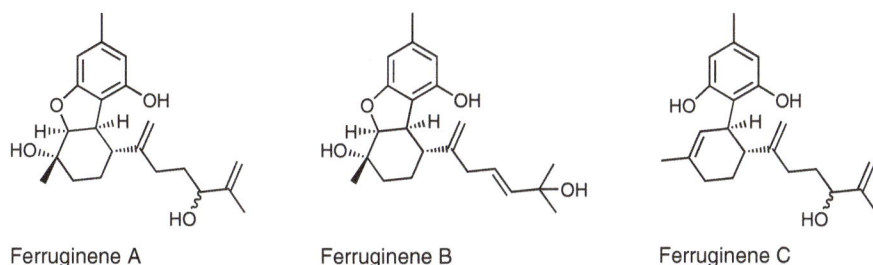

Ferruginene A          Ferruginene B          Ferruginene C

Fig. 5.3: Ferruginenes of *Rhododendron ferrugineum*.

Toxic diterpenes are also present in products originating from the plants, such as honey, labrador tea (see Section 4.4), cigarettes, and a variety of preparations used in folk and alternative medicine [133].

## 5.2.3 Pharmacology and Toxicology

The grayanotoxins target the binding site 2 ('site 2', group II receptor site) of the α-subunit of voltage-dependent $Na^+$ channels of irritable membranes and thereby suppress their inactivation after an action potential. Long-lasting depolarization and enhancement of $Ca^{2+}$ influx into the cell occur. The compounds bind to the channel only in its open state, and thus, the activation potential of the modified $Na^+$ channel is shifted in the direction of hyperpolarization. This results in positive inotropic and long-lasting hypotensive effects [133, 171]. There is a close correlation between acute toxicity and cardiac efficacy [63]. The most active compounds are those with 3-hydroxy or 2,3-epoxy and 6-hydroxy groups and with a 10-methyl and 14-$OCOCH_3$ group [172].

Rhodomolins, rhodomolleins, and rhodojaponins from *Rh. molle* possess insecticidal activity [296]. In mice, grayanotoxin XVIII and several other showed antinociceptive effects (3 mg/kg BW, i.p.) but were toxic with signs of nausea and convulsions in higher doses (10 mg/kg BW, i.p. [297]).

The $LD_{50}$ in mice for grayanotoxin I is 1.28 mg/kg BW, i.p., [98], for grayanotoxin-III 0.908 mg/kg BW, i.p., [237], and for rhodojaponin III 0.40 mg/kg BW, i.p [119].

Poisoning by Rhododendron species can be caused in humans by ingestion of the leaves (e.g., as a tea for arthritis treatment or in the form of cigarettes), of the nectar, or of the flowers (e.g., children sucking on the showy flowers) of the plants. These cases are rare and observed mainly in Asia [133]. As an example, the dried flower of *Rh. molle* (Rhododendri Mollis Flos) is monographed in the Chinese pharmacopeia as an analgesic for the treatment of rheumatoid arthritis and bruises. Poisoning incidents with symptoms of cardio- and neurotoxicity due to improper usage occasionally occurred [63]. People in Yunnan, China, who eat the flowers as seasonal delicacy, have developed skills to avoid the toxic effects [241].

Best known is the toxicity of honey from Rhododendron species (Pontic honey, 'mad honey'). Xenophon described in his 'Anabasis' in 401 B.C. a mass poisoning of Greek soldiers in Asia Minor by the honey of *Rh. ponticum* [133]. Even today, poisonings are observed after consumption of honey, mostly obtained in the Near East (especially on the Turkish Black Sea coast) and also in Nepal or on the Reunion Island. Clinical data indicate *Rh. ponticum* and *Rh. luteum* as main sources of honey poisoning [32, 60, 61, 71, 93, 98, 133, 202, 219, 244, 245, 276]. In Turkey, mad honey is sold as an alternative medicine for self-treatment of sexual dysfunction [291]. In the United States, cases of poisoning by honey are occasionally recorded on the North Pacific coast [156].

Grayanotoxin content of honey varies widely [133]. Sibel et al. [245] determined grayanotoxin-III concentrations between 2.114 and 11.371 µg/g with an average value of 6.225 µg/g in 11 fresh honey samples from different parts of the Black Sea region of Turkey. By using different nectar sources of the bees there is usually a dilution of the toxins. In 152 consultation cases evaluated in Europe after ingestion of leaves or flowers of azaleas or their nectar, symptoms of poisoning were observed in only nine patients. These symptoms were almost mild. Only one patient showed vomiting and transient hypertension [144].

Poisoning by the grayanotoxins containing honey of *Kalmia latifolia* and *K. angustifolia* is occasionally observed in the eastern parts of the USA and Canada [156].

Severe poisonings of humans by *Andromeda polifolia* and by *Chamaedaphne calyculata* are not known to us.

♦ A 28-year-old woman was poisoned after drinking a tea made from the leaves of the Ericaceae *Agarista salicifolia* (Lam.) G. Don (*Agauria salicifolia* (Lam). Oliv., a plant endemic to the Mascarenes Islands. The symptoms of poisoning were similar to those of poisoning by grayanotoxins [178].

♣ Symptoms of poisoning by grayanotoxin-containing plants, honey, and preparations, which occur after a latency period of several minutes to several hours (most within 20 min to 3 h after ingestion) include dizziness, weakness, nausea, vomiting, flushing of the face, burning of the skin, numbness, abdominal pain, diarrhea, sweating, restlessness, and, in severe poisoning, sinus bradycardia, hypotension, loss of coordination, progressive muscle weakness, arrhythmias, seizures, and possibly cardiac failure. The symptoms remain for 1–2 days. Fatal outcomes are rare [32, 60, 98, 202, 291]. Hepatitis [62], psychiatric, and neurological phenomena can occur [244]. Pathological studies in animals showed massive hemorrhages of the heart and hypostasis of the lungs [179].

⚕ For treatment of poisoning by small amounts of plants containing grayanotoxins (not more than one flower or one small leaf in children), administration of activated charcoal is recommended, for larger amounts additionally gastric emptying. Because of possible cardiac symptoms, stationary monitoring of the patient is necessary. Administration of atropine to treat bradycardia and other symptomatic measures may be required [156,161].

🐾 Poisonings by Rhododendron species in animals have been observed, for example, in sheep, goats, and kangaroos [27, 52, 128, 179, 220]. They occur preferentially in winter when there is a shortage of other forage [27]. Poisonings have also been detected in zoo animals fed by visitors with the leaves of Rhododendron shrubs standing nearby [248]. Even bees can be poisoned by the toxic nectar [40]. Pets such as cats and dogs can be affected if they eat from toxic Ericaceae, mainly azaleas, in the home or garden [133]. In Germany, according to the analyses of five German poison centers from 2012 to 2014, Rhododendron species were among the top five plants involved in pet exposures ($n = 35$) [20, 181].

Poisonings by Andromeda species are less dangerous in grazing animals than those caused by Rhododendron species [90].

Poisoning by Pieris species has been described in miniature pigs [214], goats [25, 126], and tortoises [216]. After eating the leaves of *P. formosa*, the goats exhibited teeth grinding, vomiting, convulsions, and died after about 24 h. Pathologically, hemorrhages were detectable in the small intestine [126]. The miniature pigs that were kept as pets showed sudden onset of pale oral mucosa, tachycardia, tachypnea, hypersalivation, tremor, and ataxia that progressed to lateral recumbency after ingestion of *P. japonica*. Grayanotoxin I was identified in the ingested plant, in gastric content, and organs of the involved pigs [214].

Extracts of *Rhododendron molle* and *Pieris formosa* are used as insect repellents in China. Unregistered Rhododendron extracts containing grayanotoxins are sold online and recommended for treatment of rheumatoid arthritis, skin infections, and other ailments. Because of insufficient knowledge, uncontrolled quality, and possible risk of poisoning, consumption of such products should be avoided. However, grayanotoxins and Rhododendron species gain growing attention as possible hypotensive and antimicrobial drugs [102, 133].

## 5.3 Tigliane, Ingenane, Daphnane Derivatives, and Macrocyclic Diterpenes

### 5.3.1 Chemistry

Esters of polyalcohols with the polycyclic basic bodies tigliane, ingenane, and daphnane (Fig. 5.4) are proinflammatory and tumor-promoting, but also partly antineoplastic substances of plants of the spurge family, Euphorbiaceae. They are grouped under the name phorboids. In Thymelaeaceae, daphne family, daphnane, and tigliane derivatives occur preferentially (Fig. 5.5). In addition, diterpenes with macrocyclic diterpene basic bodies, such as lathyrane, jatrophane, jatropholane, rhamnofolane, or crotofolane derivatives, are found (Fig. 5.6). They are referred to as nonphorboids and are considered precursors of polycyclic diterpenes [73, 125].

The phorboids carry a high number of hydroxy and oxo groups (4–9 O atoms per molecule). They exist unesterified, as mono-, di-, tri-, and more rarely as tetraesters.

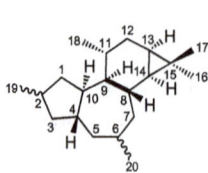

Tigliane

Tigliane derivatives

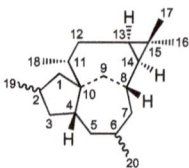

Ingenane

Ingenane derivatives

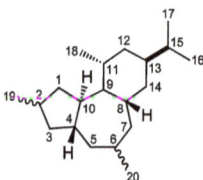

Daphnane

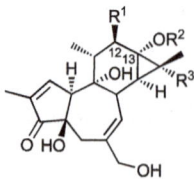

Phorbol

12-Tetradecanoyl-phorbol-13-acetate

Prostratin

$R^1$=OH, $R^2$=H, $R^3$=CH$_3$

$R^1$=OOC–(CH$_2$)$_{12}$CH$_3$
$R^2$=COCH$_3$, $R^3$=CH$_3$

$R^1$=H, $R^2$=COCH$_3$, $R^3$=CH$_3$

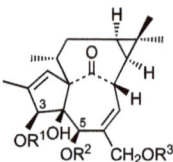

Ingenol $R^1$=$R^2$=$R^3$=H

Ingenol-3,5-dibenzoate
$R^1$=$R^2$= CO–⬡, $R^3$=H

Ingenol-3-hexadecanoate
$R_1$= CO–(CH$_2$)$_{14}$CH$_3$, $R^2$=$R^3$=H

Milliamine
$R^1$=P, $R^2$=$R^3$=H

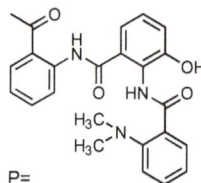

P=

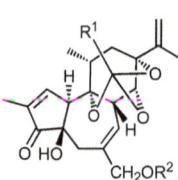

Resiniferatoxin
$R^1$= CH$_2$–C$_6$H$_5$  OCH$_3$
$R^2$= CO–CH$_2$–⬡–OH

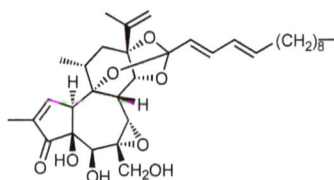

Huratoxin

Daphnane derivatives

**Fig. 5.4:** Tigliane-, ingenane-, and daphnane derivatives of Euphorbiaceae.

In the case of daphnane derivatives, three hydroxy groups form with an acid an orthoester. Epoxy groupings may also be present.

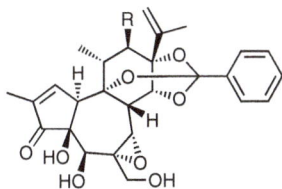

Daphnetoxin    R=H

Mezerein       R= OCO−(CH=CH)$_2$—⬡

Daphnetoxin type

Fig. 5.5: Daphnane derivatives of Thymelaeaceae.

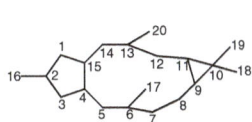

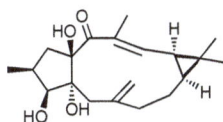

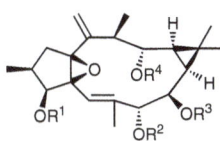

Lathyrane

Lathyrane derivatives

Lathyrol

Ingol  R$^1$=R$^2$=R$^3$=R$^4$=H

3,7,8-Triacetylingol-12-tigliate

R$^1$=R$^2$=R$^3$= CO−CH$_3$

R$^4$=CO

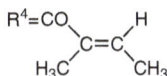

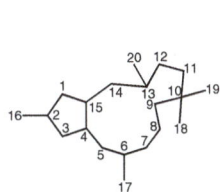

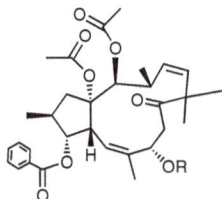

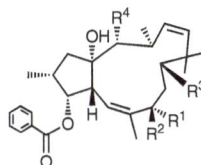

Jatrophane A

(Jatrophane B C-9 u. C-13

linked)

Jatrophane derivatives

Euphoscopin A R=H

Euphoscopin B R= CO−CH$_3$

Euphornin   R$^1$=R$^3$=R$^4$=OOC−CH$_3$, R$^2$=H

Euphornin A  R$^1$=OH, R$^2$=H

R$^3$=R$^4$= OOC−CH$_3$

Fig. 5.6: Macrocyclic diterpenes of Euphorbiaceae.

Important diterpene alcohols of the **tigliane** group are phorbol, 5-hydroxy-phorbol, 16-hydroxy-phorbol, 4-deoxy-phorbol, 12-deoxy-phorbol, 12,20-disdeoxy-phorbol, and 12-deoxy-16-hydroxy-phorbol.

Of the **ingenane**-type diterpene alcohols, e.g., ingenol, 13-hydroxy-ingenol, 16-hydroxy-ingenol, 13,19-dihydroxy-ingenol, 5-deoxy-ingenol, and 20-deoxy-ingenol are known [12].

Up to now, nearly 200 **daphnane**-type diterpenoids have been isolated and identified from plants of the Euphorbiaceae and Thymelaeaceae families [135]. They can be classified into 6-epoxy daphnane diterpenoids, e.g., daphnetoxin, resiniferonoids, e.g., daphneresiniferin A, genkwanines, e.g., genkwanine A, 1-alkyldaphnanes, e.g., gnidimacrin, and rediocides, e.g., rediocide A, based on the oxygen-containing functions at rings B and C as well as the substitution pattern of ring A.

The main representatives of the **macrocyclic** alcohols are lathyrol, 6,17-epoxy-lathyrol, 7-hydroxy-lathyrol, 17-hydroxy-lathyrol, ingol, jatrophone, and jatropholone A and B [73, 79].

Over 30 acids were detected as acid components of the esters, including linear-saturated or mono- to penta-unsaturated fatty acids ($C_2$–$C_{18}$), monomethyl or dimethyl fatty acids ($C_4$–$C_{11}$), benzoic acid, and other. In some esters, N-methylanthranilic acid or a tripep-tide of anthranilic acid residues occur as acid components. In most cases, the diesters carry a long-chain fatty acid or an aromatic acid and a short-chain fatty acid [22, 73, 76, 114].

As the first active substance of this group, 12-O-tetradecanoyl-phorbol-13-acetate (croton factor $A_1$, TPA, myristoylphorbol acetate, MPA) was isolated from the seed oil of *Croton tiglium* L., croton oil tree (Ph. 5.6, Euphorbiaceae). Its structure was elucidated by Hecker [112] in the years 1963–1967. The constitution of the main representatives of the ingenane derivatives [197] and macrocyclic diterpenes [87] was also determined by Hecker and co-workers [112].

## 5.3.2 Pharmacology and Toxicology

Main target of the **phorbol esters** is the protein kinase C (PKC), a family of serine/threonine protein kinases. Their isoforms have been classified into the subfamilies: conventional (α, $β_I$, $β_{II}$, and γ), novel (δ, ε, η, θ), and atypical (ζ, λ, ι) subfamily. The activation of conventional PKCs is dependent on 1,2-diacylglycerol (DAG) and $Ca^{2+}$, novel PKCs are only dependent on DAG, and atypical PKCs are independent on DAG and $Ca^{2+}$. PKC-α and PKC-β have been linked to increased cancer progression whereas PKC-δ increases apoptosis [260].

Because of their stereochemical similarity to DAG the phorbol esters are potent and long-lasting DAG mimetics. They activate conventional and novel PKCs by attacking C1b domains, the receptors for DAG. C1b domains are zinc finger proteins first identified in PKC. They are also present in other protein families involved in cell division and differentiation, e.g., chimerins and RasGRP3 (Ras guanylnucleotide-releasing protein). Phorbol derivatives esterified at C-12 and C-13 are slower to degrade than DAG so that their attack results in long-lasting stimulation of PKC leading to disruption of cellular processes regulated by PKC. Other proteins with C1b domains can also be affected [140].

Most striking are the tumor-promoting and inflammatory effects of phorbol esters. They are among the most potent of the known tumor promoters. They promote the promotion phase of tumorigenesis and in some cases also the propagation phase.

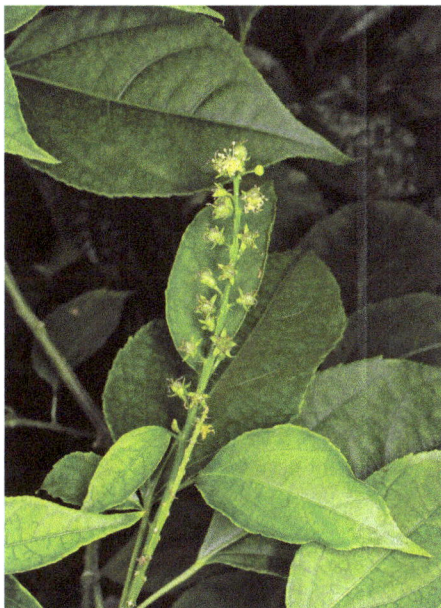

Ph. 5.6: *Croton tiglium* (source: Peter Schönfelder).

Their effects have been particularly well studied on mouse skin. They cause direct switching of cells from G1 phase to DNA synthesis and mitosis after prior exposure of the skin to low concentrations of a solitary carcinogen [26, 31, 73, 115, 136]. Also worth mentioning is the activation of Epstein–Barr viruses by phorbol esters [170].

There are considerable differences between the individual active substances of the Euphorbiaceae and the Thymelaeaceae in their pattern of action. Not only tumor-promoting but also antineoplastic effects have been found, e.g., for prostratin (13-*O*-acetyl -12-deoxyphorbol) from *Pimelea prostrata* (J.R. Forst & G. Forst) Willd. (Thymelaeaceae), Strathmore weed, a small endemic New Zealand shrub.

Possible reasons for these contrasting effects of closely related substances are
- the differential attack on the individual PKC isoenzymes, which each perform specific cellular functions [91],
- the differential attack on receptors that do not belong to the PKC family, e.g., chimerins,
- the differential influence on other mechanisms, e.g., the disruption of the mito-chondrial electron transport chain [21, 39].

TPA has the strongest tumor-promoting activity and the most extensive spectrum of activity. The following structural requirements are essential for tumor promotion:
- the position of certain hydrophilic groups in the molecule (free OH group at C-20), which are responsible for the interaction with the receptor,

–  the presence of hydrophobic groups, especially acidic residues at C-12 and C-13, whose chain length influences the strength of activity, and which influence the interaction with membrane lipids in the vicinity of the receptor.

With these two prerequisites the greatest possible structural similarity to DAG is achieved [159].

A structure-activity comparison with tumor promoters of completely different substance classes, e.g., with the microbial teleocidins, shows that in all compounds a hydrogen acceptor, two hydrogen donors, a large lipophilic group, and a corresponding spatial arrangement are prerequisites for the cocarcinogenic effect [131]. Only diesters of 4α-phorbol have cocarcinogenic activity; unesterified phorbol as well as its unesterified derivatives and diesters of 4α-phorbol are inactive. 12-Deoxy-phorbol -13,20-diester and C-13-monoester have only inflammatory effects. Triester in which the OH group at C-20 is also acylated are so-called 'hidden' or 'cryptic' tumor promoters that become active only after cleavage of the third ester bond [115, 222].

Among the **ingenol derivatives**, the di-O-acylates of 13-hydroxy-ingenol are the most potent skin irritants [199].

In the **daphnane group**, the 6,7-epoxy-daphnane orthoesters show only weak cocarcinogenic effects. However, when a double bond between C-6 and C-7 is present instead of the 6,7-epoxy group, the effect increases considerably [75]. Daphnetoxin, unsubstituted at position 12, causes uncoupling of oxidative phosphorylation and permeability disturbances of the mitochondrial membrane in addition to PKC activation [208]. The daphnane derivative mezerein, unlike TPA, attacks only in the propagation phase of tumor promotion (second-stage promoter) and is less toxic than TPA. Crucial for this is the unsaturated ester side chain of mezerein in position 12 [240]. Mezerein has less cocarcinogenic activity than daphnetoxin and, unlike the latter, possesses antileukemic properties [113]. The differences in effect are explained by differences in selectivity for PKC isozymes. Resiniferatoxin attacks the vanilloid receptor VR1 and it has effects similar to those of capsaicin [271].

The **macrocyclic diterpenes**, e.g., those of the jatrophane type, have no significant skin-irritating effect. Instead, many of them possess antineoplastic activity. Mechanisms of action are the inhibition of the mitochondrial electron transport chain, interactions with microtubules, and induction of oxidative stress and apoptosis. The compounds inhibit transport proteins of tumor cells that transport drugs out of the cell. By inhibition of the transport protein P-glycoprotein (P-gp) in cancer cells that overexpress P-gp they reverse the multidrug resistance (MDR) of cancer cells to cytostatic drugs. Antiviral (against HIV, HSV, Chikungunya virus), antifungal, and antiinflammatory activities have been reported [79, 271].

The first step to detoxify the mentioned diterpenes in the animal or human organism is hydrolytic cleavage of the long-chain acyl groups. The further steps are not fully understood.

Phorbol esters are used extensively in research to elucidate mechanisms of tumorigenesis, cell division, cell differentiation, and the role of PKC and its isoenzymes.

Because of their high toxicity, they must be used with special caution. Any contact with skin or eyes must be avoided.

Otherwise, nontumor promoting diterpenes such as prostratin or gnidimacrin (see Section 5.3.4) are potential lead substances for development of anticancer and antiviral drugs. Prostratin and gnidimacrin inhibit HIV-1 infections through activation of PKC pathway and reduce the HIV-1 latency [155, 184, 198]. Of immense interest is the ability of nontumor-promoting compounds to reduce the resistance of tumor cells to cytostatic drugs and to overcome the multi drug resistance (MDR) of human cancers. Substances with capsaicin-like effects such as resiniferatoxin are possible analgesics. Resiniferatoxin is scheduled for a Phase III clinical trial [271].

### 5.3.3  Tigliane, Ingenane, Daphnane Derivatives, and Macrocyclic Diterpenes as Toxins of Spurge Family (Euphorbiaceae)

The spurge family, Euphorbiaceae, is one of the largest families of the Angiospermae, with more than 300 genera and more than 8,000 species. Their diversity, habit, and range of distribution are very large. They are annual or perennial herbs, cactus-like succulents, shrubs, or trees. They are found in almost all regions of the world. Most species carry milky sap (latex) in unarticulated milky tubes. The most extensive genera of the family are Euphorbia, spurge, with about 2,500 species, and Croton, croton, with about 700 species [73, 136, 173].

Genera of toxicological and pharmacological importance are Aleurites, Baliospermum, Croton, Euphorbia, Excoecaria, Hippomane, Homalanthus, Hura, Jatropha, Ostodes, Sapium, Stillingia, Trigonostemon, and Vernicia. Phorbol ester activity may be present in additional genera and families, e.g., in Antidesma species (Phyllanthaceae, [22,73]).

Many Euphorbiaceae that certainly or probably contain toxic diterpenes are still used in their native countries
–   as foods and spices, e.g., *Croton lobatus* L. (*Astraea lobata* (L.) Klotzsch), lobed croton, and *Euphorbia lathyris* L., caper spurge,
–   as ingredient in beverages, e.g., the bark of *Croton eluteria* (L.) W. Wright, cascarilla, in Campari, an Italian liqueur,
–   in folk medicine, e.g., *Croton tiglium* L., purging croton, *C. lechleri* Müll. Arg., dragon's blood, *Jatropha curcas* L., physic nut, *J. gossypiifolia* L., cotton-leaf physic nut, *Hura crepitans* L., sandbox tree, *Euphorbia antiquorum* L., triangular spurge, *Eu. lathyris*, *Eu. megalatlantica* Ball, *Eu. milii* Des Moul, Christ's thorn, *Eu. resinifera* Berg, resin spurge, and *Eu. tirucalli* L., pencil tree [73, 141, 232].

Some Euphorbiaceae with proinflammatory diterpenes served as fish poisons for fishing in Southeast Asia, e.g., *Hura crepitans*, *J. curcas*, *J. multifida* L., St. Vincent physic nut, *Euphorbia balsamifera* Aiton, *Eu. buxoides* Radcl.-Sm., *Eu. piscatoria* Aiton, *Eu. tirucalli* and *Eu. trigona* Mill., African milk tree, or for making arrow poisons, e.g.,

*Croton tiglium, Eu. cereiformis* L., milk barrel, *Eu. heptagona* L., *Eu. kamerunica* Pax, *Eu. laterifolia* Schum. et Thonn., *Eu. mammillaris* L., corn cob cactus [73], and *Hippomane mancinella* L., manchinell, beach apple (in Spain manzanilla de la Muerte = little apple of death).

In Central Europe, only the genera **Euphorbia**, spurge, and Mercurialis, mercury, of the family Euphorbiaceae are represented. Hegi names 32 species of Euphorbia as native, introduced, or naturalized in Central Europe. Especially common in Europe are *Eu. cyparissias* L., cypress spurge, *Eu. helioscopia* L., sun spurge (Ph. 5.7), *Eu. peplus* L., petty spurge, *Eu. esula* L., leafy spurge, and *Eu. exigua* L., dwarf spurge.

The widespread and common cypress spurge, *Eu. cyparissias*, is a perennial that occurs in dry meadows, field margins, roadsides, and dry ruderal sites. *Eu. helioscopia, Eu. peplus, Eu. esula,* and *Eu. exigua* are field weeds.

Some spurge species are grown in gardens as ornamentals and some occur feral, e.g., *Eu. lathyris* (Ph. 5.8), *Eu. epithymoides* L. (*Eu. polychroma* A. Kern.), cushion spurge, and *Eu. myrsinites* L., blue spurge. Other species are popular houseplants. These include the shrubby succulent *Eu. milii* (Ph. 5.9), some leafless stem succulents with a cactus-like appearance, e.g., *Eu. ingens* E. Mey. ex Boiss., cactus spurge ('Candelabra tree'), *Eu. resinifera* (Ph. 5.10), and other species native to Africa. Popular as a houseplant and cut flower is also *Eu. pulcherrima* (Ph. 5.11) Willd. ex Klotzsch, poinsettia, Christmas star, native to Central America, and characterized by blood-red bracts.

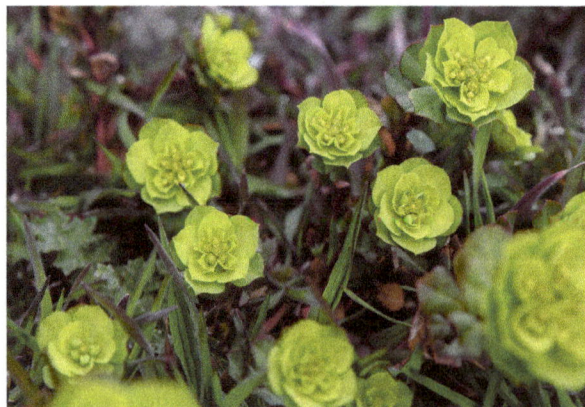

Ph. 5.7: *Euphorbia helioscopia* (source: Elena Odareeva/iStock/Getty Images Plus).

With few exceptions, e.g., *Eu. pulcherrima*, tigliane, ingenane, and/or daphnane derivatives as well as derivatives of macrocyclic diterpenes could be detected in almost all Euphorbia species examined [22, 73–75]. The active substances are contained in the milk sap and seed oil, but furthermore also in other tissues of the plants. Thus, 3.8% ingenane esters were obtained from the milk sap of *Eu. segueiriana* Neck., 1.5% from

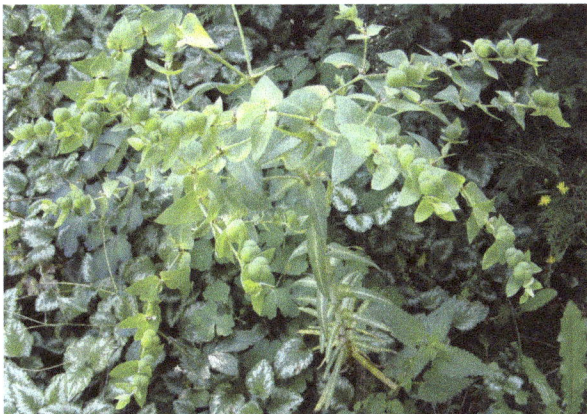

Ph. 5.8: *Euphorbia lathyris* (source: Eberhard Teuscher).

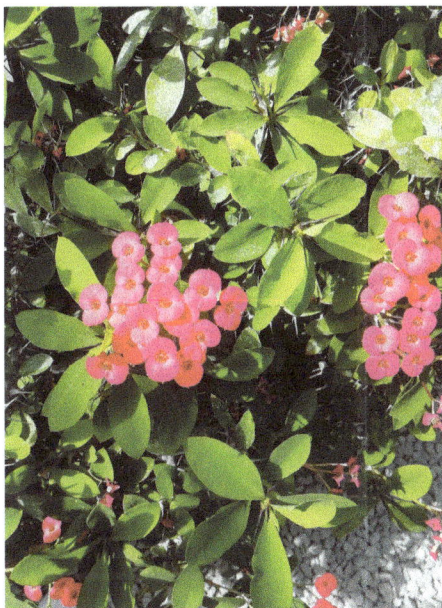

Ph. 5.9: *Euphorbia milii* (source: Ulrike Lindequist).

*Eu. serrulata*, 0.3% from *Eu. esula*, and 6.5, 5.1, and 1.0% from the whole plants, respectively [263]. Between 2008 and 2012, more than 200 novel diterpenes have been isolated from different Euphorbia species. Many of the investigated species contain constituents with two or more different cores [271].

Ph. 5.10: *Euphorbia resinifera* (source: cmspic/iStock/Getty Images Plus).

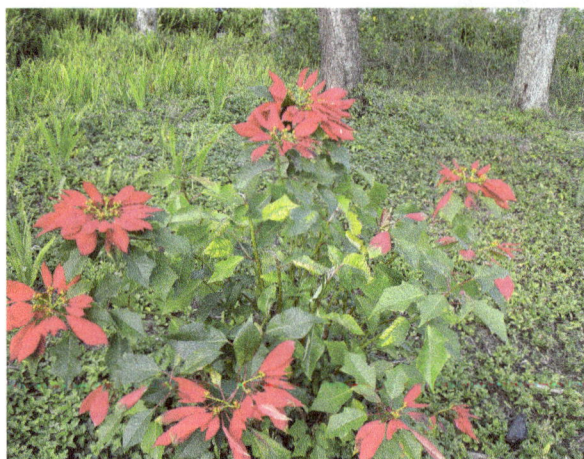

Ph. 5.11: *Euphorbia pulcherrima* (source: Margit Bukowski).

For *Eu. helioscopia* the most important constituents known so far have been summarized in tabular form (Tab. 5.1).

From *Eu. helioscopia*, were further isolated: 12-deoxy-phorbol-13-tigliate and further phorbolesters [77]; abietane derivatives with a α-unsaturated five-membered lactone ring, the helioscopinolides A–C [242]; nonirritating 20-deoxyingenol derivatives, e.g., Euphorbia factor HS 4, and diterpene esters with jatrophan or lathyrane bases, e.g., the euphoscopins A and D and the euphornins A–C [147, 234, 243, 292].

Tab. 5.1: Toxic diterpenes of *Euphorbia helioscopia* [100, 235, 290].

| Euphorbia factor | Structure | $ID_{50}$ nMol/ear |
|---|---|---|
| H1 | Ingenol-3-angelate | 0.04 |
| H2, H3 | Ingenol-3-decatrienoate | 0.06 |
| H5 | Ingenol-3-decadienoate | 0.03 |
| H4 | 20-Desoxyingenol-5-angelate | 10 |
| H7 | 20-Desoxyingenol-5-decatrienoate | 9 |

Abbreviation: ID, irritative dose

From the multitude of other Euphorbia species with toxic diterpenes here only *Eu. lathyris*, *Eu. esula*, *Eu. peplus*, and the popular houseplants *Eu. resinifera* and *Eu. milii* can be discussed in more detail.

In *Eu. lathyris* were found, among others: ingenol monoesters, e.g., ingenol-3-hexadecanoate (euphorbia factor L5, 3-HI), and ingenol-3-tetradeca-2,4,6,8,10-pentaenoate (euphorbia factor L6 [4]) and especially in the seed oil esters of lathyrol and lathyrol derivatives (SL3, SL2, SL1, SL8, and SL9 [5, 6, 132]). Lathyric acid A, isolated from seeds, was the first diterpene found in nature with a *seco*-lathyrane backbone [165].

*Eu. esula* contains ingenol-3-deca-2,4-dienoate (E1, $ID_{50}$ 0.07 µg/ear), further ingenol esters and several lathyrane and jatrophane derivatives, including esulones A–C and esulatins A–E [105, 151, 166, 239, 264, 271].

From *Eu. peplus* were obtained: ingenol-20-acetate-3-angelate (Euphorbia factor Pe1), 20-deoxy-ingenol-3-angelate (Euphorbia factor Pe2, $ID_{50}$ 0.18 µg), and ingenol-20-octanoate ($ID_{50}$ 1.0 µg, [100, 223]); jatrophane derivatives, including the pepluanines [51, 124] and pepluane and pepluanone, diterpenes with the rare tetracyclic pepluane backbone.

From the resin spurge, *Eu. resinifera*, native to Morocco, the drug Euphorbium was obtained. Euphorbium is the dried milk sap. It was used in the form of plasters or ointments for skin irritation therapy. Today it is used in homeopathy. About 20 diterpenes (RL1–RL20) have been detected in its milky sap, among them toxic ingenol and phorbol esters and resiniferatoxin [22, 116].

*Eu. milii*, Christ's star, native to Madagascar, contains, additionally to phorbol and ingenol esters [19], toxic diterpenes of unusual structure also: the potent milliamines A–G [176]. These compounds are ingenol esters bearing in positions 3 or 20 an anthraniloyl radical linked in a peptide-like manner to a 3-hydroxy-anthraniloyl radical whose amino group bears a further anthraniloyl or a *N,N*-dimethylanthraniloyl radical [121, 176].

Concomitants of the proinflammatory diterpenes in Euphorbia species are *all-cis*-1,4-polyisoprenes occurring as latex components, acetone-soluble resins, triterpene alcohols, triterpene ketones as well as polyfunctional triterpenes (e.g., euphol, tirucallol, and euphorbol), nortriterpenes (e.g., cycloeuphordenol), and small amounts of ester waxes and alkanes [141, 265]. Furthermore, unusual amino acids, e.g., *m*-hydroxyphenylglycine

[189], and the unusual peptide-*O*-acetyl-*N*-(*N*'-benzoyl-L-phenylalanyl)-L-phenylalaninol [262] were detected.

Consistent with the distribution of diterpene esters among the Euphorbia species studied, Kinghorn and Evans [142] found irritant effects on the mouse ear in 53 of 60 species studied. Only representatives of the Anisophyllum and Poinsettia sections were inactive in animal assays [142, 228].

Poisonings by spurge species are common worldwide. They occur when contact is made with the plants during gardening, care of houseplants, or when children play with the plants [14, 111, 187, 213, 250, 277, 283, 287]. Spraying the latex into the eye is particularly dangerous [65, 68, 84, 97, 129, 183].

❢ An 8-year-old girl who was struck by a boy with a specimen of *Eu. marginata* Pursh, snow on the mountain, ghostweed (often the cause of poisoning in Australia) while playing very quickly developed conjunctivitis, swelling of the face as well as eyelids, and skin manifestations [213].

❢ A 75-year-old woman had undergone cataract operations on both eyes. While gardening and breaking a plant stem a few month later, she was splashed into the eye with a whitish substance that was identified as the latex of *Eu. rigida* M. Bieb., gopher spurge. Toxic endophthalmitis developed with marked intraocular involvement and visual loss. Treatment was systemic and local with antibiotics and glucocorticoids [183].

The toxicity of the poinsettia, *Eu. pulcherrima*, is assessed differently. The absence of toxic diterpene esters and the lack of efficacy in animal experiments allow the plant to be considered harmless [273, 49], whereas individual case reports with, however, harmless symptoms, speak against this [66]. In more than 1,000 cases of consultation at the Berlin Poison Information Center, in which children had eaten mostly leaves, occasionally flower parts, only occasional abdominal pain, vomiting, and diarrhea were observed, and in one case blistering of the tongue and refusal to eat (**161**).

In tropical countries, the honey of some native Euphorbia species, e.g., *Eu. ledienii* A. Berger, 'noors honey', can cause irritation of the mouth and throat [249].

⊗ Folk medicinal use of Euphorbia species can also be the cause of poisonings. *Eu. tirucalli*, which plays a major role in folk medicine, e.g., in India and Africa, is held partly responsible for the development of endemic Burkitt's lymphoma that is caused by Epstein–Barr virus infection. An activation of Epstein–Barr viruses and an induction of characteristic chromosomal changes (8–14 translocation) in virus-infected B-lymphocytes by the diterpene esters of several Euphorbiaceae was demonstrated. In Africa, a significant association between case of Burkitt's lymphoma of persons in a household and the presence of *Eu. tirucalli* and *J. curcas* in or around the yard could be shown [268].

Ingenol mebutate, isolated from *Eu. peplus*, was approved for topical therapy of actinic keratosis, a precancerous skin condition. Due to concerns about a possible link between the application of the drug and the development of skin cancer, approval in the EU was suspended. In the USA it is yet available.

Of representatives of other genera *Hura crepitans* (Ph. 5.12), the Jatropha species (Ph. 5.13) and *Hippomane mancinella* (Ph. 5.14) should be mentioned.

**Hura crepitans**, sand bush tree, is native to tropical South America and cultivated in many tropical countries. Its fruits have been used as sand sprinkles (previously used in the cleaning of the floorbords). All parts of the plant, mainly the dried latex, were used in folk medicine, e.g., as purgative. The latex was in former times used as a fish and arrow poison. Huratoxin (Fig. 5.4) is the main active ingredient [231]. Contact with the latex can lead to skin burns [10]. Otherwise, a hydroethanolic extract of the stem bark was considered as safe in doses up to 1,000 mg/kg BW, p.o., estimated in rats and mice. Diterpenes have not been detected in the investigated extract [201].

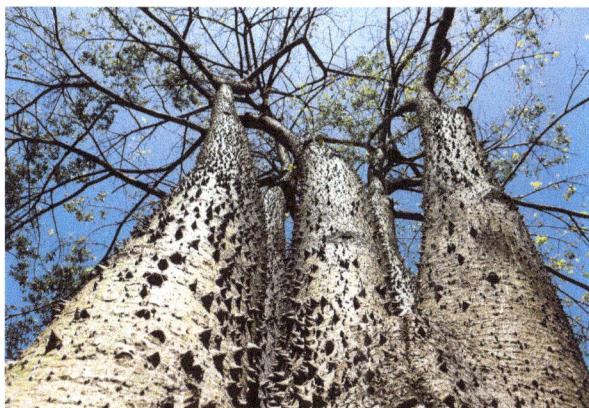

Ph. 5.12: *Hura crepitans* (source: William Rodriguez des Santos/iStock/Getty Images Plus).

**Jatropha** species are widespread in tropical and subtropical areas. *J. curcas* is the best known [86, 173]. The oil pressed from the seeds of the shrub or tree, which has originated from Mexico to Chile, is used for technical purposes, e.g., as biofuel, but also as a drastic laxative. Among other active ingredients, an intramolecular 13,16-diester of 12-deoxy-16-hydroxy-phorbol has been shown to be present in *J. gossipiifolia* [123]. Nonesterified macrocyclic diterpenes containing jatrophane A, jatrophane B, or rhamnofolane backbones, have also been isolated, for example: jatrophone (7,14-dioxo-12,15-epoxy-jatropha-3,5,8,12-tetraene A [267], Figs. 5.4 and 5.6). Wang et al. [279] hold the opinion that the toxicity of kernel cake of *J. curcas* is rather caused by hydroxy-octadecenoic acids than by phorbol esters.

Jatropha species are a frequent cause of poisonings in tropical countries [90, 150, 247]. *J. curcas* is the poisonous plant that was ingested the most by children in rural

Ph. 5.13: *Jatropha curcas* (source: Tanes Ngamson/iStock/Getty Images Plus).

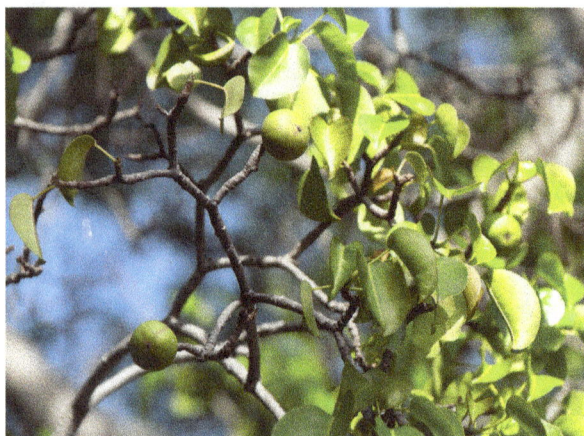

Ph. 5.14: *Hippomane mancinella* (source: CircleEyes/iStock/Getty Images Plus).

Sri Lanka. Poisoning events happen mostly in home garden [58]. The $LD_{50}$ of its oil, containing phorbol esters, is 6 ml/kg BW., p.o., in rats. The latex of *J. gossypiifolia* was genotoxic in *Allium cepa* test system [59].

† Two Nigerian children who ate ripe seeds of *J. curcas* showed vomiting and central agitation [1]. The toxic dose of seeds is not known.

Because of the promising potential of Jatropha species as source of new drugs, as bio-fuel, and as animal feed detoxification methods to remove toxic diterpenes have been developed [2, 80].

**Hippomane mancinella** (Ph. 5.14) is native to Central and northern South America, Florida, and to the Caribbean. It can be found on coastal beaches often among mangroves. All parts of the plant including the latex contain complex mixtures of esters of the tigliane and the daphnane type, among them the strongly tumor-promoting 13-(hexa-deca-2,4,6-trienoic acid) ester of 12-deoxy-5-hydroxyphorbol-6α,7α-oxide and hura-toxin [5]. The tree should be the most dangerous tree of the world [192, 289]. Poisonings are caused by eating the fruits ('beach-apples') in ignorance of the toxicity, or by contact with the latex. Even standing under the tree in the rain gave rise of dermatitis and ophthalmitis. Burning the wood would result in toxic smoke [192, 255].

🏃 The highly skin-irritating latex of most plants of the spurge family causes severe inflammation with blistering and necrosis after external contact with skin and mucous membranes, further eyelid edema, severe conjunctivitis, keratitis, and in severe cases blindness. Internal ingestion (due to the burning taste, ingestion of larger amounts is unlikely) results in redness and burning of the mouth and throat and severe gastrointestinal disturbances. Resorptive signs of acute poisoning include mydriasis, dizziness, delirium, convulsions, and collapse [215, 255, **127, 161**].

✱ Allergic reactions have been observed. Contact allergies triggered by poinsettia occur only very rarely **(65)**.

📖 Symptoms of poisoning by the latex of diterpene-containing Euphorbiaceae can, after poison removal, only be treated symptomatically, for example, with mucilaginosa and antihistamines (**49, 127, 161**). Necessary are in case of external contact thorough washing off the latex, in case of eye contact abundant rinsing of the eye with lukewarm water and presentation to an ophthalmologist, in the case of ingestion administration of activated charcoal and possibly gastric emptying.

⊗ The frequent occurrence of esophageal cancer among the population of Curaçao is explained by the fact that the roots of *Croton flavens* L., yellow croton, copahu, are chewed there as an alleged stimulant or used to prepare a bush tea (Welensali tea, [285]). The cocarcinogenic and virus-activating effects of diterpene esters may also play a role in the development of other carcinomas, such as the endemic Burkitt's lymphoma in Africa (see above) and the frequent occurrence of nasopharyngeal tumors in southern China, where tung oil from the seeds of *Vernicia fordii* (Hemsl.) Airy Shaw (*Aleurites fordii* Hemsl.), tung tree, is used as an additive to paints and varnishes [115]. The cooked seeds and seed oil of the related tree *Aleurites moluccanus* (L.) Willd., candlenut tree, are consumed in Southeast Asia as an additive to curry dishes or cooking oil. Whether toxic diterpenes are also contained in this species, which is considered toxic raw used, is not known to us.

🐾 Animals are usually discouraged from eating spurge species by their unpleasant taste. However, intoxications of livestock following ingestion of plants from the genera Euphorbia, Aleurites, and Jatropha have been reported [96]. The risk is increased if the feed contains large amounts of the plants. Symptoms include salivation, vomiting, diarrhea, hypothermia, convulsions, and paralysis. The skin-irritating substances have also been detected in the milk of animals [194, 199, **60**]. In the United States, *Eu. milii* causes many animal poisonings. The African milk bush *Eu. grantii* Oliv., African milk bush, is a common hedge plant in tropical Africa and India and a cause of animal poisoning with skin and mucous membrane damage [16, 272]. Even *Eu. pulcherrima* is considered of low toxicity, the plants are frequently implicated in pet poisonings, especially in Europe, where poinsettias are very common indoor ornamental plants during Christmas season. Generally, just mild gastrointestinal symptoms are reported [20]. According to the data collected by five German poison centers from 2012 to 2014, Euphorbia species were among the top five plant species responsible for domestic animal exposures [181].

### 5.3.4 Daphnane and Tigliane Derivatives as Toxins of Daphne Family (Thymelaeaceae)

The daphne family, Thymelaeaceae, includes about 50 genera and nearly 900 species. The plants are mainly shrubs or trees of temperate and tropical regions. A high number of the species are toxic. Some have been used to obtain fish poisons (e.g., Daphne and Gnidia species) or arrow poisons (e.g., Gnidia species). Many are used in folk medicine. With respect to their toxins, the genera Aquilaria, Daphne, Daphnopsis, Dirca, Gnidia, Lasiosiphon (now integrated into the genus Gnidia), Pimelea, Stellera, Thymelaea, and Wikstroemia have been studied [28, 29].

In Central Europe, only the genera Daphne, daphne, and Thymelaea, sparrowworts, are represented. The genus Daphne includes about 50 species, distributed in Europe and Asia. In Central Europe are found *Daphne mezereum* L., February daphne, mezereon (Ph. 5.15), *D. cneorum* L., garland flower, rose daphne, *D. laureola* L., spurge-laurel, *Daphne alpina* L., *D. striata* Tratt., striped daphne and *D. blagayana* Freyer, Balkan daphne.

*D. mezereum*, more rarely *D. cneorum*, are also grown in gardens as ornamental plants. *D. pontica* L., twin-flowered daphne, *D. caucasica* Pall., Caucasian daphne, and *D. arbuscula* Čelak, shrubby daphne, are also popular ornamental plants.

Daphne species are small shrubs, rarely small trees. The color of the flowers is mostly pink, yellowish green to white in *D. laureola* and *D. blagayana*. The ripe fruits, about the size of a pea, are red to reddish brown in *D. mezereum, D. alpina, D. cneorum*, orange-yellow in *D. striata*, whitish-yellow in *D. blagayana*, and black in *D. laureola*.

The main active ingredients of *Daphne mezereum* and *D. laureola* are the daphnane derivatives mezerein and daphnetoxin (Fig. 5.5, [73, 22, 225]). In the seeds, 0.1% mezerein and 0.02% daphnetoxin and in the whole fruits, 0.04% mezerein were found. The pulp (!) is free from toxic diterpenes.

Numerous publications deal with investigation of the ingredients of the East Asian species. In *D. genkwa* Siebold et Zucc., genkwa daphne, occurring in Southeast Asia, e.g., genkwanine and yuanhuafin were detected [139, 293], in *D. odora* Thunb. ex Murray, winter daphne, besides mezerein gniditrin, odoracin (gnidilatidin), and daphneodorins, in *D. tangutica* Maxim. (*D. retusa* Hemsl.), Tangu daphne, gniditrin [298]. These are daphnane mono- or diesters, differing from mezerein especially in the nature of the acid(s)

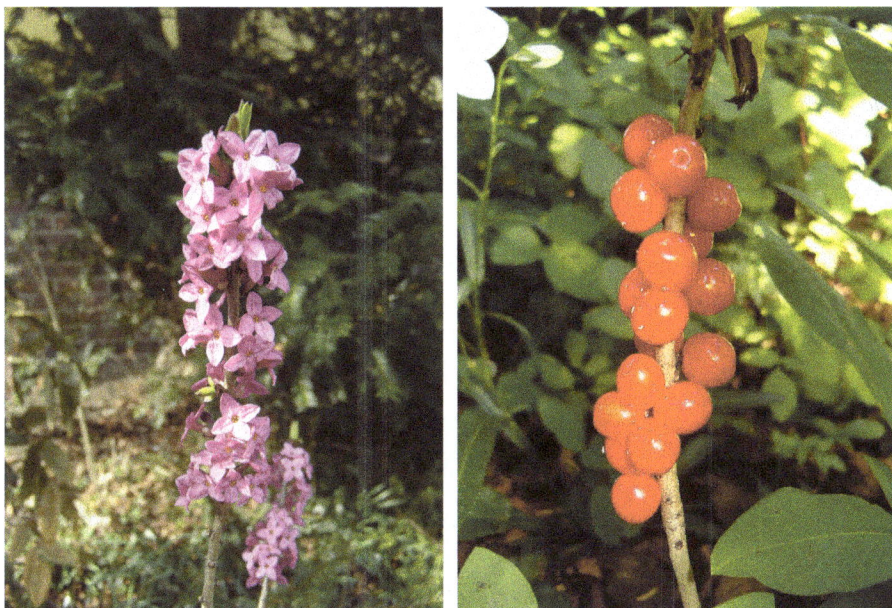

Ph. 5.15: *Daphne mezereum* (source: Eberhard Teuscher).

bound. Such esters also occur in Gnidia (in addition to 1α-alkyldaphnane derivatives) and Pimelea species. Tigliane derivatives have also been found in the genera Pimelea (e.g., prostratin, Fig. 5.4), Daphnopsis, and Aquilaria [29, 122]. Gnidiamacrin has been first isolated from *Gnidia subcordata* [152].

The $ID_{50}$ is 0.016 µM/ear for daphnetoxin and 0.03 µM/ear for mezerein [113]; the $LD_{50}$ for daphnetoxin is approximately 275 µg/kg BW, mice [254].

Poisoning by daphne occurs most frequently when children play with the plants, chew berries or flowers, or chew the twigs. Deaths have been reported after eating as few as 10–12 berries, and in other cases, poisonings involving more than 60 berries have been survived [195]. The extent of symptoms depends largely on whether the seeds are bitten or not. Gardener handling of the plants can cause dermatitis [187].

In the past, poisoning also played a role in folk medicinal use of the moist bark of the plants, especially as a laxative and blistering agent [233].

Gnidia, distributed in Africa and Asia, and Pimelea species, occurring in Australia and New Zealand, are the cause of poisoning of humans and animals in their native countries [120, 143, 196, 253, 112].

☠ The skin contact with parts of daphne plants causes swelling of the skin, blistering, rejection of the epidermis, erysipelas-like redness, and pustule formation. Prolonged exposure leads to necrotic tissue breakdown. Eye contact leads to conjunctivitis. Ingestion results in redness and swelling of the mucous membranes of the mouth, thirst sensation, salivation, stomach pain, vomiting, and severe diarrhea. Symptoms after resorption may include headache, dizziness, lightheadedness, tachycardia, convulsions, and possibly death from circulatory collapse. The kidneys as excretory organs are particularly severely damaged (**53**).

📕 After external contact, the skin must be thoroughly cleansed. After ingestion, abundant fluids should be given. Administration of activated charcoal and, if two or more two seeds have been bitten open and swallowed, induction of vomiting is recommended. Treatment of poisoning can only be symptomatic (**161**).

🐃 Although animals mostly avoid the plants poisoning has been observed, e.g., in cattle [209]. Some birds eat the flesh of the fruits apparently without suffering any damage.

## 5.4 Taxane Derivatives as Toxins of Yew Species (Taxus Species)

The genus Taxus, yew, Taxaceae, comprises about 10 species. In Central Europe, only *Taxus baccata* L., common yew, English yew (Ph. 5.16), occurs. Other species are native to North America and East Asia, e.g., *T. cuspidata* Sieb. & Zucc., Japanese yew. Worth mentioning is *T. brevifolia* Nutt. the Pacific yew, as the first identified source of the cytostatic drug taxol A (paclitaxel).

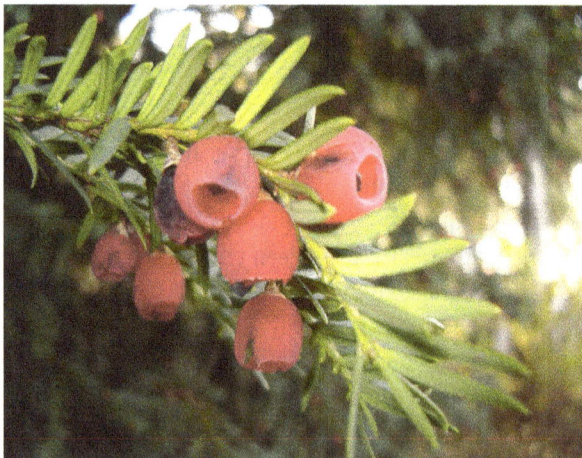

Ph. 5.16: *Taxus baccata* (source: Eberhard Teuscher).

Yew is a dioecious evergreen shrub or tree that thrives in beech slope forests, in places also as undergrowth in coniferous or deciduous forests. The female flowers consist of a single ovule, which develops into a blackish-brown seed surrounded by an initially green, later red seed coat in the shape of a cup, named aril. The length of this 'berry' is about 15 mm. Several varieties of yew exist, differing in growth habit (columnar trees to

dwarf shrubs), shape and color of needles (green, golden yellow, or white variegated needles), and color of arils (red or yellow). Yews are often grown as ornamental plants.

Probably all parts of the yew contain toxic taxane derivatives. The total fraction
of these compounds isolated by Lucas [168] in 1856 was initially declared as the alkaloid 'taxin' because of the basic character of some components. Later, it was shown
that 'taxin' is a mixture of at least 11 basic substances [101]. The content of 'taxin' was
determined to be 0.6% (in May) to 1.97% (in January [148]) in needles and 0.92% in
seeds [83]. Endophytic fungi of yew also contain taxane derivatives.

The trivial names of taxane derivatives are confusing and do not allow assignment to a specific type. Some compounds bear only semisystematic names. All the
known diterpenoids of Taxus species (taxoids) are based upon the unique taxane skeleton or rearrangement products. The pathway to taxol is considered to involve 10
steps from the diterpenoid progenitor geranylgeranyl diphosphate [53].

Currently, more than 40 taxane derivatives have been isolated from the needles,
stem and root bark, and heartwood of *Taxus baccata* [88, 106]. The compounds
(Fig. 5.7) are almost all esters of polyalcohols with tax-11-ene backbones. Hydroxy
groups may be located at C-atoms 1, 2, 5, 7, 9, 10, and 13. An oxo group can also be
attached to C-9 or C-13 (rarely also to C-10) instead of a hydroxy group. The methylene
group at C-4 can be converted into an oxirane ring (e.g., in baccatin I) or form an ox-
etane ring with the O-atom at C-5 (e.g., in taxol A and B).

Acetic acid, tiglic acid, 2-methylbutyric acid, 2-hydroxy-3-amino-3-phenylpropionic
acid, benzoic acid, and other occur as acid components of the esters. The aromatic acids
are mostly ester-linked to the hydroxy group at C-5 or C-13. The 2-hydroxy-3-amino-3-
phenylpropionic acid may in turn carry a benzoic or tiglic acid residue attached in
amide fashion. The hydroxy group at C-7 may be glycosidically linked to a D-xylose
residue.

Examples of taxane derivatives found in *T. baccata* include taxine A, taxine B,
taxol A (taxol, paclitaxel), taxol B (cephalomannine), taxol C, baccatins I–VII,
10-deacetylbaccatin III, and the taxinins [186]. According to Wilson et al. [284], 1 g of
yew needles contains an average of 5 mg of taxins, with taxin B reported to be the largest component of the *T. baccata* drug mixture and to have the greatest pharmacological
potency. Baccatin III, taxol A, and 10-deacetyl-baccatin III were found on the surface of
the twigs at concentrations ranging from 8 to 26 µg/1,000 g FW. The content varied
widely depending on the time of collection with highest amount found in January and
the plant part examined [95]. In contrast to the long-accepted assumption that the mature seed coat is free of taxine derivatives, recent studies have shown that the red arils
contain small amounts of 10-deacetylbaccatin III, baccatin III, and traces of cephalomannine, taxol A, and taxinine whereby the concentrations in the arils vary strongly
between the yews and the month of sampling [246].

The diterpenes of the other Taxus species resemble those of *Taxus baccata* [182].
Lignans [153] have also been identified in the wood of *T. baccata*. In *T. mairei* (Lemee

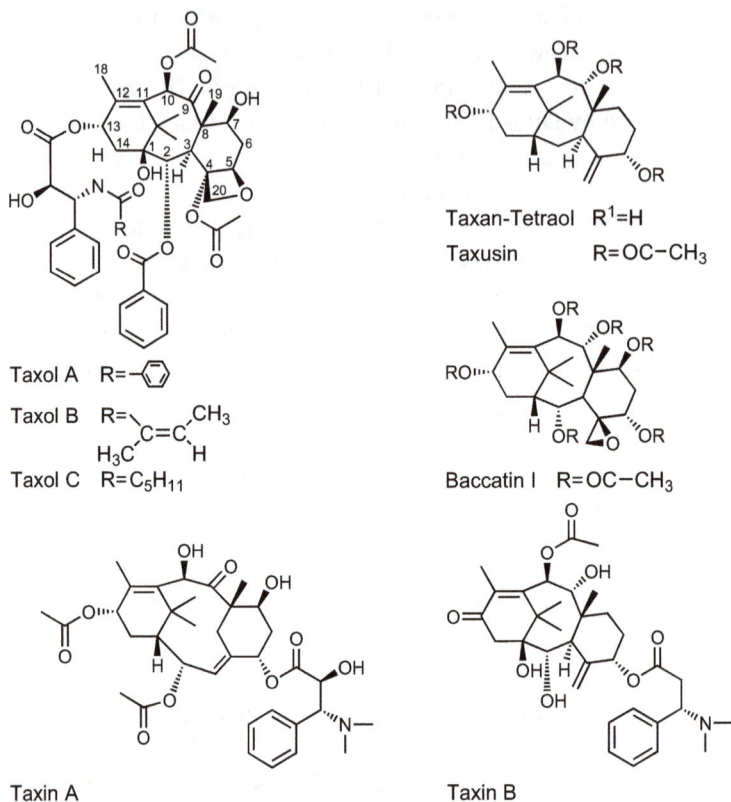

Taxol A    R=—⬡

Taxol B    R=\ /CH₃
              C=C
          H₃C'   'H

Taxol C    R=C₅H₁₁

Taxan-Tetraol   R¹=H

Taxusin         R=OC–CH₃

Baccatin I   R=OC–CH₃

Taxin A

Taxin B

**Fig. 5.7:** Taxane derivatives of Taxus species.

et Levl.) S.Y.Hu, Maire yew, native to China, the tropolone derivatives taxamairin A and B were found [164].

Because of its antitumor activity, in particular taxol A has been studied for its pharmacological properties. It binds specifically and reversibly to the microtubules of animal or human cells, rendering them nonfunctional. Thereby cell division in the late G2 phase is blocked [127].

Structural requirement for the proliferation inhibitory and thus antineoplastic effect (ED$_{50}$ of Taxol A for KB cells 0.01 ng/ml) are the intact taxane basic body, an ester side chain at C-13, and a hydroxy group at C-2 of the side chain [127, 174]. Taxines, especially taxine B, have negative inotropic and negative chronotropic effects. Cardiotoxicity is due to antagonistic attack at calcium and sodium ion channels of the heart [9, 284].

Taxamairins have also a cytotoxic effect [164]. The lignans show anti-inflammatory and, like some taxanes, antinociceptive activity [153].

Total extracts from yew leaves stimulate smooth and striated muscles but have a paralyzing effect on the heart muscles. Blood pressure is lowered, respiration is initially stimulated, and later paralyzed. Hypersensitivity has been observed [34].

The lethal dose of needles after p.o. administration is 0.2–0.4 g/kg BW for horses, 0.7 mg/kg BW for pigs and rabbits, 2.0 mg/kg BW for cattle, 2.3 mg/kg BW for dogs, 12.0 mg/kg BW for goats, and 16.5 mg/kg BW for chicken. Ingestion of needles in an amount of 0.6–1.3 g/kg BW (equivalent to 3.0–6.5 mg of taxine) is considered as lethal to humans [284].

The toxicity of yew was already known in ancient times. The plants played a particularly important role in the cult of the dead (death tree) and were supposed to protect against demons and lightning. Extracts of the needles were used as murder, suicide, and arrow poisons. Even a longer stay under the trees was considered dangerous [83, 149].

Today, the seeds with the attractive red seed coat are often a cause for toxicological consultation cases. However, since the aril contains no (or only very low amounts) toxins and the seeds are usually swallowed unchewed, serious poisoning rarely occurs. Chewing the yew needles, on the other hand, is dangerous. An infusion made from 50 to 100 g of needles may be fatal. Even in more recent times, the needles or the bark or extracts made from *T. baccata* or *T. cuspidata* have been used in suicide attempts, suicides, and for other reasons [13, 85, 103, 117, 134, 145, 193, 210–212, 258, 266, 269, 274].

† A 14-year-old boy died after following instructions from the Internet to commit suicide with yew needles. For this purpose, he had cut off the needles in his parents' garden, chopped them up, and ate them. When the corpse was opened by forensic medicine, the stomach contained partly crushed and partly completely preserved needles. Histopathological findings were nonspecific, such as a clear general blood congestion of the internal organs and a pronounced brain and lung edema [280].

† A 40-year-old man who had ingested an extract of about 120 g of yew needles with suicidal intent developed severe dizziness after about 1 h and a little later severe nausea. He was admitted to the hospital in a dazed state. ECG showed bradyarrhythmia (40 beats/min) changing to tachycardia and bizarre, very wide ventricular complexes. Blood pressure was decreased. Treatment with lidocaine substantially improved the condition, and the patient was discharged after four days without further arrhythmias [274].

🜊 Symptoms of poisoning after ingestion of yew needles or chewed seeds begin about one hour (to 24 h) after ingestion with vomiting, abdominal pain, diarrhea, and colic. This is accompanied by dizziness, unconsciousness, and a red coloration of the lips. After initial acceleration of respiration and pulse, cardiac and respiratory activity slow down, and respiration becomes more and more superficial until death finally occurs from respiratory paralysis and/or cardiac failure [117, 145, 193, 211, 266, 284].

📠 Ingestion of about three chewed seeds or more requires gastric emptying and administration of activated charcoal. If the needles are ingested and remain in the stomach for a long time, gastric emptying is necessary, even after several hours. Endoscopic removal of plant residues can be helpful.

Glauber's salt must be given in addition to activated charcoal. Inpatient monitoring (ECG for at least 24 h) and symptom-oriented therapy, e.g., atropine for bradycardia, and digitalis antidote or i.v. calcium gluconate for severe arrhythmias, are necessary (**161**). P.e. administration of lidocaine [274] or lipid resuscitation (i.v. infusion of a high dose lipid emulsion to reduce the absorption of the lipophilic toxins) may improve prognosis. The early use of extracorporeal life support is recommended in patients presenting with polymorphic ventricular tachycardia and cardiogenic shock [117, 258].

🐾 Taxus needles are among the most dangerous plants for animals. Cause of poisoning is often careless handling of hedge cuttings. For example, poisoning has been reported in cattle [203], goats [47], bears [15], emus [81], moose, reindeer [108], and turtles [282]. Horses react particularly sensitive. While death occurs within about one hour in horses, the course of disease is much slower in cattle. Symptoms of poisoning in cattle include restlessness, roaring, convulsions, staggering, diarrhea, bradycardia, dyspnea, and collapse. Deaths may occur even after several days [60]. The fruits are eaten in large quantities by birds, especially blackbirds and greenfinches, without harm. The seeds are apparently not bitten.

In folk medicine, yew preparations were used as anthelmintics or abortives. Taxol A (paclitaxel) and its derivatives docetaxel and cabazitaxel are important cytostatics in tumor therapy today. Important side effects are myelosuppression and peripheral neuropathy.

Taxol A, 10-deacetylbaccatin III, baccatin III, paclitaxel C, and 7-epipaclitaxel have also been found in shells and leaves of hazel plants, *Corylus avellanus* L. [200]. Since these plant parts are not eaten, there is no risks for humans. The taxane derivatives are probably produced by endophytic fungi, predominantly ascomycetes. Extracts of *C. avellana* reduce the viability of human tumor cells [89].

## 5.5 Diterpenes as Hallucinogenic Compounds of Sage of the Diviners (*Salvia divinorum*)

*Salvia divinorum* Epling et Játiva, sage of the diviners, herb of the virgin, Lamiaceae, is a hallucinogenic sage species native to the mountain forests of the Sierra Mazateka in the state of Oaxaca of Mexico. Mazatecs who live there use *S. divinorum* in low doses for medicinal purposes and in higher doses for divination ceremonies. They chew the leaves or brew them into tea to facilitate spiritual encounters. Shamans use only the fresh leaves. Cultivation and use were long concealed from strangers.

The plant (Ph. 5.17) is a perennial that grows to over 1 m tall with conspicuous winged shoot axils. The plant can grow up to 2.50 m high in the greenhouse. The active substances are excreted via trichomes.

It was not until the early 1960s that R. Gordon Wasson and A. Hofmann became aware of the plant during ethnopharmacological expeditions in Mexico. The first

Ph. 5.17: *Salvia divinorum* (source: nndanko/iStock/Getty Images Plus).

botanical description was in 1962, after which cultivation and use spread first to North America and in the 1980s to Europe and other parts of the world.

S. *divinorum* contains as main active ingredient the neoclerodane diterpene salvinorin A (0.089–0.37%, with an average of 0.245%, first named divinorin A, Fig. 5.8), which has not yet been found in other Salvia species (see Section 3.2.1). Other constituents are the salvinorins B–F, the salvinicins A and B, which also have a neoclerodane basic body and the divinatorins A–C (derivatives of the antibiotically active hardwickiic acid [24, 54, 104, 110, 190]).

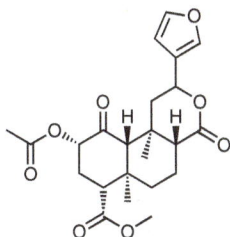

Salvinorin A

Fig. 5.8: A diterpene of *Salvia divinorum*.

Salvinorin A is a highly selective κ-opiod receptor agonist and an allosteric modulator of cannabinoid type 1 (CB$_1$) receptors. It is responsible for the hallucinogenic effects of sage of diviners [227]. It shows no action at serotoninergic receptors, which are the main targets for classical hallucinogens like LSD or psilocybin. Moreover, salvinorin A inhibits biosynthesis of leukotrienes related with inflammatory processes [35, 48, 54]. Acute or chronic toxic effects of salvinorin or the leaves of S. *divinorum* beyond psychic effects have not been clearly demonstrated. However, the number of conclusive studies is still small [48, 104].

The dried leaves of *S. divinorum* are smoked (usually in hookahs known as 'bongs'), or the fresh leaves are rolled up and chewed for prolonged period or leached by remaining under the tongue. Absorption occurs through the lungs or oral mucosa. When smoked, the effect begins within 30 s, peaks after 2–5 min, and dissipates within 30 min. With oral use, it begins after 10 min and subsides after 1–2 h. Effects in humans are expected from about 0.25 mg salvinorin A [54, 104, 226, **125**].

*Salvia divinorum* and its use as recreational drug to achieve a euphoric effect are propagated worldwide today via the Internet. Preparations are also available in scene stores. In 2008, the US Drug Enforcement Administration estimated that nationally 1.8 million people, mainly of them being young, with ages 18–25 years, have used *S. divinorum* in their lifetime [226]. In the United States, *S. divinorum* use is prevalent among recent or active drug users who have used other hallucinogens or stimulants and associated [288].

⚕ Hallucinations caused by the leaves of *Salvia divinorum* are short and extremely intense. They include the appearance of bizarre effects, for example, curvatures of space or rolling body sensations are perceived. Unlike many other hallucinogens, mainly perception and logical thinking are altered, but hardly any emotion. Some users describe hearing colors or smelling sounds (synesthesia) while others endorse an 'out of body experience'. The high is often felt as a drastic to distressing rather than a euphoric experience. Complete loss of reality and numerous rash actions may be triggered. Drowsiness/lethargy and tachycardia may occur, but vital signs were largely normal. Effects can last more than 24 h. Fatal cases have not been reported. Of 500 users surveyed, less than 1% reported feelings of addiction or dependence [23, 54, 221, 226, **125**]. Long-term damage has not yet been researched.

⚕ There are no clear recommendations for clinical management of *S. divinorum* expositions. Supportive care including sedation with benzodiazepines for severe agitation is prudent [226].

The prohibitive degree of *S. divinorum* varies widely from country to country. In Australia and Denmark, its use has been banned already since 2002 and 2003, resp. In the United States, the regulation of *S. divinorum* varies between several federal states. Control exists presently (2021) in 33 states and territories of the US. It is restricted in many European countries including the United Kingdom and Germany (here included in Schedule I of the Narcotics Act. that means not marketable), in Canada, Japan, South Korea, Brazil, and Armenia. In many other countries *S. divinorum* is legal.

Salvinorin A is a useful tool to study the role of the κ-opiod receptor system that has become a therapeutic target for the treatment of neurological and psychiatric diseases. At low doses, salvinorin A shows anxiolytic, antidepressant, neuroprotective, analgesic, and anti-inflammatory activities. The clinical usefulness of salvinorin A, its analogues (able to produce long-lasting analgetic and anti-inflammatory effects but without the hallucinatory component) and the leaves of *S. divinorum* in chronic pain and mental disorders is in investigation. Salvinorin A analogues hold promise for the development of anti-addiction drugs, e.g., as treatments for cocaine abuse [35, 48, 54, 104].

## 5.6 Forskolin as Toxin of Coleus and Plectranthus Species

Forskolin (coleonol, colforsin, Fig. 7.9) is a labdane-type diterpene, found in *Coleus for-skohlii* (Willd.) Briq. (*Plectranthus forskohlii* Willd., Ph. 5.18), Indian coleus, and *Plectran-thus barbatus* Andrews (*Coleus barbatus* (Andrews) Benth. ex G. Don), false boldo, both Lamiaceae. The species are distributed from tropical Africa to India, Pakistan, and China. *Coleus forskohlii* is used in traditional medicine of India (Hindu and Ayurveda) to treat various diseases of the cardiovascular, respiratory, gastrointestinal, and central nervous systems.

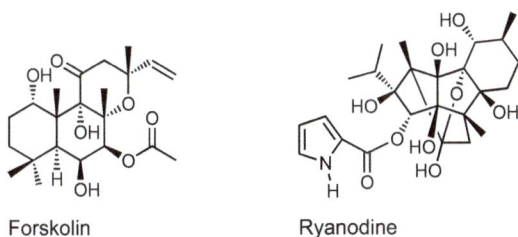

Forskolin        Ryanodine

Fig. 5.9: Diterpenes of Coleus and Ryania species.

Ph. 5.18: *Coleus forskohlii* (source: Barbara Liepke).

Forskolin is a direct, rapid, and reversible activator of adenylate cyclase and promotes the formation of cAMP. The compound directly targets the catalytic subunit of adeny-late cyclase and affects numerous cAMP-dependent processes. Other targets include membrane transport proteins, e.g., glucose transport proteins, and channel proteins, e.g., of the nicotinic acetylcholine receptor. Forskolin exhibits cardiotonic, spasmolytic, and hypotensive effects and inhibits platelet aggregation. It increases the chemosensi-tivity of tumor cells. In animal assays, forskolin stimulates lipolysis, thermogenesis, and basal metabolic rate and leads to reduction of body weight. However, related studies in human are inconclusive [8, 18]. In addition, forskolin leads to blood pressure drop, car-diovascular problems, nausea, headache, and metabolic disturbances [18].

Reports about intoxications caused by the forskolin containing plants are not known to us.

Forskolin and its better water-soluble derivatives are used as model substances to study processes mediated by adenylate cyclase and the role of cAMP [7].

Forskolin containing products have received popular attention as a weight loss agent. Because of possible toxic effects it is necessary to warn against taking it. Interactions with antihypertensive and anticoagulant drugs can be expected [18].

## 5.7 Ryanodine as Toxin of Ryania Species

Ryanodine (Fig. 7.9) and 9,21-didehydro-ryanodine are esters of the diterpenes ryanodol and dehydroryanodol, resp., with a pyrrole-2-carboxylic acid. They are formed by Ryania species native to tropical America, especially *Ryania angustifolia* (Turcz.) Monach, narrow leaved ryania, and *R. speciosa* Vahl, Salicaceae. The plants are long been recognized for their insecticidal properties. In 1948, ryanodine was identified as the compound responsible for the action [224].

Ryanodine was subsequently found to induce paralysis of cardiac and skeletal muscle and used to track the purification of the relevant receptors from sarcoplasmatic reticulum preparations, the so-called ryanodine receptors (RyRs). RyRs are huge ion channels that are responsible for the release of $Ca^{2+}$ from the sarco/endoplasmatic reticulum and thus control many $Ca^{2+}$-dependent processes within the cell [270]. Ryanodine binds RyRs preferentially in the open state. At concentrations > 100 μM it inhibits $Ca^{2+}$ release. $LD_{50}$ values for the rat are 750 mg/kg BW, p.o., and 750 mg/kg BW, s.c. [154, 99, 140]. The RyR is also activated by cardioactive steroid glycosides (see Section 8.3.5).

In humans, gastrointestinal and central symptoms develop after ingestion of plant parts of Ryania species. External contact results in severe irritation of the skin.

The powdered stem wood and extracts of Ryania species are used as selective feeding and contact poisons against various pests (140). In their native habitat, the plants serve to poison alligators (113).

## 5.8 Diterpenes as Hepatotoxins of Germander Species (Teucrium Species) and Bitter Grape (*Tinospora crispa*)

Teucrin A (Fig. 5.10) and teuchamaedryn A are furano-clerodane diterpenes responsible for the hepatotoxicity of the wall germander, *Teucrium chamaedrys* L. (Ph. 5.19), Lamiaceae. Teucrium species are found all over the world, most commonly in Mediterranean climates. *Teucrium chamaedrys* is native to Mediterranean region of Europe and North

Africa and to Asia. Extracts of the plant have been used, e.g., for treatment of gout, digestive disorders, and overweight.

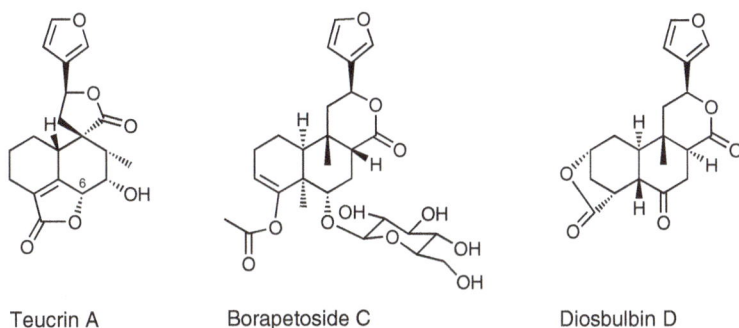

Teucrin A            Borapetoside C            Diosbulbin D

**Fig. 5.10:** Furano diterpenes of Teucrium species and *Tinospora crispa.*

**Ph. 5.19:** *Teucrium chamaedrys* (source: Orest Lyzhechka/iStock/Getty Images Plus).

Teucrin A and teuchamaedryn A came to the attention first in 1992 in France after the poisoning of patients which had ingested Teucrium extracts for treatment of obesity [207]. Since then, other cases of poisoning by this or other Teucrium species, e.g., *Teucrium capitatum* L. and *T. polium* L., golden germander, have been observed [55, 64, 167, 218, 251, 252]. A prerequisite for toxicity is CYP450 3A-dependent activation of teucrin A to electrophilic metabolites. These bind to hepatic proteins, deplete hepatic stores of glutathione and cytoskeleton-associated protein thiols, form plasma membrane blebs and cause apoptosis of liver cells [30, 82, 158].

In the past, poisonings occurred during medical application of commercial germander extracts. Today, the use of germander as a food additive or as a medicinal plant is no longer approved or is restricted. Nevertheless, intoxication cases have been due to drinking tea from leaves of locally collected germander plants [99, 167].

> ☠ Typical symptoms of poisoning, which begin after a latency period of about 2 months, include vomiting, abdominal pain, jaundice, and weight loss. Most patients recovered rapidly upon withdrawal of germander, but few fatal cases were recorded [167].

> 🚂 Hepatotoxicity is usually self-limiting once the ingestion is discontinued. Rechallenge should be avoided. In acute cases of hepatotoxicity administration of *N*-acetylcysteine is recommended [167]. Liver transplantation may become necessary [180].

***Tinospora crispa*** (L.) Hook. F. & Thomson (Menispermaceae), bitter grape, is a popular traditional medicinal plant in Asia, used for the treatment of diabetes. Several cases of hepatotoxicity have been observed and led to the identification of 18 furano-diterpenoids structurally related to the major compound borapetoside C (Fig. 5.10). A hepatotoxic mechanism similar to that of Teucrium is hypothesized [38].

Diosbulbins B and D (Fig. 5.10) are hepatotoxic furano norclerodane diterpenoids isolated from *Dioscorea bulbifera* L. (Dioscoreaceae), coralroot, widely distributed in tropical and subtropical regions [162].

## 5.9 Diterpene Glycosides as Toxins of Mastix Thistle (*Carlina gummifera*), Cocklebur (*Xanthium strumarium*), Ox-Eye Daisy (*Callilepis laureola*), and Wedelia Species

Norditerpenes of the kaurane type are the aglycones (atractyligenin) of the highly toxic atractylosides (Fig. 5.11) and of related glycosides. They are obtained together with carboxy-atractyloside (gummiferin) from the rhizome of **Carlina gummifera** (L.) Less. (*Chamaeleon macrophyllus* (Desf.) D.P. Petit, *Atractylis gummifera* L. (Ph. 5.20), mastix thistle, glue thistle, Asteraceae [56]. *C. gummifera* is common in the Mediterranean region, where it is used in folk medicine. It was also used for suicides and as a murder poison.

Carboxyatractyloside is the main toxin in seeds and seedlings of **Xanthium strumarium** L., common cocklebur, California burr and *X. strumarium* ssp. *sibiricum* (Patrin ex Widder) Greuter, Asteraceae. The toxins are preferentially present in the cotyledon stage of *X. strumarium* and are diminished after germination of the seeds [49, 78, 256]. *X. strumarium* is native to North America and invasive worldwide. It

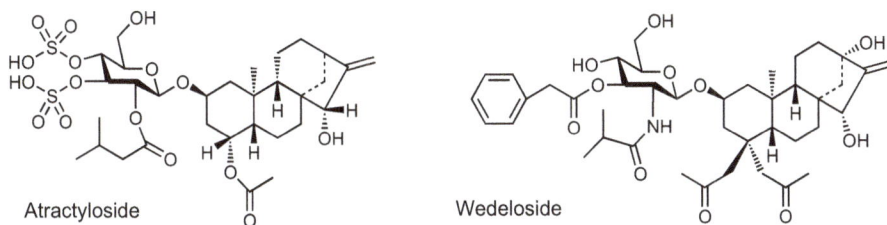

**Fig. 5.11:** Diterpene glycosides of Carlina, Xanthium, Callilepis, and Wedelia species.

**Ph. 5.20:** *Carlina gummifera* (source: Peter Schönfelder).

is used in TCM and in traditional medicine in South Asia and of some North American Indians [78].

Atractyloside is also present in the rhizome of **Callilepis laureola** DC., the ox-eye-daisy, Asteraceae. Its extracts are used rectally or perorally by the Zulu in South Africa to protect against 'evil spirits' but also to treat stomach problems and impotence. The preparations, called 'Impila', cause liver and kidney necrosis. The plant was identified as the cause of fatal poisonings in children [36, 252].

**Wedelia glauca** (Ortega) Hoffm. ex Hicken (Asteraceae), indigenous to South America, is also reported to contain atractyloside [236]. In *W. asperrima* Benth., cause of animal poisoning in Australia and South America, the related diterpene aminoglycoside wedeloside and its 4'-*O*-rhamnosyl analog have been identified (Fig. 5.11, [160, 169]).

Analogues of atractyloside can be found in small amounts in coffee beans (34.5–624 mg/kg in green beans, 17.5–32.5 mg/kg in roasted beans). Considering the low content and the fact that some of the analogues are nontoxic, the risk related to coffee consumption is presumably negligible [36, 56]. Mention should also be made of the kaurane-type diterpenes kahweol and cafestol. Esters of these alcohols with fatty acids can be found in the oil of Coffeae semen.

Atractyloside is a mitochondrial uncoupler and inhibits the oxidative phosphorylation of ADP. It binds specifically to the adenine nucleotide carrier in the inner mitochondrial membrane. This translocator is responsible for the ATP/ADP antiport and involved in mitochondrial membrane permeabilization. The action of atractyloside is exerted especially in cells rich in mitochondria such as hepatocytes and in proximal tubular epithelial cells, which contain carriers that allow atractyloside to cross cell membranes. Consequently, competitive inhibition of ADP and ATP transport by atractyloside leads to hepatic necrosis and renal failure [36, 56, 140].

The $LD_{50}$ of atractyloside is 143 mg/kg BW, i.p., in rats, and 15 mg/kg BW, i.v., in dogs and of carboxyatractyloside 2.9 mg/kg BW, i.p., in rats [157]. In mice, the $LD_{50}$ of an infusion from the root of *Carlina gummifera* is 11.4 mg/kg BW, p.o [177]. Comparably to atractyloside wedeloside inhibits the mitochondrial ADP/ATP transport. The $LD_{50}$ of wedeloside is 1 mg/kg BW, i.p., mouse. Experimental feeding of *Wedelia glauca* to sheep and cows (4–10 g/kg BW) caused primarily liver damage. The minimum toxic dose of *W. glauca* is 4–5 g fresh plant/kg BW of animals [50].

Causes of poisonings by *Carlina gummifera* are accidental ingestion of the plants, overdosage during folk medicinal use, or intended use for suicides or as murder poison. In regions where *C. gummifera* grows it can be easily confused with wild artichoke. Children chew the root like chewing gum because of its sweet tasting juice [56, 261].

🚹 A 7-year-old boy who had been given an extract from the roots of mastic thistle as a vermifuge was admitted to a hospital two days later in an unresponsive state, with abdominal pain and vomiting. The EEG showed generalized cerebral edema, respiratory activity was impaired, and the liver was enlarged. Intense jaundice developed. Despite intensive symptomatic treatment, the boy died on day 8 after receiving the root extract [92].

🚹 The consumption of *Xanthium strumarium* (in Bangladesh named Ghagra shak) seedlings in large amounts, due to inaccessibility of other foods during destructive monsoon floods in November 2007, caused fatal outbreak in Bangladesh. Seventy-six people in nine villages were affected. Nineteen persons (25%) died [107].

🧎 Symptoms of poisoning with plant parts containing atractyloside include nausea, vomiting, elevated liver enzymes, epigastric and abdominal pain, diarrhea, anxiety, headache, altered mental status, and convulsions, often followed by coma. Poisoning is often fatal due to renal or hepatic failure [56, 107, 252].

💊 Therapeutic approaches are only symptomatic and include fluid and electrolyte replacement, cardiovascular and respiratory support, seizure control, and conventional therapeutic methods for severe hepatic and renal failure. A new approach could be the development of specific Fab fragments against the toxic components [56].

In TCM, the toxicity of atractyloside containing plants is reduced by hydrothermal processing [43].

🐂 Poisonings in livestock have been reported after consumption of seeds and seedlings of *Xanthium strumarium* [256, 286]. *Wedelia glauca* causes livestock poisonings in South America and *W. asperrima* in Australia and in South America [160, 169, 236].

## 5.10 Diterpenes of Seaweeds (Marine Macroalgae)

Among the **green algae**, Chlorophyta, diterpenes are found preferentially in representatives of the families Caulerpaceae, Codiaceae, and Udotaceae (Caulerpales) [206]. They are almost of aliphatic or of monocyclic structure and often possess a terminal 1,4-diacetoxybutadiene group (masked dialdehyde grouping), e.g., in a diterpene from *Penicillus dumetosus* [205]) or two 1,4-diacetoxybutadiene groups in chlorodesmin from *Chlorodesmis fastigiata* (Fig. 5.12 [281]).

In the **brown algae**, Phaeophyta, aliphatic diterpenes occur, e.g., in the genera Dictyota and Cystoseira. There is a tendency to form macrocycles (e.g., pachylactone from *Pachydiction coriaceum* [130]), but these can also be secondarily contracted by bridges, e.g., in the case of spatol from *Spatoglossum schmittii* (Fig. 5.12 [94]). Reactive groups include:

- epoxy groups, e.g., in spatol,
- dialdehyde groupings, e.g., in the dictyodial from Dictyota species, [70]),
- terminal furan rings (masked dialdehydes), e.g., in the taonianon from *Taonia australasica*, [191]) or
- hydroquinone residues that can be transformed to *p*-quinones, e.g., in mediterraneol A from *Cystoseira stricta* [11].

However, compounds without reactive groups also have pronounced biological effects, apparently due to their membrane activity. Because of their unpleasant taste they repel herbivorous fishes. Species of the genus Dictyota have been particularly intensively studied and are a promising source of compounds with antiviral, cytotoxic, and chemical defensive activities. About 200 diterpenoids, belonging to 15 chemical classes, have been isolated from this genus, among them the pachydictyols A–C ([3, 42], Fig. 5.12). Dollabelladienol A from the Brazilian marine species *Dictyota pfaffii* [204] and 13-deacetoxyamijdictyol from *D. plectens*, isolated from the South China sea [45], are potential anti-HIV agents.

The **red algae**, Rhodophyta, preferentially form carbocyclic six-rings (more rarely five-rings), which are often fused and brominated. The brominated diterpenes have 20

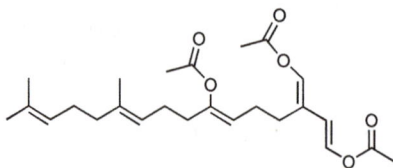

Diterpene from *Penicillus dumetosus*

<u>Chlorophyta</u>

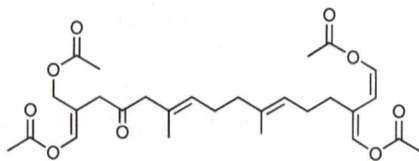

Chlorodesmin

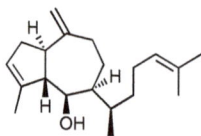

Pachydictyol A

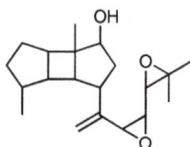

Spatol

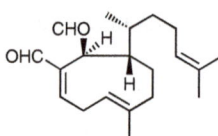

Dictyodial

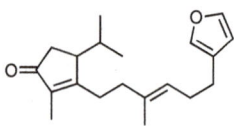

Taonianon

<u>Phaeophyta</u>

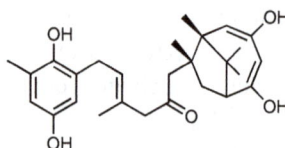

Mediterraneol A

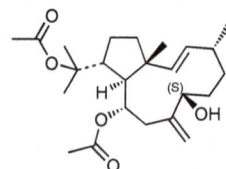

Dolabelladienol A

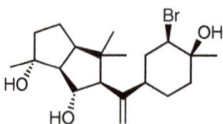

Neorogioltriol

<u>Rhodophyta</u>

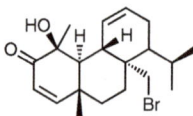

Sphaerococcenol A

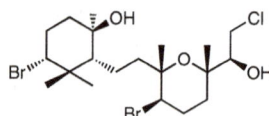

Laurencianol

**Fig. 5.12:** Diterpenes of macroalgae.

different basic bodies, which are especially often parguerane, isoparguerane, irieane, labdane, and sphaerane. About 70% of the brominated compounds found to date are from various Laurencia species and about 30% are from *Sphaerococcus coronopifolius*. A labdane-type brominated diterpene was obtained from *Chondria tenuissima*. Some compounds were also found in Aplysia species and were probably derived from the diet of sea hares [146].

Examples of diterpenes from red algae include neorogioltriol, a tricyclic brominated diterpenoid, and related compounds from *Laurencia glandulifera*, sphaerococcenol A from the red alga *Sphaerococcus coronopifolius* [72], which occurs on the Atlantic coast of Morocco, and laurencianol, obtusadiol, rogioldiol A, and prevezoles A and B from *Laurencia obtusa* ([37, 185], Fig. 5.12).

Biological tests have shown antimicrobial, cytotoxic, and ichthyotoxic effects [37, 146, 205]. Sphaerococcenol A has antimalarial activity [72]. Neorogioltriol exhibits analgesic and anti-inflammatory properties, e.g., in a mouse colitis model [41, 57].

# References

(for numbers in bold see cross chapter literature)

[1]    Abdu-Aguye I et al. (1986) Hum Toxicol 5: 269
[2]    Abou-Arab AA et al. (2019) Heliyon 5: e01689
[3]    Abou-El-Wafa GSE et al. (2013) Marine Drugs 11: 3109
[4]    Adolf W, Hecker E (1975) Z Krebsforsch 84: 325
[5]    Adolf W, Hecker E (1984) J Nat Prod 47(3): 482
[6]    Adolf W, et al. (1984) Phytochemistry 23(7): 1461
[7]    Alasbahi RH, Melzig MF (2012) Pharmazie 67(1): 5
[8]    Alasbahi RH, et al. (2010) Part 1: Planta Med 76(7): 653. Part 2: Planta med 76 (8):753
[9]    Alloatti G et al. (1996) Life Sci 58: 845
[10]   Alonso-Castro AJ et al. (2017) eCAM 2017: 9439868
[11]   Amico V et al. (1989) J Nat Prod 52(5): 962
[12]   Appendino G (2016) Ingenane Diterpenoids. In: Kinghorn A et al. ed. Progr Chem Org Nat Prod, vol. 102. Springer, cham, https://doi.org/10.1007/978-3-319-33172-0-1
[13]   Arens AM et al. (2016) Clin Toxicol 54: 878
[14]   Asilian A, Faghihi G (2004) Contact Dermat 51(1): 37
[15]   Bacciarini LN et al. (1999) Eur J Pathol 5: 29
[16]   Bagavathi R et al. (1988) Planta Med 54(6): 506
[17]   Bao GH et al. (2003) Planta Med 69(5): 434
[18]   Barrea L et al. (2019) Int J Obesity Suppl 9: 32
[19]   Baslas RK, Gupta NC (1984) Herba Hung 23: 67
[20]   Bertero A et al. (2020) Front Vet Sci 7: 487
[21]   Betancur-Galvis L et al. (2003) J Ethnopharmacol 85(2-3): 279
[22]   Beutler JA et al. (1989) Phytother Res 3(5): 188
[23]   Biggs JM et al. (2017) J Pediatr Pharmacol Ther 22(6): 385
[24]   Bigham AK et al. (2003) J Nat Prod 66(9): 1242
[25]   Bischoff K et al. (2014) J Med Toxicol 10: 411
[26]   Blumberg PM, Boutwell RK (1980/81) Crit Rev Tox 8: 199
[27]   Bolton JF (1955) Vet Rec 67: 138
[28]   Borris RP, Cordell GA (1984) J Nat Prod 47(2): 270
[29]   Borris RP et al. (1988) J Ethnopharmacol 24: 41
[30]   Brewer CT et al. (2017) Int J Mol Sci 18: 2353
[31]   Brooks G et al. (1989) Carcinogenesis 10: 283
[32]   Broscaru L et al. (2017) Eur J Case Rep Int Med (EJCRIM) 2017 (4): 10.12890/2017_000742
[33]   Burke JW et al. (1989) J Am Chem Soc 111: 5831
[34]   Burke MJ et al. (1979) N Y State J Med 79: 1576
[35]   Butelman ER, Kreek MJ (2015) Front Pharmacol 6: 190
[36]   Bye SN et al. (1990) Toxicon 28: 997
[37]   Caccamese S et al. (1982) Tetrahedron Lett 23: 3415
[38]   Cachet X et al. (2018) Scient Rep 8: 13520

[39]  Caloca MJ et al. (2001) J Biol Chem 276: 18303

[40]  Carey FM et al. (1959) J Pharm Pharmacol 11: 269T

[41]  Chatter R et al. (2011) Mar Drugs 9: 1293

[42]  Chen J et al. (2018) Mar Drugs 16: 159

[43]  Chen LY et al. (2013) Molecules 18: 2018

[44]  Chen SN et al. (2004) J Nat Prod 67(11): 1903

[45]  Cheng S et al. (2014) J Nat Prod 77(12): 2685

[46]  Chung S et al. (1980) Planta Med 38(3): 269

[47]  Coenen M, Bahrs F (1994) Dtsch Tierärztl Wochenschr 101: 364

[48]  Coffeen U, Pellicer F (2019) J Pain Res 12: 1069

[49]  Cole RJ et al. (1980) J Agric Food Chem 28: 1330

[50]  Collazo L, Riet-Correa F (1996) Vet Hum Toxicol 38: 200

[51]  Corea G et al. (2004) J Med Chem 47: 988

[52]  Crawford JE (1999) Vet Rec 144: 680

[53]  Croteau R et al. (2006) Phytochem Rev 5(1): 75

[54]  Cunningham CW et al. (2011) Pharmacol Rev 63: 316

[55]  Dağ M et al. (2014) Ann Saudi Med 34(6): 541

[56]  Daniele C et al. (2005) J Ethnopharmacol 97: 175

[57]  Daskalaki MG et al. (2019) Mar Drugs 17(2): 97

[58]  Dayasiri MBKC et al. (2017) Int J Pediatr 2017: 6187481

[59]  de Almeida PM et al. (2015) Gen Mol Biol 38(1): 93

[60]  Demir H et al. (2011) Int Scholar Res Network ISRN Toxicol 2011: 526426

[61]  Desel H, Neurath H (1998) Toxichem Krimtech 65: 63

[62]  Dogan FS et al. (2015) Turk J Emerg Med 15: 185

[63]  Dong LC et al. (2014) J Ethnopharmacol 157: 69

[64]  Dourakis SP et al. (2002) Eur J Gastroenterol Hepatol 16: 693

[65]  Eberle MM et al. (1999) Klin Monatsbl Augenheilkd 215: 203

[66]  Edwards N (1983) J Pediat 102: 404

[67]  Eijkman IF (1882) Recl Trav Chim Pays Bas 1: 224

[68]  Eke T et al. (2000) Arch Ophthalmol 118: 13

[69]  El-Naggar SF et al. (1980) J Nat Prod 43(5): 617

[70]  Enoki N et al. (1983) Chem Lett 1749 and 1837

[71]  Ergun K et al. (2005) Int J Cardiol 99: 347

[72]  Etahiri S et al. (2001) J Nat Prod 64(8): 1024

[73]  Evans FJ (1986) Naturally Occurring Phorbol Esters. Evans FJ (ed.) CRC Press, Boca Raton, p. 139

[74]  Evans FJ, Kinghorn AD (1977) Botan J Lin Soc 74: 23

[75]  Evans FJ, Soper CJ (1978) J Nat Prod 41: 193

[76]  Evans FJ, Taylor SE (1983) Prog Chem Org Nat Prod 44: 1

[77]  Evans FJ et al. (1975) Acta Pharmacol Toxicol 37: 250

[78]  Fan W et al. (2019) Molecules 24: 359

[79]  Fattahian M et al. (2020) Phytochem Rev 19(2): 265

[80]  Félix-Silva J et al. (2014) eCAM 2014: 369204

[81]  Fiedler HH, Perron RM (1994) Berl Münchn Tierärztl Wochenschr 107: 50

[82]  Frenzel C, Teschke R (2016) Int J Mol Sci 17: 588

[83]  Friese W (1951) Pharm Zentralhalle 90: 259 and 289

[84]  Frohn A et al. (1993) Ophthalmologe 90: 58

[85]  Frohne D, Pribilla O (1965\66) Arch Toxicol 21: 150

[86]  Fujiki H et al. (2017) J Cancer Res Clin Oncol 143: 1359

[87]  Fürstenberger G, Hecker E (1972) Planta Med 22(7): 241

[88]   Gabetta B et al. (1998) Phytochemistry 47(7): 1325
[89]   Gallego A et al. (2017) Biomed Pharmacother 89: 565
[90]   Gandhi VM et al. (1995) Food Chem Toxicol 33: 39
[91]   Geiges D et al. (1997) Biochem Pharmacol 53: 865
[92]   Georgiou M et al. (1988) J Toxicol Clin Toxicol 26: 487
[93]   Gerke R et al. (2003) Internist 44: 1308
[94]   Gerwick WH et al. (1980) J Am Chem Soc 102: 7991
[95]   Glowniak K et al. (1999) Phytomedicine 6(2): 135
[96]   Goel G et al. (2007) Int J Toxicol 26: 279
[97]   Gómez-Valcárcel M, Fuentes-Páez G (2016) Case Rep in Ophthalmol 7: 125
[98]   Gössinger H et al. (1983) Dtsch Med Wochenschr 108: 1555
[99]   Gori L et al. (2011) Basic Clin Pharmacol Toxicol 109: 521
[100]  Gotta H et al. (1984) Z Naturforsch B: Anorg Chem Org Chem 39 B 683
[101]  Graf LE et al. (1982) Liebigs Ann Chem 2: 376
[102]  Grimbs A et al. (2017) Front Plant Sci 8: 551
[103]  Grobosch T et al. (2013) Forensic Sci Int 227: 118
[104]  Grundmann O (2007) Planta Med 73(10): 1039
[105]  Günther G et al. (1998) Phytochemistry 47(7): 1309
[106]  Guo Y et al. (1995) J Nat Prod 58(12): 1906. Erratum in J Nat Prod 59 (10,1996): 1002
[107]  Gurley ES et al. (2010) PLoS ONE 5(3): e9756
[108]  Handeland K et al. (2017) PLoS ONE 12(12): e0188961
[109]  Hanson JR (2006) Nat Prod Rep 23: 875
[110]  Harding WW et al. (2005) Org Lett 7(14): 3017
[111]  Hausen BM (2005) Aktuelle Derm 31: 167
[112]  Hecker E (1968) Cancer Res 28: 2338
[113]  Hecker E (1971) In: Pharmacognosy and Phytochemistry. Wagner H, Hörhammer L (eds.) Springer, Berlin
[114]  Hecker E (1977) Pure Appl Chem 49: 1423
[115]  Hecker E (1987) Bot J Linnean Soc 94: 197
[116]  Hergenhahn M et al. (1984) J Cancer Res Clin Oncol 108: 98
[117]  Hermes-Laufer J et al. (2021) ESC Heart Fail 8: 705
[118]  Hikino H et al. (1972) Chem Pharm Bull (Tokyo) 20: 1090
[119]  Hikino H, et al. (1976) Toxicol Appl Pharmacol 35: 303
[120]  Hill MWM (1970) Aust Vet J 46(6): 287
[121]  Hirata Y (1975) Pure Appl Chem 41: 175
[122]  Hirata Y (1987) Bioact Mol 2: 181
[123]  Hirota M et al. (1988) Cancer Res 48: 5800
[124]  Hohmann J et al. (1999) Phytochemistry 51(5): 673
[125]  Hohmann J et al. (2004) In: Poisonous Plants and Related Toxins. Acamovic T (ed.), et al. CABI Publ, Wallingford, p. 96
[126]  Hollands RD, Hughes MC (1986) Vet Rec 118: 407
[127]  Horwitz SB et al. (1986) Ann New York Acad Sci 466: 733
[128]  Hough I (1997) Aust Vet J 75(3): 174
[129]  Hsueh KF et al. (2004) J Chin Med Assoc 67: 93
[130]  Ishitsuka M et al. (1983) Tetrahedron Lett 24(46): 5117
[131]  Itai A et al. (1988) Proc Natl Acad Sci USA 85: 3688
[132]  Itokawa H et al. (1990) Phytochemistry 29(6): 2025
[133]  Jansen SA et al. (2012) Cardiovasc Toxicol 12: 208
[134]  Janssen J, Peltenburg H (1985) Ned Tijschr Geneeskad 129: 603

[135] Jin YX et al. (2019) Molecules 24: 1842

[136] Jury SL et al. (1987) The Euphorbiales, Chemistry, Taxonomy and Economic Botany. Acad Press, London

[137] Kaiya T, Sakakibara J (1985) Chem Pharm Bull (Tokyo) 33: 4637

[138] Karisawa H (1962) Ref C A 57: 12335

[139] Kasai R et al. (1981) Phytochemistry 20(11): 2592

[140] Kazanietz MG et al. (2000) Biochem Pharmacol 60: 1417

[141] Kemboi D et al. (2020) Molecules 25: 4019

[142] Kinghorn AD, Evans FJ (1975) Planta Med 28(8): 325

[143] Kiptoon JC et al. (1982) Toxicology 25: 129

[144] Klein-Schwartz W, Litovitz T (1985) Clin Toxicol 23: 91

[145] Kobusiak-Prokopowicz M et al. (2016) BMC Pharmacol Toxicol 17: 41

[146] Kornprobst JM, Al-Easa HS (2003) Curr Org Chem 7: 1181

[147] Kosemura S et al. (1985) Bull Chem Soc Jpn 58: 3112

[148] Kuhn A, Schäfer G (1937) Dtsch Apoth Ztg 52: 1265

[149] Kukowka A (1969) Heimatbote, Kr. Greiz, XV, p. 187

[150] Kulkarni ML et al. (2005) Indian J Pediatr 72: 75

[151] Kupchan SM et al. (1976) Science 191: 571

[152] Kupchan SM et al. (1976) J Amer Chem Soc 98(18): 5719

[153] Küpeli E et al. (2003) J Ethnopharmacol 89: 265

[154] Kushnir A, Marks AR (2010) Adv Pharmacol 59: 1

[155] Lai W et al., (2015) J Med Chem 58(21): 8638

[156] Lampe KF (1988) J Am Med Assoc 259: 2009

[157] Lang R et al. (2013) Phytochemistry 93: 124

[158] Larrey D, Faure S (2011) J Hepatol 54: 599

[159] Leli U et al. (1990) Mol Pharmacol 37: 286

[160] Lewis IAS et al. (1981) Tetrahedron 37(24): 4305

[161] Li CH et al. (2017) Molecules 22: 1431

[162] Li R et al. (2016) Nat Prod Rep 33(10): 1166

[163] Li Y et al. (2016) Scient Rep 6: 36752

[164] Liang J et al. (1987) Chem Pharm Bull (Tokyo) 35: 2613

[165] Liao SG et al. (2005) Org Lett 7: 1379

[166] Liu LG et al. (2002) Planta Med 68(3): 244

[167] Livertox: clinical and research information on drug-induced liver injury [internet]. Bethesda (MD): national Institute of Diabetes and Digestive and Kidney Diseases; 2012-. Germander. [Updated 2018 Mar 12]. https://www.ncbi.nlm.nih.gov/books/NBK548282

[168] Lucas H (1856) Arch Pharm 95: 145

[169] MacLeod JK et al. (1990) J Nat Prod 53(5): 1256

[170] MacNeil A et al. (2003) Br J Cancer 88: 1566

[171] Maejima H et al. (2003) J Biol Chem 278: 9464

[172] Mager PP et al. (1981) Pharmazie 36: 381

[173] Maghuly F et al. (2015) Appl Plant Genomics Biotechnol 2015: 87

[174] Magri NF, Kingston DGI (1988) J Nat Prod 51(2): 298

[175] Mancini SD, Edwards JM (1979) J Nat Prod 42(5): 483

[176] Marston A, Hecker E (1983) Planta Med 47(3): 141

[177] Martin ML et al. (1985) Ann Rev Acad Farm 51: 751

[178] Martinet O et al. (2005) Presse Med 34: 797

[179] Matschullat G (1974) Prakt Tierarzt 55: 624

[180] Mattei A et al. (1995) J Hepatol 22: 597

[181] McFarland SE et al. (2017) Vet Rec 180: 327

[182] McLaughlin JL et al. (1981) J Nat Prod 44(3): 312

[183] Mennel S et al. (2005) Ophthalmologe 102: 1099

[184] Miana GA et al. (2015) Mini Rev Med Chem 15(13): 1122

[185] Mihopoulos N et al. (2001) Tetrahedron Lett 42: 3749

[186] Miller RW (1980) J Nat Prod 43(4): 425

[187] Mitchell J, Rook A (1979) Botanical Dermatology, Plants Injurious to the Skin, Green Grass. Vancouver

[188] Morales P et al. (2017) Front Pharmacol 8: 422

[189] Müller P, Schütte HR (1968) Z Naturforsch 23B: 659

[190] Munro TA, Rizzacasa MA (2003) J Nat Prod 66(5): 703

[191] Murphy PT et al. (1981) Tetrahedron Lett 22: 1555

[192] Muscat M (k/a M) (2019) eJIFCC 30(3): 346

[193] Natascha G et al. (2020) Eur Heart J–Case Rep 4: 1

[194] Nawito M et al. (2001) J Cancer Res Clin Oncol 127: 34

[195] Nöller HG (1955) Monatsschr Kinderheilkd 103: 327

[196] Nwude N (1981) J Anim Prod Res 1: 109

[197] Opferkuch HJ, Hecker E (1973) Tetrahedron Lett 3611

[198] Otsuki K et al. (2020) Org Lett 22(1): 11

[199] Ott HH, Hecker E (1981) Experientia 37: 88

[200] Ottaggio L, et al. (2008) J Nat Prod 71(1): 58

[201] Owojuyigbe OS et al. (2020) Eur J Med Plants 31(8): 1

[202] Özhan H et al. (2004) Emerg Med J 21: 742

[203] Panter KEet al. (1993) J Am Vet Med Assoc 202: 1476

[204] Pardo-Vargas A et al. (2014) Marine Drugs 12: 4247

[205] Paul VJ, Fenical W (1984) Tetrahedron 40: 2913

[206] Paul VJ, Fenical W (1987) Biorg Marine Chem 1: 1

[207] Pauwels A et al. (1992) Gastroenterol Clin Biol 16: 92

[208] Peixoto F et al. (2004) Planta Med 70(11): 1064

[209] Pernthaner A, Langer T (1993) Wien Tierärztl Monatsschr 80: 138

[210] Persico A et al. (2011) J Anal Toxicol 35: 238

[211] Pierog JE et al. (2009) J Med Toxicol 5(2): 84

[212] Pietsch J et al. (2007) Int J Leg Med 121: 417

[213] Pinedo JM, et al. (1985) Contact Dermat 13(1): 44

[214] Pischon H et al. (2018) Vet Pathol 55(6): 896

[215] Pitts JF et al. (1993) Br J Ophthalmol 77: 284

[216] Pizzi R et al. (2005) Vet Rec 156: 487

[217] Plugge PC (1983) Arch Pharm 221: 813

[218] Polymeros D et al. (2002) J Clin Gastroenterol 34: 100

[219] Popescu R, Kopp B (2013) J Ethnopharmacol 147: 42

[220] Puschner B et al. (2001) J Am Vet Med Assoc 218: 573

[221] Ranganathan M et al. (2012) Biol Psychiatry 72(10): 871

[222] Rippmann F (1990) Quantit Struct Activity Relat 9: 1

[223] Rizk AM et al. (1988) Phytochemistry 27(6): 1605

[224] Rogers EF et al. (1948) J Amer Chem Soc 70(9): 3086

[225] Ronlán A, Wickberg B (1970) Tetrahedron Lett 11: 4261

[226] Rosenbaum CD et al. (2012) J Med Toxicol 8: 15

[227] Roth BL et al. (2002) Proc Natl Acad Sci USA 99: 11934

[228] Runyon R (1980) Clin Toxicol 16: 167

[229]  Sakakibara J, Kaiya T (1983) Phytochemistry 22(10): 2547
[230]  Sakakibara J et al. (1981) Phytochemistry 20(7): 1744
[231]  Sakata K et al. (1971) Tetrahedron Lett 1141
[232]  Salehi R et al. (2019) Biomolecules 9: 337
[233]  Schindler H (1962) Planta Med 10(2): 232
[234]  Schmidt RJ (1986) Natural Occurring Phorbol Esters. Evans FJ (ed.) CRC Press, Boca Raton, p. 87
[235]  Schmidt RJ, Evans FJ (1980) Contact Dermat 6(3): 204
[236]  Schteingart CD, Pomilio AB (1984) J Nat Prod 47: 1046
[237]  Scott PM et al. (1971) Food Cosmetics Toxicol 9: 179
[238]  Seephonkai P et al. (2011) J Nat Prod 74(4): 712
[239]  Seip EH, Hecker E (1982) Planta Med 46(12): 215
[240]  Sharkey NA et al. (1989) Carcinogenesis 10: 1037
[241]  Shi Y et al. (2021) J Ethnopharmacol 265: 113320
[242]  Shizuri Y et al. (1983) Chem Lett 12: 65
[243]  Shizuri Y et al. (1984) Tetrahedron Lett 25: 1155
[244]  Shrestha TM, et al. (2018) Clin Case Rep 6: 2355
[245]  Sibel S et al. (2014) J Ethnopharmacol 156: 155
[246]  Siegle L, Pietsch J (2018) Phytochem Anal 29: 446
[247]  Singh RK et al. (2010) Mjafi 66(1): 80
[248]  Smith MC (1978) J Am Vet Med Assoc 173: 78
[249]  Sosath S et al. (1988) J Nat Prod 51(6): 1062
[250]  Spoerke DG, Temple AR (1979) Am J Disease Childhood 133: 28
[251]  Starakis I et al. (2006) Eur J Gastroenterol Hepatol 18: 681
[252]  Stickel F, Shouval D (2015) Arch Toxicol 89: 851
[253]  Storie GJ et al. (1986) Aust Vet J 63(3): 135
[254]  Stout GH et al. (1970) J Am Chem Soc 92: 1070
[255]  Strickland NH (2000) Bmj 321(7258): 428
[256]  Stuart BP et al. (1981) Vet Pathol 18: 368
[257]  Suffness M, Cordell GA (1985) Manske RHF, The Alkaloids, vol. XXV: p. 1 Acad Press, New York
[258]  Swinnen J et al. (2020) Ann Case Rep 14: 413
[259]  Terai T et al. (2000) Chem Pharm Bull (Tokyo) 48: 142
[260]  Tsai JY et al. (2020) Int J Mol Sci 21: 7579
[261]  Turgut M et al. (2005) Ann Trop Paed 25: 125
[262]  Tyler MI, Howden MEH (1985) Plant Toxicol Proc Aust-USA Poisonous Plants Symp 1984. Seawright
       AA (ed.). Queensland Poisonous Plants, Commun Yeerongpilly. Australia, p. 367
[263]  Uemura D et al. (1975) Chem Lett 6: 537
[264]  Upadhyay RR et al. (1977) Indian J Chem 15B: 294
[265]  Uzabakiliho B et al. (1987) Phytochemistry 26(11): 3041
[266]  Vališ M et al. (2014) J Med Case Rep 8: 4
[267]  Vandenbark GR, Niedel JE (1984) J Nat Cancer Inst 73: 1013
[268]  Van den Bosch C et al. (1993) Br J Cancer 68: 1232
[269]  Van Ingen G et al. (1992) Forens Sci Int 56: 81
[270]  Van Petegem F (2012) J Biol Chem 287(38): 31624
[271]  Vasas A, Hohmann J (2014) Chem Rev 114(17): 8579
[272]  Verdourt B, et al. (1969) Common Poisonous Plants of East Africa. Collins, London, p. 254
[273]  Vogg G, et al. (1999) Environ Health Perspect 107: 753
[274]  von Dach B, Streuli RA (1988) Schweiz Med Wochenschr 118: 1113
[275]  von Kürten S et al. (1971) Arch Pharm 304: 753
[276]  von Malottki K, Wiechmann HW (1996) Dtsch Med Wochenschr 121: 936

[277]  von Mühlendahl KE (2005) Monatsschr Kinderheilkunde 153: 1103
[278]  Wang LQ et al. (2001) Fitoterapia 72: 779
[279]  Wang XH et al. (2020) Comm Biol 2020(3): 228
[280]  Wehner F, Gawatz O (2003) Arch Kriminol 211: 19
[281]  Wells RJ, Barrow KD (1979) Experientia 35: 1544
[282]  Wiechert JM et al. (2001) Prakt Tierarzt 82: 260
[283]  Wilken K, Schempp CM (2005) Hautarzt 56: 955
[284]  Wilson CR et al. (2001) Toxicon 39: 175
[285]  Wink M (2015) Medicines 2: 251
[286]  Witte ST et al. (1990) J Vet Diagn Invest 2: 263
[287]  Wolf-Abdolvahab S et al.. (1997) Z Hautkrankheiten 8: 596
[288]  Wu LT et al. (2011) Substance Abuse Rehabilitation 2: 53
[289]  www.floraqueen.com
[290]  Yamamura S et al. (1989) Phytochemistry 28(12): 3421
[291]  Yarlioglues M et al. (2011) Texas Heart Inst J 38(5): 577
[292]  Zayed SMAD et al. (2001) J Cancer Res Clin Oncol 127: 40
[293]  Zhan ZJ et al. (2005) Bioorg Med Chem 13: 645
[294]  Zhang HP et al. (2005) J Asian Nat Prod Res 7: 87
[295]  Zhang HP et al. (2005) Bioorg Med Chem 13: 5289
[296]  Zhong G et al. (2005) J Nat Prod 68(6): 924
[297]  Zhu Y et al. (2019) RSC Adv 9: 18439
[298]  Zhuang L et al. (1982) Planta Med 45(7): 172

# 6 Triterpenes

## 6.1 General

The basic bodies of the triterpenes are composed of six isoprene units, i.e., they are $C_{30}$ compounds. More than 14,000 structures basing on more than 100 different basic bodies are known [26].

In addition to the aliphatic squalene frequently found in the unsaponifiable fraction of plant and animal fats, aliphatic triterpenes or polyethers derived from them have so far been found mainly in marine organisms [43].

The majority of triterpenes possess tetracyclic or pentacyclic ring systems. Tetracyclic triterpenes include the derivatives of the hypothetical basic compounds lanostane, dammarane, euphane, and cucurbitane. Pentacyclic triterpenes are derived from oleanane, ursane, hopane, and lupane, among others. Mono-, bi-, and tricyclic triterpenes occur rarely. In higher plants, 3,4-*seco* derivatives of polycyclic triterpenes are also commonly found [10]. To them belong the iridal derivatives, which can be considered 3,4-*seco* derivatives of bi- or tricyclic triterpenes.

The triterpenes often serve as aglycones of the triterpene saponins (see Chapter 9). Also of toxicological interest are the icterogenically active triterpene esters of various plants (see Section 6.2), limonoids of Meliaceae (see Section 6.3), the cucurbitacins (see Section 6.4), iridals and cycloiridals in Iris species (see Section 6.5), and gossypol, which occurs in the cotton plant and is biogenetically a dimeric sesquiterpene of the cadinane type (see Section 6.6).

Triterpenes are also found in macroalgae (see Section 6.7), basidiomycetes, and sponges.

Review of pharmacological activities of triterpenes: [38].

## 6.2 Icterogenic Triterpene Esters

Some triterpenes and steroid derivatives have a hepatotoxic effect and lead, presumably via disturbance of bile excretion, to accumulation of degradation products of porphyrin derivatives (chlorophyll and hemoglobin) in the blood and thus to secondary photosensitization.

Icterogenic plants are *Narthecium ossifragum* (L.) Huds., bog asphodel, Nartheciaceae, Tribulus species, e.g., *Tribulus terrestris* L., cat's head, Zygophyllaceae, Lippia species, and *Lantana camara* L., lantana, both Verbenaceae.

In *N. ossifragum* and *T. terrestris*, steroid saponins seem to be the main active ingredients so that they are described there (see Section 9.4).

Main active ingredients of **Lippia** species, Verbenaceae, are the triterpene esters lantadene A, icterogenin, and the corresponding 3α-hydroxy analogs [25].

https://doi.org/10.1515/9783110724738-006

The genus Lippia contains around 200 species of tropical shrubs. Because of their pleasant fragrance and their pharmacological activities due to essential oil, some species, e.g., *L. citriodora* (Palau) Kunt. (*Aloysia citriodora* Palau), lemon verbena, are used as culinary plants and herbal drugs. Icterogenic Lippia species, especially *L. rehmannii* H. Pearson, occur in Central and South America, and to a lesser extent in Africa.

🐾 Photodermatoses and jaundice caused by Lippia species in animals are described in the old literature [101].

Of greater importance than the Lippia species is **Lantana camara** (Ph. 6.1), Verbenaceae. The plant is native to tropical America and is now common in many subtropical and tropical areas of the world. It is kept in Central Europe as an ornamental plant indoors or in summertime outdoors also. It occurs in several varieties. The cultivated representatives of the genus Lantana, comprising about 150 species, are mostly hybrids.

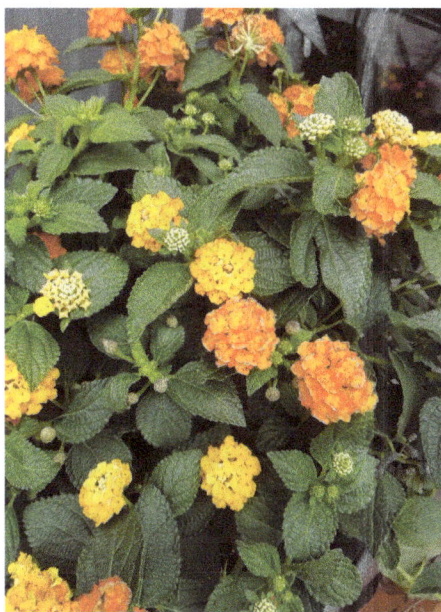

Ph. 6.1: *Lantana camara* (source: Ulrike Lindequist).

Lantana is a sparse-branching plant that can form a shrub up to 1 m high. The flowers change the color from white or yellow to red or purple. There are also white, yellow, pink, orange, and reddish-brown hybrids.

The icterogenic agents of *Lantana camara* are triterpenic acid esters, which are preferentially derived from oleanolic acid (Fig. 6.1). In addition, ursane derivatives, e.g., 3-keto-ursolic acid and lantinic acid, lupane derivatives, e.g., betulinic acid and

Oleanolic acid
3β-OH

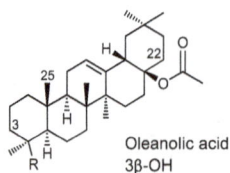

| Name | Structure, Substituents | | | |
|---|---|---|---|---|
| | 3 | 3/25 | 22β | R |
| Lantadene A (Rehmannic acid) | =O | | OAng | $CH_3$ |
| Lantadene B | =O | | ODma | $CH_3$ |
| Lantadene D | =O | | OiBu | $CH_3$ |
| 22β-2-Methylbutyryloxy-3-oxo-oleanolic acid | =O | | OMBu | $CH_3$ |
| Icterogenin | =O | | OAng | $CH_2OH$ |
| 22β-Angeloyloxy-oleanolic acid | βOH | | OAng | $CH_3$ |
| 22β-Angeloyloxy-23-hydroxy-oleanolic acid | βOH | | OAng | $CH_2OH$ |
| 22β-Dimethylacryloyl-oleanolic acid | βOH | | ODma | $CH_3$ |
| Lantanolic acid | βOH | -O- | | $CH_3$ |
| Lantanilic acid | βOH | -O- | ODma | $CH_3$ |
| Camaric acid | βOH | -O- | $OCOCH_3$ | $CH_3$ |

OAng, Angeloyloxy-; OiBu, Isobutyryloxy-; ODma, 3,3-Dimethylacryloyloxy-;
OMBu, 2-Methylbutyryloxy-

Fig. 6.1: Oleanolic acid derivatives of *Lantana camara*.

lantabetulinic acid, and euphane-type triterpenes that contain a cyclic lactone ring have been found [15, 17, 50, 86, 90, 104].

The isolation of the main active ingredients, lantadenes, was carried out by Louw in 1948 [77], and their structure was elucidated by Barton and de Mayo in 1954 [14].

The main active ingredient is considered to be lantadene A. The parent form and the hybrids with colored flowers preferentially contain lantadene A and lantadene B, and the white-flowered varieties also contain betulinic acid. The orange ones contain little lantadene but 22-angeloyloxy-23-hydroxy-oleanolic acid, and the pink ones no lantadene but icterogenin and lantanolic acid [50].

After peroral intake, the active ingredients of *Lantana camara* are rapidly absorbed and transported to the liver via the portal vein. They are metabolized to polar compounds and excreted into the bile. Damage to the membranes of the bile ducts and cholestasis result. The activity of many enzymes is altered, the protein content of liver mitochondria and microsomes is reduced, and the content of bilirubin and other degradation products of porphyrin derivatives in serum is increased. The motility of the ruminant stomach is decreased, so that the toxins remain in the stomach for a longer time [111, 121]. The mechanism of toxicity is not known but may be related to the effects of lantadene A on mitochondrial energetics [34].

There is conflicting information on the toxicity of the lantadenes. Crystal polymorphisms may play a role [110]. In sheep, severe signs of poisoning were induced by peroral administration of 65–75 mg/kg BW lantadene A or 200–300 mg/kg BW lantadene B [107]. In guinea pigs, lantadenes (6–24 mg/kg BW, p.o., daily for 90 days) led to a dose-dependent decrease in body weight, nephrotoxicity, and alterations in liver enzymes and oxidative stress markers [71]. About 6 g/kg BW leaf powder, p.o., induced cholestasis in guinea pigs [111]. Embryotoxic effects of aqueous alcoholic leaf extracts have been shown in rats [82].

Poisoning has been described only in children, caused by consumption of the immature fruits [128]. The ripe fruits are even eaten in tropical countries [**49**].

☠ The symptoms of poisonings by presumably unripe berries of common lantana are said to be similar to those of poisoning by *Atropa belladonna*. For example, pupil dilation has been described. A case of death is reported [**49**].

▣ Ingestion of the immature fruits of *Lantana camara* may necessitate gastric emptying and symptomatic treatment [128].

🐏 *Lantana camara* is a dangerous plant for grazing animals in many countries, e.g., Australia, India, South Africa, and South America [82, 110]. Poisonings, including fatal ones, are most common in cattle, but rarely in sheep, goat, and horses. It has also been observed in kangaroos [100]. The outbreaks of Lantana poisonings are seen during fodder scarcity, drought, and flood where animals may consume small quantities of leaves either while grazing or due to mixing with regular fodder. Therefore, often a subclinical disease develops [71].

Poisoning of animals is manifested as subacute or chronic cholestasis. Symptoms include vomiting, anorexia, constipation, photodermatitis, and severe icterus. Death occurs from liver and kidney failure in a period up to 2 weeks. Histopathologic evidence includes hepatocellular enlargement and fine cytoplasmic vacuolation, dilatation of the gallbladder, biliary congestion in the liver, and necrosis of liver tissue and renal damage [34, 58, 61, 81, 110].

Animal poisoning should be treated by administering large amounts of activated charcoal and supplying electrolyte solution [81, 94]. Antibodies to the triterpenes provide some protection against the symptoms of poisoning [113]. To prevent photodermatosis, affected animals must be kept in the dark. Further treatment is symptomatic [110].

In tropical and subtropical countries, decoctions of the roots of Lantana species are used for treatment of wounds, rheumatism, and malaria. The essential oil of the leaves is of interest because of its leishmanicidal and trypanocidal activities [13].

## 6.3  Limonoids as Toxins of *Azadirachta indica* and *Melia azedarach*

Limonoids are complex, highly oxygenated tetranortriterpenes with a reactive furanolactone core structure (Fig. 6.2). Limonoids possess, among others, cytotoxic, insecticidal, antimicrobial, and antimalarial activities [103].

Of interest is their occurrence in ***Azadirachta indica*** A. Juss. (*Antelaea azadirachta* (L.) Adelb.), the neem tree, margosa (Ph. 6.2), and ***Melia azedarach*** L., China berry (Ph. 6.3), both Meliaceae.

Ph. 6.2: *Azadirachta indica* (source: Wolfgang Kiefer).

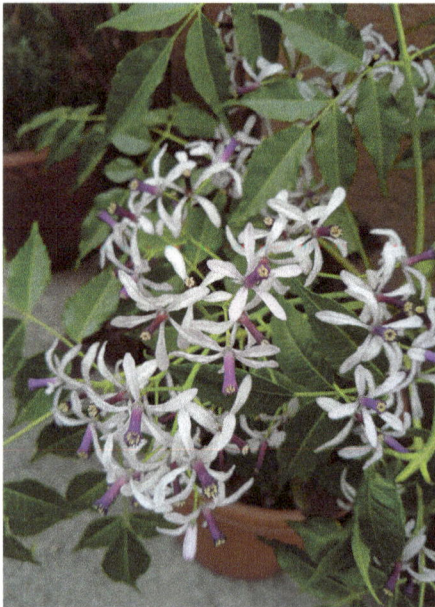

Ph. 6.3: *Melia azedarach* (source: Meena Rajbhandari).

*Azadirachta indica* is an evergreen, fast-growing, draft-resistant tree that thrives in a subhumid to subarid climate. It is native to South Asia and naturalized in the Arabian region. It has been used for thousands of years in traditional Indian medicine such as Ayurveda [22]. Seeds comprise 40% of oil (neem oil) with the limonoid azadirachtin (Fig. 6.2) as the major active ingredient. Azadirachtin and other triterpenes such as nimbin (Fig. 6.2), nimbolide, gedunin, and azadirone are contained in all parts of the plants [66]. 'Margosa oil' means different preparations from *A. indica*.

Limonoids such as toosendanin (Fig. 6.2) from the bark and meliatoxins from the fruit are the bioactive components of *Melia azedarach*. The evergreen tree is native to Asia, but also found in Australia, tropical South America, and Africa and widely used for a variety of ailments in Iranian traditional medicine and in TCM [96, 124].

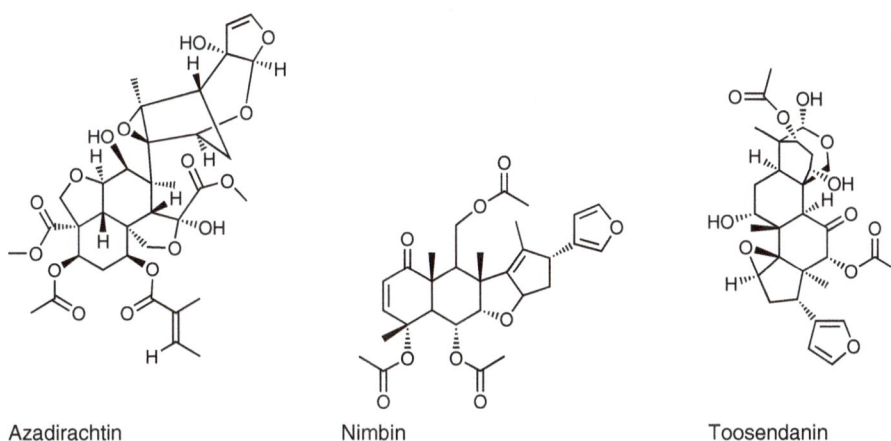

| Azadirachtin | Nimbin | Toosendanin |

Fig. 6.2: Limonoids of *Azadirachta indica* and *Melia azedarach*.

For insects, azadirachtin acts as an antifeedant and repellent agent. It induces sterility by preventing oviposition and suppresses sperm production in males. It is structurally similar to the ecdysones, insect hormones responsible for metamorphosis in insects, and disturbs hormonal and metabolic processes [30].

For rodents, toxicity of extracts of *A. indica* or neem oil after p.o. administration is low. The estimated safe dose for azadirachtin is 15 mg/kg BW/day, for the aqueous extract is 0.3 mg/kg BW/day, and for the seed oil is 0.26 mg/kg BW/day. The negative effects on reproduction of both male and female mammals were reversible [19]. The 24 h oral median lethal dose was determined as 14 mL/kg BW in rats and 24 mL/kg BW in rabbits. Lungs and CNS were the target organs for toxic effects. Intragastric administration of the oil to mice was not toxic at a dose of 2 mL [66]. Fruits and leaves of *M. azedarach* cause central nervous paralysis and death in mice and rat. The $LD_{50}$ of toosendanin in mice ranges between 250 and 500 mg/kg BW, p.o. [96].

🜸 Neem oil poisoning has been observed in children. Symptoms were vomiting, hepatic toxicity, metabolic acidosis, and encephalopathy [6].

🜸 Human poisonings by *Melia azedarach* are rarely reported. Poisoning occurred with application within TCM. Symptoms, ranging from a few hours to up to 3 weeks, were weakness, myalgia, generalized numbness, and ptosis. In most cases, patients recovered without consequences. According to Chinese medical literature, human poisoning can occur if 6–9 fruits, 30–40 seeds, or 400 g of the bark are consumed. Besides neurological, gastrointestinal, cardiovascular, and respiratory effects, in severe cases, death is possible [96].

🜂 Treatment is symptomatic [96].

🐾 The fruits of *Melia azedarach* were the cause of poisoning of dogs, cattle, pigs, and ostriches with predominant neurotoxic symptoms [96].

*Azadirachta indica, Melia azedarach*, and isolated compounds such as azadirachtin are useful biopesticides [19]. Because of their anticancer and antimalarial activities, limonoids are candidates for drug development [43, 103].

## 6.4 Cucurbitacins

### 6.4.1 Chemistry and Distribution

Cucurbitacins are derivatives of the hypothetical triterpene hydrocarbon cucurbitane (19(10→9)*abeo*-5α-lanostane, Fig. 6.3). They carry a large number of hydroxy groups (especially at C-2, C-16, C-20, and C-25, the latter often acetylated) and oxo groups (especially at C-3, C-11, and C-22 or C-24). The aliphatic side chain is sometimes joined by an oxygen bridge to form a furan ring. Double bonds in the tetracyclic ring system ($\Delta^{1,2}$), but also in the side chain ($\Delta^{23}$), are common. Momordicins (Fig. 6.5) have a formyl group at C-9. Cucurbitacins B and E have been identified to be the primary cucurbitacins. Other cucurbitacin types can be generated by enzymatic conversion of them [125].

More than 50 naturally occurring cucurbitane derivatives are known [32, 60, 92].

Cucurbitacins have been isolated in free or glycosidical forms. The monosaccharide components are D-glucose and L-rhamnose. In addition to monodesmosides, e.g., bryoamaroside, bisdesmosides, e.g., bryoside, are found (Tab. 6.1). Since almost all plants containing cucurbitacins investigated so far contain a very active β-glucosidase (elaterase), which rapidly releases the cucurbitacins from the glycosides when the plant tissues are destroyed, it is still unclear whether and to what extent free cucurbitacins occur in the intact plant.

The first cucurbitacin, called elaterin, was obtained as early as 1831 from the drug Elaterium, the thickened juice of *Ecballium elaterium* (L.) A. Rich., squirting cucumber (Ph. 6.4). The exact structural formula (elaterin = cucurbitacin E) was established 1961 by Lavie and coworkers [73]. The first isolated cucurbitacin glycoside was probably bryonoside. It was isolated, probably in an impure form, from *Bryonia cretica* ssp. *dioica* by Vauquelin in 1806. The structure of the aglycone was determined by Biglino and coworkers in 1963 [18].

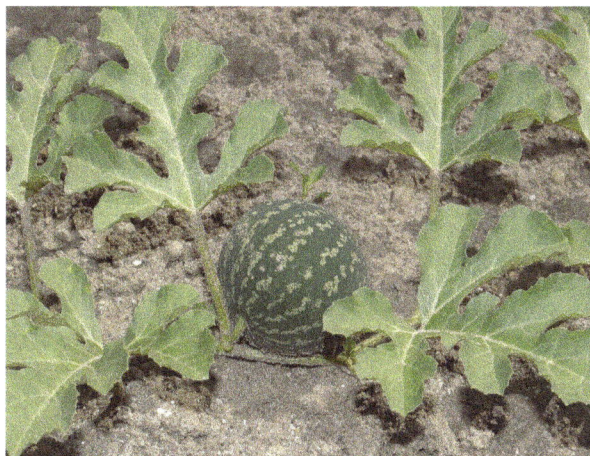

Ph. 6.4: *Citrullus colocynthis* (source: Wolfgang Kiefer).

Cucurbitacins are widely distributed in the Cucurbitaceae family, cucurbits, gourd family. So far, they have been detected in about 100 of the more than 900 species of Cucurbitaceae. Special mention should be made of their occurrence in Bryonia species, bryony (see Section 6.4.3), in *Cucumis sativus* L., cucumber, in *Cucumis melo* L., melon, in *Cucurbita pepo* L., pumpkin, *Cucurbita pepo* ssp. *pepo*, zucchini, *Cucurbita maxima* Duchesne, squash, in *Cayaponica racemosa* (Mill.) Cogn., mountain melon leaf, Luffa species, sponge cucumber, Coccinia species, scarlet gourds, Echinocystis species (*Echinocystis lobata* (Michx.) Torr. and A. Gray, wild balsam apple, prickly cucumber), Lagenaria species (*Lagenaria siceraria* (Molina) Standl., bottle gourd, calabash), Citrullus species (*Citrullus lanatus* (Thunb.) Matsum. and Nakai, watermelon, *Citrullus colocynthis* (L.) Schrad., coloquinte [47]), *Ecbalium elaterium* [51], Trichosanthes species (*Trichosanthes cucumerina* L., serpent guard), and Hemsleya species. The cucurbitacins are responsible for the bitter taste of these plants which are widely used for culinary purposes and in traditional medicine of diverse regions of the globe [105].

In Cucurbitaceae, cucurbitacins occur in high concentrations in the roots, and in lower concentrations in leaves and stems. In fruits, the highest concentration is reached at maturity. However, in some species, e.g., in the nara plant, *Acanthosicyos horridus* Welw. ex Benth. and Hook.f. (Ph. 6.5), endemic to Namib desert, the concentration decreases at fruit maturity. The seeds of this plant, known as butternuts or butterpips, are poor in or free from cucurbitacins. Therefore, the melon-like fruits (weight between 1 and 2.5 kg) can be eaten [31].

**Ph. 6.5:** *Acanthosicyos horridus* (source: Christian Edelmann/iStock/Getty Images Plus).

The fruits of the cultivated Cucurbitaceae (cucumber, squash, melon, and watermelon) used are free of cucurbitacins as a result of breeding measures. Reverse mutations, which are relatively common, result in plants with bitter, toxic fruits.

Cucurbitacin-containing plants of other taxa include Gratiola species (Plantaginaceae, see Section 6.4.4), *Momordica charantia* (Cucurbitaceae, see Section 6.4.5), *Begonia × tuberhybrida* Voss., tuberous begonia (Begoniaceae [37]), *Anagallis arvensis* L. (*Lysimachia arvensis* (L.) U. Manns and Anderb.), scarlet pimpernel (Primulaceae [130]), *Phormium tenax* J.R. Forst and G. Forst, New Zealand flax (Xanthorrhoeaceae [72]), *Tropaeolum majus* L., garden nasturtium (Tropaeolaceae [127]), and *Purshia tridentata* (Pursh) DC., bitterbrush (Rosaceae, native to Western North America [27]). Cucurbitacins also occur in the seeds of some Brassicaceae, e.g., of *Iberis sempervirens* L., evergreen candytuft (Ph. 6.6), *I. gibraltarica* L., Gibraltar candytuft, *I. amara* L., wild candytuft, *I. umbellata* L., annual candytuft, and the garden cress, *Lepidium sativum* L. The plants are grown as ornamentals and cress for culinary purposes [60].

Among the basidiomycetes, cucurbitacins are found in Hebeloma and Russula species. Some marine gastropods also contain cucurbitacins [118].

## 6.4.2 Pharmacology and Toxicology

The cucurbitacins have strong cytotoxic and skin- and mucosa-irritant effects. Their cytotoxic and antiproliferative effects are thought to be due to inhibition of nucleic acid and protein synthesis, cell cycle arrest, induction of apoptosis, and inhibition of cell migration and invasion and of angiogenesis. Selective inhibition of oncogenic signaling pathways such as the Janus kinase/signal transducer and activator of transcription pathway (JAK/STAT pathway) could be shown [23, 117]. A mechanistic study demonstrated that cucurbitacin C inhibited the Akt/PKB (Akt = protein kinase B, PKB) signaling by inhibition of Akt

phosphorylation at Ser473 [129]. Some cucurbitacins, e.g., bryoniosides 2 and 3, have anti-inflammatory and hepatoprotective activities and inhibit TPA-triggered Epstein–Barr virus activation in animal experiments [120]. Antimicrobial, anthelmintic, and insecticidal effects have also been demonstrated (reviews: [32, 33, 63, 85, 125, 133]).

The $LD_{50}$ values of cucurbitacin A are 1.2 mg/kg BW, i.p., in mice and 2 mg/kg BW, i.p., in rats [57], and those of cucurbitacin B are 10.9 mg/kg BW, p.o., and 1.0 mg/kg BW, i.p., in the mice [84]. A double bond from C-23 to C-24 and an acetyl group at C-25 augment the toxicity of cucurbitacins [63]. Cucurbitacin B has a low oral bioavailability (approximately 10%) but could be distributed extensively into internal organs with a high volume of distribution and tissue to plasma ratio [54].

Poisonings by plants containing cucurbitacins are caused by Bryonia species (see Section 6.4.3). They are also possible, when one eats unexpectedly bitter, toxic fruits that occur among the usually eatable fruits of Cucurbitaceae (pumpkin, melon, cucumber, watermelon, and zucchini [11, 36, 65, 114]). The bitter taste is often not immediately noticed, especially by children.

Other possibilities for poisoning include accidental ingestion, e.g., of colocynths [4, 95], improper folk medicinal use, e.g., of the squirting cucumber *Ecbalium elaterium* [9, 99, 106] or the bottle gourd, *Lagenaria siceraria* [52, 64], and ingestion of plants to which cucurbitacin-containing pesticides have been applied [36].

If *E. elaterium* fruit bursts and its juice splashes into eyes, severe keratoconjunctivitis can result [24].

⚑ Two independent intoxication cases occurred in Southern Germany in 2015 after the ingestion of zucchini. Two couples had consumed meals of vegetable stew containing homegrown, bitter-tasting zucchini: 118 µg/g cucurbitacin B, 681 µg/g cucurbitacin E, 127 µg/g 74 cucurbitacin I, and 127 µg/g E-glucoside were found in a cooked zucchini sample involved in the poisoning. The patients developed abdominal cramps and bloody diarrhea. One patient died a couple of days later [11].

✝ In the USA, five adult patients presented with nausea, vomiting, and diarrhea within 5–25 min after ingestion of cooked bitter bottle gourd. One patient required hospitalization because of severe diarrhea, hematemesis, and hypotension. All patients improved within a few days after i.v. application of fluids and proton pump inhibitors [52].

🐾 Animals are at risk from consumption of wild Cucurbitaceae. *Phormium tenax*, an important fiber and ornamental plant, causes poisonings of animals in New Zealand and Australia [76].

Increasing attention is being paid to the antitumor effects of cucurbitacins [33, 105, 129, 133]. Cucurbitacin G 2-glucoside and cucurbitacin H are candidates for development of drugs against SARS-CoV-2 [62].

### 6.4.3 Cucurbitacins as Toxins of Bryony Species (Bryonia Species)

The genus Bryonia, bryony, Cucurbitaceae, includes 12 species distributed from the Canary Islands through the Mediterranean region to Iran and South Asia. In Central Europe, only *Bryonia alba* L., white bryony (Ph. 6.7), and *B. cretica* ssp. *dioica* (Jacq.) Tutin, Cretan bryony (*B. dioica* Jacq., red bryony) are found. *B. alba* has been valued as a medicinal plant since ancient times. In northern and central-eastern Europe, the root was often sold and used as a substitute for mandrake (*Mandragora officinarum* L. [70]).

Ph. 6.7: *Bryonia alba* (source: Witheway/iStock/Getty Images Plus).

Bryonia species are perennial plants with beet-shaped roots. *B. alba* is monoecious. The ripe berries, about the size of a pea, are black. *B. cretica* ssp. *dioica* is dioecious.

The ripe berries are red. *B. alba* is found from Europe to Central Asia, and the other species mainly in Europe and Northwest Africa. Locations are hedges, fences, and riparian forests.

Selected compounds found in the roots of Bryonia species are shown in Fig. 6.3 and Tab. 6.1 [28, 55, 56, 57, 89, 92, 93, 97, 102, 119, 120].

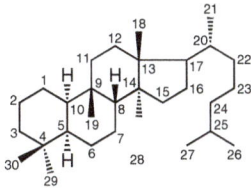

Cucurbitane

| Name | Substituents | | | | | | | | Double bonds |
|------|----|----|----|-----|-----|----|----|----|--------------|
| | 2β | 3 | 11 | 16α | 20β | 22 | 24 | 25 | |
| Cucurbitacin B | HO | =O | =O | OH | OH | =O | | OAc | $\Delta^5$ $\Delta^{23}$ |
| Cucurbitacin D | HO | =O | =O | OH | OH | =O | | OH | $\Delta^5$ $\Delta^{23}$ |
| Cucurbitacin E | HO | =O | =O | OH | OH | =O | | OAc | $\Delta^1$ $\Delta^5$ $\Delta^{23}$ |
| Cucurbitacin I | HO | =O | =O | OH | OH | =O | | OH | $\Delta^1$ $\Delta^5$ $\Delta^{23}$ |
| Cucurbitacin J | HO | =O | =O | OH | OH | =O | OHα | OH | $\Delta^1$ $\Delta^5$ |
| Cucurbitacin K | HO | =O | =O | OH | OH | =O | OHβ | OH | $\Delta^1$ $\Delta^5$ |
| Cucurbitacin L | HO | =O | =O | OH | OH | =O | | OH | $\Delta^1$ $\Delta^5$ |
| Cucurbitacin S | HO | =O | =O | | —O— | OCH₃ | ⎡—O—⎤ | | $\Delta^1$ $\Delta^5$ |
| Bryodulcosigenin | | OHβ | =O | | | | OHβ | OH | $\Delta^5$ |
| Bryosigenin | | OHβ | =O | | | | =O | OH | $\Delta^5$ |

Fig. 6.3: Cucurbitacins of Bryonia species.

Tab. 6.1: Glycosides of Bryonia species.

| Name | Aglycone | Monosaccharides |
|------|----------|-----------------|
| Elaterinide | Cucurbitacin E | ← 2-β1Glc |
| Bryoamaride | Cucurbitacin L | ← 2-β1Glc |
| Bryonoside | Bryodulcosigenin | ← 3-β1Glc2 ← 1Rha<br>← 25-β1Glc6 ← 1βGlc |
| Bryodulcoside | Bryodulcosigenin | ← 3-β1Glc6 ← 1βGlc2<br>← 25-β1Glc6 ← 1βGlc2 |
| Bryoside 2 | Bryodulcosigenin | ← 3-1Rha2 ← 1βGlc<br>← 25-β1Glc |
| Bryonioside B | Bryosigenin | ← 3-1Rha2 ← 1βGlc |

Other interesting substances that occur in Bryonia species include triterpenes with multiflorane backbones, e.g., bryonolic acid (3-hydroxy-multiflor-8-ene-29-acid) and its p-hydroxycinnamoyl ester bryocoumaric acid [56, 57]. Cycloartane-type triterpenes have been detected in the berries [3]. Of particular interest are the unsaturated polyhydroxy fatty acids similar to eicosanoids in structure and action, such as 9,12,13-trihydroxy-octadeca-10($E$)-15($Z$)-dienoic acid. In addition, cholest-7-en-3α-ols, methylated and ethylated at C-4 and C-24, were found [92]. Furthermore, the occurrence of the ribosome-inactivating proteins bryodin 1 and 2 is noteworthy. The unusual amino acids $N$-ethyl-asparagine and $N$-hydroxyethyl-asparagine were detected in the herb of *B. cretica* ssp. *dioica* [44].

Poisonings with Bryonia species occur through consumption of berries, especially by children, through contact with the juice of fresh plants or through folk medical use of the plants as a drastically acting laxative, abortifacient, or diuretic. According to older data, for children already 15 berries and for adults 40 berries are potentially fatal [**161**].

🐾 Symptoms of poisoning after ingestion of plant parts of Bryonia species are vomiting (observed in children already after six to eight berries), bloody diarrhea, colic, kidney irritation, anuria, collapse, convulsions, paralysis, and during pregnancy abortion. Fatal outcome due to central respiratory paralysis is possible. External contact causes skin inflammation with formation of blisters and ulcers [69].

💊 After ingestion of small amounts (up to three berries in children), activated charcoal along with plenty of fluids should be given, for larger amounts (more than three berries in children) gastric emptying is recommended. Further treatment is symptomatic [**161**].

🐐 Bryony is avoided as food by animals due to its unpleasant smell with the exception of goat which can eat small amounts of it. Birds can safely eat of the berries [70]. Fatal poisonings have been recorded in dogs, pigs, chickens, ducks, and cattle. Symptoms included vomiting, diarrhea, dehydration, fever, respiratory distress, tachycardia, convulsions, renal dysfunction, and collapse. Pathologically, numerous hemorrhages were detected [126, **60**].

### 6.4.4 Cucurbitacins as Toxins of Hedge Hyssop (*Gratiola officinalis*)

The genus Gratiola includes about 25 species. *Gratiola officinalis* L., hedge hyssop (Ph. 6.8, Plantaginaceae), is the only species native from Europe to middle Asia. In Canada and the USA, *Gratiola neglecta* Torr., clammy hedge hyssop, occurs.

*Gratiola officinalis* is a perennial plant growing to 40 cm tall. The pale pink to white flowers are solitary in the leaf axils. The fruit is a four-valved capsule. The plant occurs in wet places, preferably in marshy meadows, in the plain.

In the herb of *G. officinalis*, the cucurbitacins gratiogenin, 16-hydroxy-gratiogenin (Fig. 6.4), cucurbitacin L, E, and I (Fig. 6.3), and the cucurbitacin glycosides gratiogenin-3α-D-glucopyranoside, gratioside (gratiogenin diglucoside), elaterinide, and desacetyl elaterinide were found [49]. The content of cucurbitacins is about 1%.

Ph. 6.8: *Gratiola officinalis* (source: Wolfgang Kiefer).

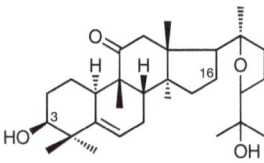

Gratiogenin

Fig. 6.4: A cucurbitacin of *Gratiola officinalis*.

Saponins and alkaloids occur as minor compounds. Attempts to declare the cardio-toxic effect of the plant by the presence of cardioactive steroid glycosides yielded different results. Presumably, the cardiac activity is due to the cucurbitacins [87].

The irritant effect of cucurbitacins seems to be responsible for the toxicity of the plant.

*G. officinalis* was formerly used as a drastic, anthelmintic, abortifacient, and diuretic. Today, poisoning due to erroneous use of gratiole in herbal teas is possible, but very rare.

☠ Symptoms of poisoning after ingestion of tea made from the leaves of *Gratiola officinalis* were vomiting, bloody diarrhea, cramps, kidney irritation, and disturbances of heart activity and respiratory function. Fatal cases occurred [87, 127].

🚑 The treatment of poisoning is symptomatic.

### 6.4.5 Cucurbitacins in Bitter Melon (*Momordica charantia*)

*Momordica charantia* L., bitter melon, balsam pear (Ph. 6.9), Cucurbitaceae, is a climbing plant with edible fleshy, oblong, prickly fruits. Two varieties of the plant exist: the Indian bitter melon and the Chinese bitter melon. They are widespread in tropical and subtropical regions in Asia, Africa, and the Caribbean and cultivated as a vegetable, spice, and medicinal plant.

During the last decades, *M. charantia* obtained worldwide growing attention as dietary supplement. The karela powder and karela juice, manufactured from the fruits, are used in many countries in the form of ready-made products as dietetic, e.g., as vitamin source and by diabetic patients. The powder is a component of the curry spice, which is widely used in the cuisines of East, South, and Southeast Asia. Bitter gourd tea, made from dried fruit slices, is used in folk medicines of many regions for a wide range of medical applications, including gastrointestinal complaints, viral diseases, and to prevent and treat type 2 diabetes [21, 59].

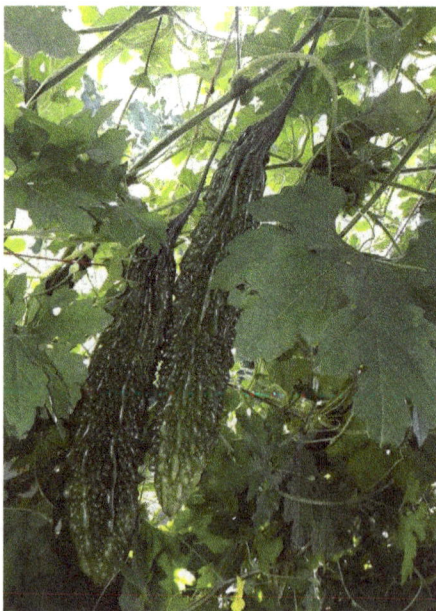

Ph. 6.9: *Momordica charantia* (source: Meena Rajbhandari).

The plants contain as cucurbitacins momordicins I–III (Fig. 6.5), momordicosides A, C, F1, I, and K, charantosides A–K, momorcharacoside A, and some further compounds. Momordicins and momodicorsides differ from the cucurbitacins of the abovementioned Cucurbitaceae in possessing a formyl group at C-9 [48, 59, 131, 134]. Distribution of cucurbitacins differs across different chemotypes [136].

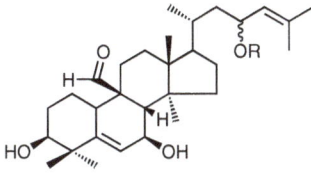

Momordicin I   R=H
Momordicin II  R=βGlc

Fig. 6.5: Cucurbitacins of *Momordica charantia*.

Other notable ingredients are peptides and proteins, among them ribosome-inactivating proteins (momocharins, *Momordica charantia* lectins), polysaccharides, saponins, and phenolic compounds [59].

Hypoglycemic and lipid-lowering effects of the plants have been demonstrated in animal and small-scale human studies [40, 91]. Cucurbitacins and polypeptides are mainly held responsible for the antidiabetic effects. Possible mechanisms of action are activation of peroxisome proliferator-activated receptor gamma PPAR-γ, inhibition of protein tyrosin phosphatase 1b (PTP1B), a negative regulator of insulin signaling pathway, and inhibition of α-glucosidase [48, 59, 134].

The $LD_{50}$ values of the pressed juice and the ethanolic extract in mice are very high (919 mg/kg BW and 3,623 mg/kg BW, respectively, both s.c.). Postmortem damage to the lungs and heart and darkening of the blood were detected [40].

Few poisoning cases during self-medication use were reported [83].

⚑ A 22-year-old man developed atrial fibrillation with rapid ventricular response after ingestion of crushed bitter melons and a drink of bitter melon juice [41]. Another one showed symptoms of acute gastric ulceration after consumption of half a liter of concentrated liquid extract [88]. In clinical studies with type 2 diabetes patients, investigating possible hypoglycemic effects, moderate adverse effects including diarrhea and abdominal pain have been reported [91].

⚑ Two small children had convulsions followed by hypoglycemic coma after drinking beverage prepared from *M. charantia* on an empty stomach [21].

Long-term consumption will lead to a glucose-6-phosphate dehydrogenase deficiency and risk of fava bean disease [21].

Because of insufficient evidence on the effects, *M. charantia* is not recommended as a substitute for insulin or hypoglycemic drugs in diabetes treatment at present [91]. Interactions with normal diabetes treatment cannot be excluded [49]. It is contradicted in pregnant woman because it caused abortion in animals [21]. Further studies are needed to explore the existing potential of *M. charantia* for prevention and/or therapy of diabetes, obesity, and other health problems [42].

## 6.5 Iridals and Cycloiridals in Iris Species

Iris species, Iridaceae, contain iridals or cycloiridals (Fig. 6.6). These are triterpenes with one or two isolated carbocyclic six-rings and a highly reactive aldehyde function. The compounds have been isolated only from fresh rhizomes, but they are likely to occur also in other parts of the plant. The cycloiridals possess as second ring an unsaturated methyl ionone ring (extramethyl group, $C_{31}$ compounds!). By oxidative degradation during prolonged storage, they yield a ketone mixture with violet scent, consisting of γ-irone and α-irone. From that originates the violet scent of the formerly used drug Rhizoma iridis, violet root. The iridals and cycloiridals are also present (in the intact plant only?) as fatty acid esters, especially as myristates and linolenates.

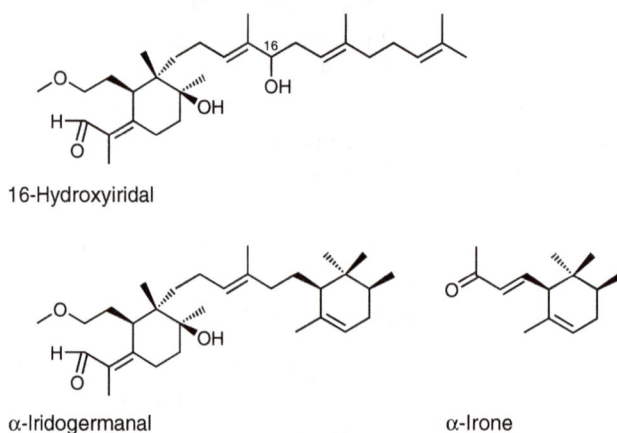

16-Hydroxyiridal

α-Iridogermanal                    α-Irone

Fig. 6.6: Triterpenes of Iris species.

The compounds detected are in *Iris germanica* L. German iris, the iridal iridogermanal and the cycloiridals α-iridogermanal, δ-iridogermanal, 10-deoxy-iridogermanal, and the noriridal irigermanone (together about 1% of FW [7, 67, 98]), in *I. versicolor* L., blue flag, the iridal iriversical [67], in *I. pseudacorus* L., flag iris (Ph. 6.10), the iridal 16-hydroxyiridal [80], in *I. missouriensis* Nutt. Missouri flag, the iridal isoiridogermanal, in *I. pallida* Lam., Dalmatian iris, the iridal isoiridogermanal, and the cycloiridals iripallidal, α-irigermanal, and deoxyiripallidal [68, 80], and in *I. pallisii* Fisch. var. *chinensis* Fisch. and *I. foetidissima* L., gladwyn, a spirobicyclic triterpene [79].

In addition to the triterpene derivatives, Iris species contain potentially toxic alkylbenzoquinones.

Iridals, like phorbol esters (see Section 5.3.2), bind to the C1b domain of protein kinase C and other proteins, e.g., the GTPase RasGRP3, and exhibit similar effects like phorbol esters [109].

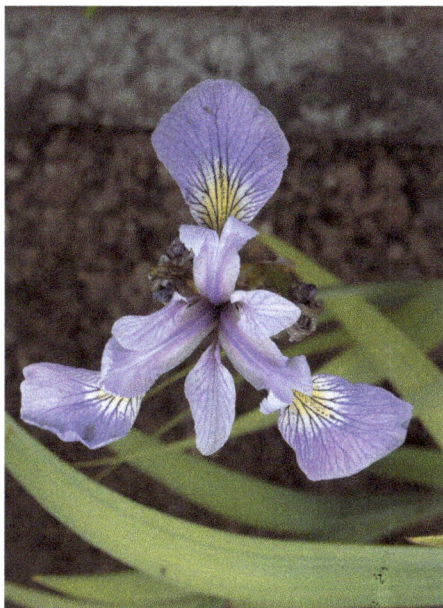

Ph. 6.10: *Iris versicolor* (source: Wolfgang Kiefer).

♟ Symptoms of poisoning after chewing the Iris rhizome are inflammation of the lips and oral mucosa, vomiting, severe diarrhea, and colic [**127**].

🛏 Treatment of poisoning is symptomatic.

🐃 Bloody diarrhea and death have been observed in animals after eating leaves of Iris species found in hay.

In an animal model, an aqueous extract of *I. germanica* rhizomes showed beneficial effects against Alzheimer's disease [20]. Methanolic extracts of *I. pallida*, *I. versicolor*, and *I. germanica*, rich in isoflavonoids, inhibited bacterial biofilms [53].

## 6.6  Gossypol as Toxin of Cotton Species (Gossypium Species)

(+)-Gossypol and (–)-gossypol (Fig. 6.7) are produced by pigment glands in all parts of Gossypium species, cotton, Malvaceae, with greatest concentration in the seeds.

Gossypium species are crops that are cultivated since thousands of years in tropical and subtropical regions and domesticated in many regions independently of each other. The original home of the wild plant is unclear. They belong to the economically most important plants of the world. *Gossypium hirsutum* L., upland cotton, *G. barbadense* L., sea

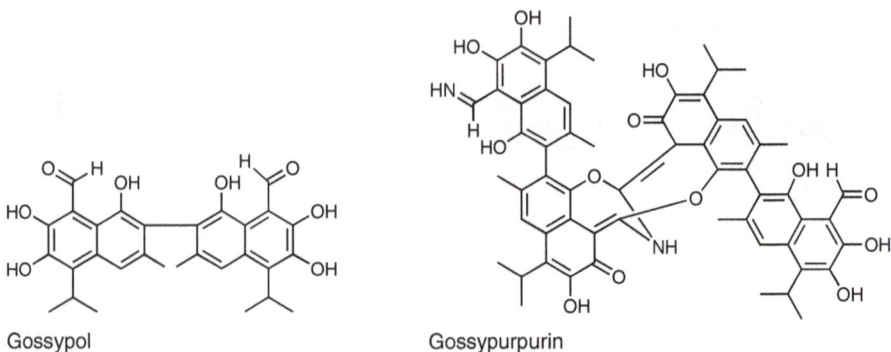

Fig. 6.7: A triterpene and a storage product of Gossypium species.

Ph. 6.11: *Gossypium herbaceum* (source: Wolfgang Kiefer).

island cotton, *G. arboreum* L., tree cotton, and *G. herbaceum* L., common cotton (Ph. 6.11) are cultivated for cotton fiber and production of seed oil and cotton seed meal.

Of toxicological interest is especially gossypol of the seeds. Their content of gossypol is 1.0–1.5% (up to 6.6% [45]). However, varieties poor in gossypol are also known (gossypol content of seeds <0.01%). The ratio of the biologically less active (+)-gossypol to the more active (–)-gossypol is 3:2 in *G. hirsutum* and 2:3 in *G. barbadense* L. [116]. Gossypol is also present in the tropical tree *Thespesia populnea* (L.) Sol. ex Corrĕa, portia tree, milo in Polynesian languages, Malvaceae [2].

The yellow colored gossypol sometimes changes to more toxic, intensely colored artifacts during processing and storage. About 15 kinds of the artifacts are known, e.g., the red gossypurpurin (Fig. 6.7, $LD_{50}$ 6.68 g/kg BW, p.o., rat) and the green gossyverdurine ($LD_{50}$ 0.66 g/kg BW, p.o., rat [78, 115]).

The gossypol absorption rate after ingestion is inversely proportional to the amount of iron in the diet. In ruminants, free gossypol is bound to proteins by microbial enzymes, but it is not known whether the bound form can be absorbed or if microorganisms can release free gossypol. The absorbed gossypol accumulates in the liver and kidneys. Excretion occurs via bile and feces [46].

Gossypol is a chemically very reactive substance which reduces the activity of various enzyme systems by reacting with amino groups through formation of Schiff bases. Oxidoreductases such as lactate dehydrogenase (LDH), transferases such as catechol-*O*-methyltransferase (COMT) as well as hydrolases such as calcineurin and lyases such as adenylate cyclase are influenced. Other targets include signal transduction mechanisms affecting protein expression and membranes. Gossypol antagonizes the antiapoptotic BCL-2 (B-cell lymphoma 2) proteins and induces apoptosis. It belongs to the 'BH3-mimetics', promising compounds for cancer therapy [12, 35]. The inhibitory effect of gossypol on spermatogenesis and sperm motility is due to the inhibition of enzyme systems involved in sperm maturation and motility, particularly a mitochondrial isoenzyme of LDH (LDH-C), which is present only in testes and spermatozoa and is essential for sperm energy production [35]. In some test systems, including rat germ cells, gossypol causes chromosomal damage, so it must be considered a potentially genotoxic compound. Embryotoxicity has been observed [16, 46, 74, 135].

While rats, hamsters, and monkeys are not very sensitive, the toxic effects of gossypol are strong in guinea pigs, rabbits, pigs, and dogs. Rabbits died within 4 min after i.v. administration of 0.05 g gossypol. The $LD_{50}$ of the enantiomeric mixture was 2.4–3.3 g/kg BW in rats and 0.55 g/kg BW in pigs [1, 39, 112, **91, 100**]. Dogs died after eating the pressed residues of cottonseeds used as a bedding material, which contained 1.6 mg/kg free gossypol [122].

Since products from cottonseed are of great importance for the nutrition of humans (especially cottonseed oil) and animals (cottonseed meal), they must be freed from gossypol by suitable treatment processes, e.g., by heating, or low-gossypol seeds must be used [46, **91, 100**]. When cottonseed meal is used for animal feed, there is a risk of gossypol transfer through the food chain to humans [108, 132].

🐾 Acute poisonings occur after ingestion of cottonseed meal with high amounts of gossypol. Cumulative effects of smaller concentrations appear following an ingestion period of 1–3 months. They have been reported in chicken, ducks, pigs, dogs, sheep, goats, and other animals. Poisoning by gossypol containing plant material is one of the most diagnosed plant-associated intoxications in cattle in California [46, 123, 132]. Monogastric animals are more susceptible than ruminants, and young ruminants more than adult ones.

Symptoms of poisoning are reduced feed intake, weight loss, diarrhea, and disturbances of heart and liver functions. Histopathological, necrotic changes can be detected in the heart, liver, lungs, and skeletal muscles. By interference with immune function, gossypol reduces animal's resistance to infections and impairs the efficiency of vaccines [46, 123].

The contraceptive effect of gossypol has been clinically tested. However, insufficient efficacy, possible genotoxic effects, possible hypokalemia, and failure to fully restore spermatogenesis after discontinuation of the drug make extensive use as a contraceptive questionable so that the research on this field was not continued. But the antineoplastic, antiviral, and antiparasitic properties of gossypol are yet of interest [35].

## 6.7 Triterpenes of Seaweeds (Marine Macroalgae)

In macroalgae, in contrast to the lower terpenes, triterpenes occur comparatively rarely. An example of a triterpene from **brown algae** is turbinaric acid (Fig. 6.8), a cytotoxic trinorsqualene derivative from *Turbinaria ornata* Agardh, Sargassaceae [8].

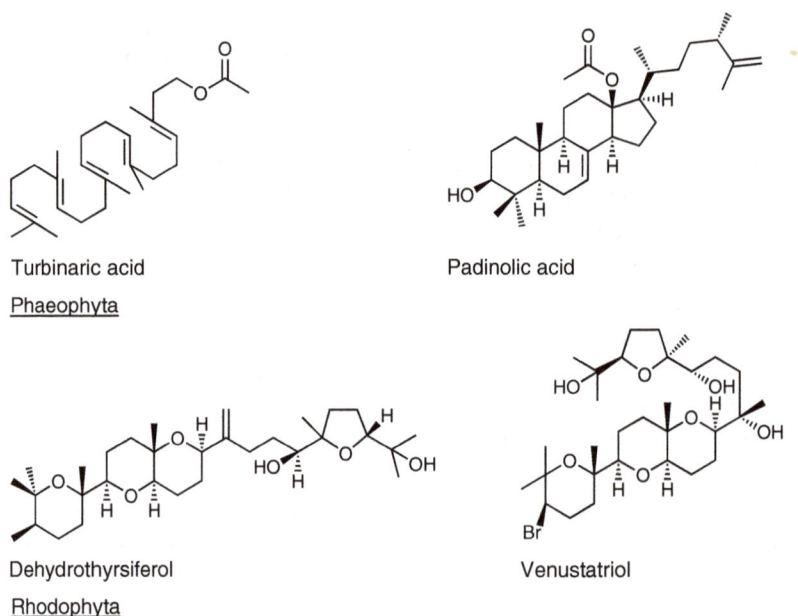

Turbinaric acid

Phaeophyta

Padinolic acid

Dehydrothyrsiferol

Rhodophyta

Venustatriol

Fig. 6.8: Triterpenes of macroalgae.

Padinolic acid, isolated from the marine brown alga *Padina boergesenii* Allender and Kraft, occurring at the coast of Oman, inhibits the activity of urease, α-glucosidase, and lipid peroxidation [5].

Among **red algae**, triterpenes have been found primarily in Laurencia species. The triterpenes from the Rhodomelaceae can be divided into compounds with a dioxabicyclo [4.4.0]decane ring system, e.g., thyrsiferol, dehydrothyrsiferol, and venustatriol (Fig. 6.8), in compounds with a dioxabicyclo[5.4.0]undecane ring system, e.g., enshuol, and compounds with symmetrical elements, e.g., longilenes and teurilene. Thyrsiferyl-23-acetate, first isolated from *Laurencia obtusa*, and longilene peroxide, isolated from *L. viridis*, inhibit similar like okadaic acid, a product of Dinophyceae, serine–threonine phosphatase type 2A (PP2A) and are therefore highly cytotoxic [29]. Dehydrothyrsiferol, identified in *Laurencia pinnatifida* and *L. viridis,* inhibits tumor cell growth without affecting the re-sistance-causing transport molecule P-gp. Extracts of *L. venusta* exhibit antiviral activity. The compounds are potential anticancer drug leads [75].

## References

(for numbers in bold, see cross-chapter literature)

[1]     Adams R et al. (1960) Chem Rev 60: 555
[2]     Akhila A, Rani K (1993) Phytochemistry 33(2): 335
[3]     Akihisa T et al. (1998) Phytochemistry 49(6): 1757
[4]     Al-Faraj S (1995) Ann Trop Med Parasitol 89: 695
[5]     Ali L et al. (2017) Mar Drugs 15: 19
[6]     Alzohairy MA ((2016)) eCAM 2016: 7382506
[7]     Amin HIM et al. (2014) Molecules 26: 264
[8]     Asari F et al. (1989) J Nat Prod 42(5): 1167
[9]     Aydogan K et al. (2012) Ann Dermatol 24(3): 358
[10]    Baas WJ (1985) Phytochemistry 24(9): 1875
[11]    Bajcsik N et al. (2017) J Chrom B 1052: 128
[12]    Balakrishnan K et al. (2008) Blood 112(5): 1971
[13]    Barros LM et al. (2016) Molecules 21: 209
[14]    Barton DHR, de Mayo P (1954) J Chem Soc 1954: 887
[15]    Barua AK et al. (1976) Phytochemistry 15(6): 987
[16]    Beaudoin AR (1985) Teratology 32: 251
[17]    Begum S et al. (2003) Chem Pharm Bull (Tokyo) 51: 134
[18]    Biglino G et al. (1963) Tetrahedron Lett 11651
[19]    Boeke SJ et al. (2004) J Ethnopharmacol 94: 25
[20]    Borhani M et al. (2017) Afr J Tradit Complement Altern Med 14(4): 140
[21]    Bortolotti M et al. (2019) Front Pharmacol 10: 485
[22]    Braga TM et al. (2021) Molecules 26: 252
[23]    Brouwer IJ et al. (2020) Biochem Biophys Rep 24: 100832
[24]    Brouzas D et al. (2012) Case Rep Ophthalmol 3: 87
[25]    Brown JMM et al. (1963) Proc Roy Soc B 157: 473
[26]    Cárdenas PD et al. (2019) Front Plant Sci 10: 1523
[27]    Carmely S et al. (1983) Tetrahedron Lett 24: 3673
[28]    Cattel L et al. (1978) Gazz Chim Ital 108: 1
[29]    Cen-Pacheco F et al. (2018) Mar Drugs 16: 131
[30]    Chaudhary S et al. (2017) Front Plant Sci 8: 610
[31]    Cheikhyoussef N et al. (2017) 3 Biotech 7: 297
[32]    Chen JC et al. (2005) Nat Prod Rep 22: 386
[33]    Chen X et al. (2012) Anti-Cancer Drugs 23: 777
[34]    Cullen JM, Stalker MJ (2016) Liver and Biliary System. Elsevier Amsterdam, p. 258
[35]    Dodou K (2005) Expert Opin Invest Drugs 14: 1419
[36]    Dolan LC et al. (2010) Naturally occurring food toxins. Toxins 2: 2289

[37]    Doskotch RW (1969) J Nat Prod 33: 115

[38]    Dzubak P (2006) Nat Prod Rep 23: 394

[39]    Eagle E (1950) Arch Biochem 26: 68

[40]    El Sattar El Batran SA et al (2006) J Ethnopharmacol 108: 236

[41]    Erden I et al. (2010) Ann Saudi Med 30(1): 86

[42]    Fan M et al. (2019) Int J Environ Res Public Health 16: 3251

[43]    Fernández JJ et al. (2000) Nat Prod Rep 17: 235

[44]    Fowden L (1961) Biochem J 81: 154

[45]    Frampton VL (1960) Econ Bot 14: 197

[46]    Gadelha ICN et al. (2014) Sci World J 2014: 231635

[47]    Gamlath CB et al. (1988) Phytochemistry 27(10): 3225

[48]    Gao Y et al. (2020) Nat Prod Bioprospect 10(3): 153

[49]    Gmelin R (1967) Arch Pharm 300: 234

[50]    Hart NK et al. (1976) Aust J Chem 29: 655

[51]    Hegnauer R (1957) Pharm Acta Helv 32: 334

[52]    Ho CH et al. (2014) J Emerg Med 46: 772

[53]    Hoang L et al. (2020) Antibiotics 9: 403

[54]    Hunsakunachai N et al. (2019) BMC Compl Altern Med 19: 157

[55]    Hylands PJ, Kosugi J (1982) Phytochemistry 21(6): 1379

[56]    Hylands PJ, Mansour ESS (1982) Phytochemistry 21(11): 2703

[57]    Hylands PJ et al. (1980) J Chem Soc, Perkin Trans 1: 2933

[58]    Ide A, Tutt CL (1998) J S Afr Vet Assoc 69: 30

[59]    Jia S et al. (2017) Int J Mol Sci 18: 2555

[60]    Jian CC et al. (2005) Nat Prod Rep 22(3): 386

[61]    Johnson JH, Jensen JM (1998) J Zoo Wildl Med 29: 203

[62]    Kapoor N et al. (2020) Inform Med Unlocked 21: 100484

[63]    Kaushik U et al. (2015) Phcog Rev 9: 12

[64]    Khatib KI, Borawake KS (2014) J Clin Diag Res 8(12): MD05

[65]    Kirschman JC, Suber RL (1989) Food Chem Toxicol 27: 555

[66]    Koriem KMM (2013) Asian Pac J Trop Biomed 3(10): 834

[67]    Krick W et al. (1983) Z Naturforsch C Biosci 38 C: 689

[68]    Krick W et al. (1984) Helv Chim Acta 67: 318

[69]    Krienke EG, von Mühlendahl KE (1978) Notfallmedizin 4: 619

[70]    Kujawska M, Svanberg I (2019) J Ethnobiol Ethnomed 15: 22

[71]    Kumar R et al. (2018) BMC Vet Res 14: 129

[72]    Kupchan SM et al. (1978) Phytochemistry 17(4): 767

[73]    Lavie D et al. (1961) Tetrahedron Lett 2(18): 615

[74]    Li YF et al. (1989) Reprod Toxicol 3: 59

[75]    Li YX et al. (2013) Molecules 18: 7886

[76]    Lindsay Lauder W (1865) Foreign Med Chir Rev 36(71): 153

[77]    Louw PGJ (1948) Onderstepoort J Vet Sci 23: 233

[78]    Lyman CL (1963) J Am Oil Chem Soc 40: 571

[79]    Marner FJ (1990) Helv Chim Acta 73: 433

[80]    Marner FJ et al. (1988) Helv Chim Acta 71: 1331

[81]    McLennon MW, Amos M (1989) Aust Vet J 66(3): 93

[82]    Mello FB et al. (2005) Toxicon 45: 459

[83]    Memorial Sloan Kettering Cancer Center. https://www.mskcc.org/cancer-care/integrative-medicine/
        herbs/bitter-melon (retrieved 7_10_2022)

[84]    Metcalf RL et al. (1980) Proc Natl Acad Sci USA 77: 3769

[85]    Miró M (1995) Phytother Res 9: 159

[86]    Misra L, Laatsch H (2000) Phytochemistry 54(8): 969

[87]    Mueller A, Wichtl M (1979) Pharm Ztg 124: 1761

[88]    Nadkarni N et al. (2010) Ind J Gastroenterol 29(19): 37

[89]    Nakano K et al. (1995) Phytochemistry 39(1): 209

[90]    O'Neill M et al. (1998) J Nat Prod 61(11): 1328

[91]    Ooi CP et al. (2012) Cochrane Database Syst Rev 2012(8): CD007845

[92]    Panosyan AG, Avetisyan GM (1985) Arm Khim Zh 38: 644

[93]    Panosyan AG et al. (1985) Khim Prir Soedin 5: 679

[94]    Pass MA (1986) Aust Vet J 63(6): 169

[95]    Pfab R (1999) MMW Fortschr Med 141: 41

[96]    Phua DH et al. (2008) Clin Toxicol 46: 1067

[97]    Pohlmann J (1975) Phytochemistry 14(7): 1587

[98]    Potterat O et al. (2014) Helv Chim Acta 97: 32

[99]    Raikhlin-Eisenkraft B, Bentur Y (2000) J Toxicol Clin Toxicol 38: 305

[100]   Reddy YR et al. (2002) Ind Vet J 79: 91

[101]   Rinnington C, Quin JI (1937) Onderstepoort J Vet Anim Ind 9: 225

[102]   Ripperger H (1976) Tetrahedron 32: 1567

[103]   Roy A, Saraf S (2006) Biol Pharm Bull 29(2): 191

[104]   Roy S, Barua AK (1985) Phytochemistry 24(7): 1607

[105]   Salehi B et al. (2019) Molecules 24: 1854

[106]   Satar S et al. (2001) Eur J Emerg Med 8: 337

[107]   Seawright AA, Hrdlicka J (1977) Aust Vet J 53(5): 230

[108]   Semon B (2012) Med Hypotheses 78: 293

[109]   Shao L et al. (2001) J Med Chem 44: 3872

[110]   Sharma OP et al. (1988) Toxicon 26: 975

[111]   Sharma S et al. (2000) Toxicon 38: 1191

[112]   Smith HA (1957) Am J Pathol 33: 353

[113]   Stewart C et al. (1988) Aust Vet J 65(11): 349

[114]   Steyn DG (1950) S Afr Med J 24: 713

[115]   Stipanovic RD et al. (1975) Phytochemistry 14(4): 1077

[116]   Stipanovic RD et al. (2005) J Agric Food Chem 53: 6266

[117]   Sun J et al. (2005) Oncogene 24: 3236

[118]   Tan JW et al. (2002) Z Naturforsch C Biosci 57: 963

[119]   Tunmann P et al. (1966) Liebigs Ann Chem 693: 158

[120]   Ukiya M et al. (2002) J Nat Prod 65(2): 179

[121]   Uppal RP, Paul BS (1982) Indian Vet J 59: 18

[122]   Uzal FA et al. (2005) J Vet Diagn Invest 17: 626

[123]   Varga A, Puschner B (2012) Vet Med: Res Rep 3: 111

[124]   Vishnukanta ACR (2008) Pharmacogn Rev 2: 173

[125]   Wang X et al. (2017) Peer J 5: e3357

[126]   Whur P (1986) Vet Rec 119: 411

[127]   Wojciechowska B, Wizner L (1983) Herba Polon 29: 97

[128]   Wolfson SL, Salomons TWG (1964) Am J Dis Child 107: 173

[129]   Wu D et al. (2019) Front Pharmacol 10: 1287

[130]   Yamada Y et al. (1978) Chem Pharm Bull 26(10): 3107

[131]  Yue J et al. (2019) Phytochemistry 157(1): 21
[132]  Zeng QF et al. (2014) Poult Sci 93: 2000
[133]  Zeng Y et al. (2021) Phytother Res 35: 4155
[134]  Zhang LJ et al. (2014) Phytomedicine 6: 564
[135]  Zhu X et al. (2020) Front Neurosci 14: 318
[136]  Zhou S et al. (2019) Metabolomics 15(8): 104

# 7 Tetraterpenes

## 7.1 General

Tetraterpenes are $C_{40}$ compounds built up from eight isoprenoid residues. They are either present in aliphatic form or form long-chain molecules that bear rings at one or both ends. Polycyclic representatives, as known from the sesqui-, di-, and triterpenes, do not occur with them.

The known tetraterpenes can be counted in total to the group of carotenoids. These aliphatic lipophilic compounds, formed by microorganisms and plants, are mostly cyclized at the end of the chain and are mostly yellow to reddish-violet-colored. They are structured in such a way that two methyl groups are located in the center of the molecule in the 1,6-position, while the remaining methyl groups, with the exception of the terminal ones, are located in the 1,5-position. Degradation products of carotenoids in which this arrangement is preserved are called apocarotenoids. Chain extensions to $C_{45}$ or $C_{50}$ compounds also occur in nature.

While carotenoids are generally important as flower pigments, accessory pigments in photosynthesis, photoprotectors, and precursors of vitamin A, toxic effects are attributed to the degradation products of protocrocin, a carotenoid in *Crocus sativus* (see Section 7.2).

## 7.2 Poisonous Cleavage Products of Carotenoids in Croci Species (Crocus Species)

The genus Crocus, croci, crocuses, Iridaceae, comprises about 90 species. Most of them are native to the Mediterranean region and across Central Asia. In Central Europe the main species is *Crocus vernus* (L.) Hill, spring crocus. This perennial tuberous plant with mostly white, but rarely light to dark purple flowers is especially common in moist, humus-rich alpine meadows. Many species of croci, e.g., *C. vernus* and *C. chrysanthus* (Herb.) Herb., golden crocus, are spring bloomers, *C. speciosus* M. Bieb., large purple crocus, blooms in autumn.

*Crocus sativus* L., saffron crocus, autumn crocus (Ph. 7.1), a triploid and therefore sterile plant with light purple veined flowers is cultivated. The vivid orange stigmata of the flowers are used in dried form as the spice saffron. This plant is cultivated, for example, in Iran, Greece, Spain, Switzerland, Marocco, India, and Pakistan in continental, not tropical climate [148].

The spice saffron, the most expensive spice of the world, contains, when fresh harvested acylglycosides of the brown-red colored, water-soluble apocarotenoid crocetin (8,8'-diapo-ψ,ψ'-carotenoid dicarboxylic acid) with β-D-glucose, with the disaccharide β-D-gentiobiose, or the trisaccharide β-D-neapolitanose. By cleavage these

https://doi.org/10.1515/9783110724738-007

Ph. 7.1: *Crocus sativus* (source: Jean-Luc Farges/iStock/Getty Images Plus).

compounds arise crocin 1 (= crocin A, crocetin-di-(β-D-gentiobiosyl)-ester), crocetin-mono-(β-D-glucosyl)-ester, crocetin-di-(β-D-glucosyl)-ester, crocetin-mono-(β-D-gentiobiosyl)-ester, crocetin-(β-D-gentiobiosyl)-(β-D-neapolitanosyl)-ester, and crocetin-di-(β-D-neapolita-nosyl)-ester. Other important compounds are the bitter glucoside picrocrocin, which is also derived from carotenoids, carotenoids like lycopene, α-, β-, and γ-carotene and a xanthone-carotenoid conjugate, magnicrocin (magniferin-6'-*O*-(crocetyl-1'''-*O*-β-D-glucoside ester)). The native all-trans-crocins rearrange into *cis*-14-crocins when exposed to oxygen or light ([4, 5, 9, **148**], Fig. 7.1).

The constitution of crocetin was elucidated in 1932 by Karrer et al. [3] and that of picrocrocin in 1934 by Kuhn and Winterstein [4].

If the drug is stored, the odorless picrocrocin is converted by hydrolytic or non-hydrolytic glucose cleavage to 4-hydroxy-β-cyclocitral or 4,5-dehydro-β-cyclocitral (safranal, odoriferous! [12]). Other transformation products are probably isophorone, 4-oxoisophorone, and the crocusatins (Fig. 7.1, [5]). The stored drug contains 5–30% crocins and 0.4–2.5% essential oil with safranal as major constituent [**148**].

Crocin, crocetin, and other crocetin glycosides were also detected in the stigmata of Crocus species cultivated as ornamentals in Central Europe. Picrocrocin was not found in those species [11].

Saffron has been used since ancient times not only as a spice and coloring agent but also as remedy, especially for diseases of the gastrointestinal tract. Poisoning occurred in the past when the drug was misused as an abortifacient [2]. No adverse effects are known for humans up to a maximum daily dose of 1.5 g of drug. The drug is considered toxic when ingested with doses more than 5 g and could be lethal if taken about 20 g/day [7]. The abortive dose is 10 g [**140**].

We are not aware of any cases of poisoning by *C. vernus, C. chrysanthus* or *C. speciosus.*

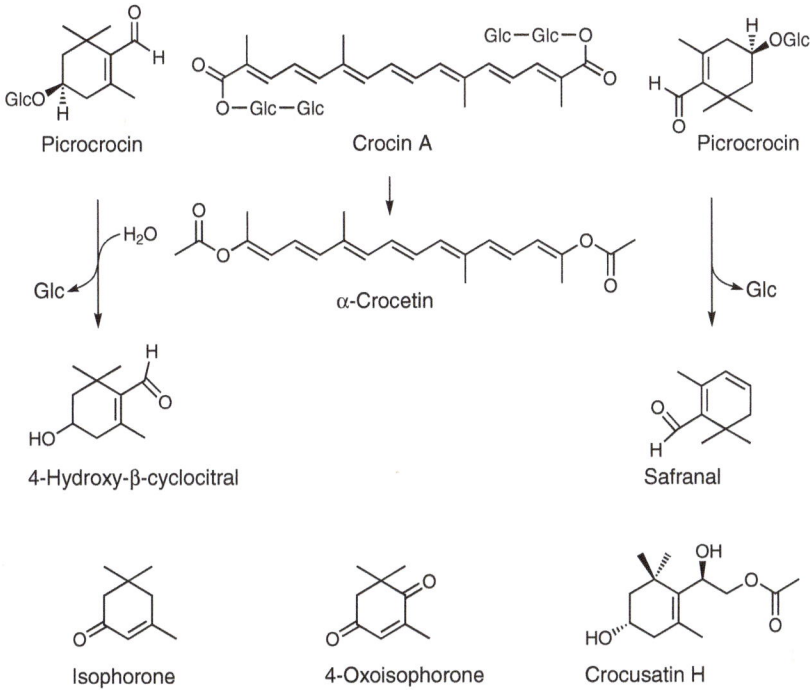

Fig. 7.1: Components of *Crocus sativus*.

♟ Predominant symptoms of poisoning after peroral ingestion of high doses of saffron are nausea, vomiting, diarrhea, and dizziness. More severe symptoms are swelling of the lips, eyelids, and joints, hemorrhages, especially of the kidneys as the main organ of excretion. In pregnant women abortion may occur [2, 7, **53, 127**].

♦ In a 28-year-old pregnant woman who had ingested about 5 g of saffron in milk together with estrogen-containing tablets, severe gastroenteritis, hematuria, circulatory collapse, and as a result of hemorrhage in the area of the nasal septum, severe purpura with jet-black necrosis of the nose occurred in addition to abortion [2].

🚑 Treatment of poisoning is symptomatic after primary toxin removal [**127**].

Extracts of saffron and crocins in nontoxic amounts show neuroprotective and antidepressant activity in animal and human studies. The effects can be explained by interactions, especially of *trans*-crocetin, with central N-methyl-D-aspartate (NMDA) receptors. Therefore, saffron could be an effective intervention for symptoms of depression and anxiety [6, 7]. Other small-scale clinical studies demonstrated anti-Alzheimer, antipruritic, and antiobesity effects. Symptoms of premenstrual syndrome have been reduced [7]. Because of the slightly bitter taste, saffron is a component of gastric tonics. Regarding insufficient investigations on safety of saffron in pregnancy, pregnant woman should avoid using high doses of saffron [1].

Crocin (gardenin) is also an ingredient of *Gardenia jasminoides* J. Ellis, cape jasmine (Rubiaceae). Ethanolic extracts of the fruits of these shrubs or trees growing in warm climates serve in Asia as a food coloring ('Gardenia yellow'). In TCM Gardenia fruits (*Gardeniae fructus*, Zhi Zi) are used to treat bleedings and reduce swellings, in Korea to lower fever. In vitro studies revealed genotoxic effects for extracts of *G. jasminoides* and the iridoid genipin contained therein, but not for crocin [8]. In dietary supplements, gardenia has been detected as an undeclared adulteration of saffron [10].

## References

(for numbers in bold see cross-chapter literature)

[1]    Bostan HB et al. (2017) Iranian J Basic Med Sci 20: 110
[2]    Frank A (1961) Dtsch Med Wochenschr 86: 1618
[3]    Karrer P et al. (1932) Helv Chim Acta 15: 1218, 1399
[4]    Kuhn R, Winterstein A (1934) Ber Dtsch Chem Ges 67: 344
[5]    Li CY, Wu TS (2002) J Nat Prod 65 (10): 1452
[6]    Marx W et al. (2019) Nutr Rev 77(8): 557
[7]    Moshiri M et al. (2015) Drug Res 65: 287
[8]    Ozaki A et al. (2002) Food Chem Toxicol 40: 1603
[9]    Pfänder W, Wittwer F (1975) Helv Chim Acta 58: 2233
[10]   Sendker J et al. (2022) Z Phytother 43 (4): 178
[11]   Wagner K (1969) Dissertation, University Saarbrücken, Germany
[12]   Zarghami NS, Heinz DE (1971) Phytochemistry 10 (11): 2755

# 8 Steroids

## 8.1 Chemistry and Biogenesis

Steroids are compounds that are formally derived from the hypothetical tetracyclic hydrocarbon sterane, a $C_{17}$ compound: 1H-perhydro-cyclopenta[a]phenanthrene (Fig. 8.1). With hypothetical replacement of the hydrogen atoms on the ring system with different substituents, a wide variety of compounds are formed.

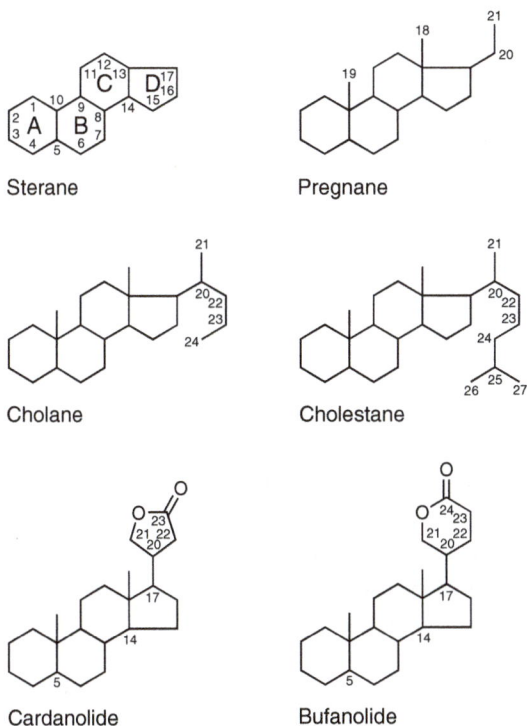

Sterane

Pregnane

Cholane

Cholestane

Cardanolide

Bufanolide

Fig. 8.1: Hypothetical base bodies of steroids.

If the three-dimensional ring systems of sterane derivatives are projected onto a plane, the substituent at C-10 is selected as a reference point. This substituent is assumed to be oriented above the ring plane. The linkages to substituents oriented in the same direction were drawn by amplified lines or long compact wedges. These substituents are called β-positioned. The substituents facing away from the observer are called α-positioned. They are linked to the ring system by dashed lines, lines of dots, or lines of transverse dashes.

https://doi.org/10.1515/9783110724738-008

To simplify the nomenclature of steroids, hypothetically substituted sterane derivatives are used as parent structures when developing the rational names, e.g., pregnane, cholane, cholestane, ergostane (24-methyl-cholestane), stigmastane (24-ethyl-cholestane), lanostane (4,4,14-trimethyl-cholestane), cardanolide, bufanolide (Fig. 8.1), spirostane, and furostane.

The widening of a ring of the sterane base body by insertion of one C atom is indicated by the prefix homo-, in case of insertion of two C atoms by dihomo-, after the capital letter denoting the respective ring, e.g., A-homo steran. The narrowing of a ring by elimination of a C atom is indicated by nor-, e.g., C-norsterane. Opened rings are given the prefix *seco*-, e.g., 14,15-*seco*-pregnane.

According to the linkages of rings A, B, C, and D, the steroids can be divided as representatives of the
– 5α-series: A/B *trans*, B/C *trans*, C/D *trans*,
– 5β-series: A/B *cis*, B/C *trans*, C/D *trans*,
– Cardenolide series: A/B *cis*, B/C *trans*, C/D *cis*.

The 5α-series includes cholestane derivatives, the 5β-series includes cholane derivatives, and the cardenolide series includes the majority of the aglycones of the cardioactive steroid glycosides. The aglycones of the steroid saponins are partly derivatives of the 5α-series and partly of the 5β-series. The transition from one series to the other occurs via unsaturated intermediates.

Almost all living organisms are capable of synthesizing steroids. Exceptions are some microorganisms, mollusks, crustaceans, and insects.

Biogenesis of steroids occurs from the aliphatic triterpene hydrocarbon squalene. At first, 2,3-oxidosqualene is formed, which, by cyclization through simultaneous migration of two methyl groups as well as hydride ions, converts in microorganisms and animals to lanosterol, and in most plants to cycloartenol. Lanosterol or cycloartenol can be transformed by elimination of three methyl groups in a multistep reaction to the cholestane derivative cholesterol.

Fungi and plants can convert cholestane derivatives by methylation in position 24 in the side chain to ergostane derivatives (24-methylcholestane derivatives, e.g., campesterol) and by a methylation of the inserted methyl group to stigmastane derivatives (24-ethylcholestane derivatives, e.g., β-sitosterol).

Some plants, rarely animals, can convert cholesterol via pregnan derivatives by connecting them with a $C_2$- or a $C_3$-body in positions C-20 and C-21 to the aglycons of cardioactive steroid glycosides.

Some plants convert cholesterol by cyclization of its side chain to aglycones of steroid saponins.

The almost always present, mostly β-positioned hydroxyl group at C-3 can be linked to fatty acids (acylsterols), to mono- or oligosaccharides (steroid glycosides and steroid saponins) or to sulfuric acid (in the case of saponins of echinoderms).

## 8.2 Distribution

Steroids are ubiquitously distributed. Sterols (sterines and 3-hydroxy-steranes with a side chain of eight to ten C atoms at C-17) are of particular importance. They are components of cell membranes and cellular endomembrane systems of all living organisms. They are precursors of steroid hormones, bile acids, vitamin D, as well as many secondary metabolites.

Of toxicological interest are the cardioactive steroid glycosides of plants (see Section 8.3) and animals, the steroid saponins (see Chapter 9), withanolides (see Section 8.4), petuniasterones (see Section 8.6), pregnane- and *seco*-pregnane glycosides (see Section 8.7), 1,25-dihydroxycholecalciferol (see Section 8.8), and toxic steroid glycosides of South African Ornithogalum species (see Section 8.5).

Structurally remarkable cytotoxic steroids have been detected in red algae, e.g., some 24-hydroperoxy-fucosterols in *Turbinaria conoides* [376].

Polyoxygenated, cytotoxic steroids have been isolated from basidiomycetes, e.g., from *Polyporus umbellatus* and *Trametes versicolor* [459].

In animals, steroids occur in the secretion of the skin glands of toads, in nuchal glands of some species of snakes, and in some insects. They serve especially as defense against microorganism, predators, and other attackers.

## 8.3 Cardioactive Steroid Glycosides

### 8.3.1 Chemistry

Cardioactive steroid glycosides are steroids that, administered in subtoxic doses, are capable to exert a positive inotropic effect on the heart, which means they strengthen the contractile force of the heart. Their negative chronotropic effect lowers the heart rate. Their negative dromotropic effect slows down the conduction of impulses in the heart. Their positive bathmotropic effect increases the cardiac excitability.

So far, more than 500 cardioactive steroid glycosides with more than 100 aglycones are known. The aglycones have a 10,13-dimethyl-sterane basic body with *cis–trans–cis*, or more rarely *trans–trans–cis* linkage of rings A/B/C/D (reviews on chemistry: [97, 139, 384]).

The first cardioactive steroid glycoside digitoxin was isolated in an approximately pure form in 1867 by Nativelle [40]. The cleavability of digitoxin to the aglycone digitoxigenin and the sugar digitoxose was demonstrated by Kiliani in 1895 [208]. The research groups of Windaus [435, 436], Jacobs [180], Tschesche [408], Elderfield [321], Reichstein [174], and Meyer [292] were involved in the elucidation of the structure of digitoxigenin. In 1933, Stoll and Kreis [400] demonstrated that digitoxin is a secondary glycoside derived from the primary glycoside purpurea glycoside A.

**Cardenolides** (derivatives of 5β,14β-card-20(22)-enolide or 5α,14β-card-20(22)-enolide, Fig. 8.2) have a β-positioned lactone ring in position 17. This lactone ring is five-membered and monounsaturated (butenolide ring, but-2-ene-4-olide ring, (5$H$)-furan-2-one ring).

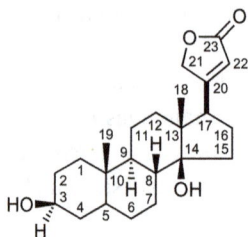

| Name | Substituents | | | | | | |
|------|---|---|---|---|---|---|---|
| | 1 | 5 | 11 | 12 | 16 | 10 | 13 |
| Digitoxigenin | | | | | | $CH_3$ | |
| Uzarigenin | | Hα | | | | $CH_3$ | |
| Thevetiogenin | | | | | | $CH_3$ | =$CH_2$ |
| Syriogenin | | Hα | | OH | | $CH_3$ | |
| Gitoxigenin | | | | | OH | $CH_3$ | |
| Gitaloxigenin | | | | | OForm | $CH_3$ | |
| Oleandrigenin | | | | | OAc | $CH_3$ | |
| Adigenin | | | | | OiVal | $CH_3$ | |
| Digoxigenin | | | | OH | | $CH_3$ | |
| Periplogenin | | OH | | | | $CH_3$ | |
| Sarmentogenin | | | OHα | | | $CH_3$ | |
| Cannogenol | | | | | | $CH_2OH$ | |
| Cannogenin | | | | | | CHO | |
| Corotoxigenin | | Hα | | | | CHO | |
| Diginatigenin | | | | OH | OH | $CH_3$ | |
| Evonogenin | OH | OH | | | | $CH_3$ | |
| Bipindogenin | | OH | OHα | | | $CH_3$ | |
| Strophanthidol | | OH | | | | $CH_2OH$ | |
| Adonitoxilogenin | | | | | OH | $CH_2OH$ | |
| Adonitoxigenin | | | | | OH | CHO | |
| Strophanthidin | | OH | | | | CHO | |
| Sarmentologenin | | OH | OHα | | | $CH_2OH$ | |
| Strophadogenin | | OH | | | OH | CHO | |
| Nigrescigenin | | OH | OHα | | | CHO | |
| Ouabagenin | OH | OH | OHα | | | $CH_2OH$ | |
| Hyrcanogenin | | $\Delta^{4(5)}$ | | | | CHO | |

β-positioned substituents are not marked separately
Abbreviations: Form, formyl residue; Ac, acetyl residue; iVal, isovaleryl residue

**Fig. 8.2:** Aglycones of cardenolide glycosides.

**Bufadienolides** (derivatives of 5β,14β-bufa-20,22-dienolides, Fig. 8.3) have a six-membered and double unsaturated ring in position 17 (pentadienolide ring, penta-2,4-diene-5-olide ring, pyran-2-one ring).

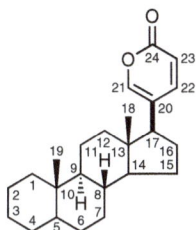

| Name | Substituents | | | | | | | | | | | Double bond (Δ) |
|------|----|----|----|----|----|----|----|----|----|----|----|------|
| | 1 | 3 | 5 | 6 | 8 | 11 | 12 | 14 | 15 | 16 | 10 | |
| Bersaldegenin | CHO | | OH | | | | | OH | | | CH$_3$ | |
| Scillarenin | | OH | | | | | | OH | | | CH$_3$ | 4 |
| Bufalin | | OH | | | | | | OH | | | CH$_3$ | |
| Helleborogenon | | O= | | | | | | OH | | | CH$_3$ | 1, 4 |
| Scilliglaucogenin | | | OH | | | | | OH | | | CHO | 3 |
| Scillicyanogenin | | | OH | | | | | OH | | OAc | CHO | 3 |
| Telocinobufagin | OH | OH | | | | | | OH | | | CH$_3$ | |
| Gamabufotalin | OH | | | | | OHα | | OH | | | CH$_3$ | |
| Scilliphaeoside | OH | | | | | | OH | OH | | | CH$_3$ | |
| Bufotalin | OH | | | | | | | OH | | OAc | CH$_3$ | |
| Hellebrigenin | OH | OH | | | | | | OH | | | CHO | |
| Scilliglaucoside | OH | | | | | | | OH | | | CHO | 4 |
| Resibufogenin | OH | | | | | | | -O- | | | CH$_3$ | |
| Scillirubroside | OH | | | | OH | | | OH | | | CH$_3$ | 4 |
| Scilliroside | OH | | | OAc | OH | | | OH | | | CH$_3$ | 4 |
| Arenobufagin | OH | | | | | OHα | O= | OH | | | CH$_3$ | |

Fig. 8.3: Aglycones of bufadienolide glycosides. Abbreviations see Fig. 8.2.

**Allocardenolide glycosides** and **allobufadienolide glycosides** differ in the structure of their aglycones from those above described. For example, compounds with α-positioned lactone rings occur, e.g., glycosides of 17α-digitoxigenin in Cerbera species [289] and 4,5-dehydrodigitoxin (canarigenin).

Many allocardenolide glycosides were detected in Nerium species. In these plants occur aglycones with secondary modifications of the carbon skeleton of the sterane basic body, e.g., the aglycone of neriaside (Fig. 8.6), or 15(14→8)*abeo*-cardenolides, e.g., oleagenin (Fig. 8.6). The aglycones differ in the substitution pattern of the sterane basic body also. Hydroxy-, epoxy-, or oxo-groups occur as substituents, like in adynerigenin (Fig. 8.6). The hydroxy groups are acylated in some cases. The aglycones may differ in the occurrence and in the position of double bonds in the basic ring system, e.g., the aglycones of neriumosides Fig. 8.6) and 2α,3β,14β-trihydroxy-card-16,20(22)-dienolide (Fig. 8.6). The methyl group at C-10 can be oxidized to a CH$_2$OH, CHO, or

COOH group. In almost all representatives, β-positioned hydroxyl groups are present at C-3 and C-14.

There are also deviations from the usual substitution pattern. Thus, there exist agly-cones unsubstituted in position 3, e.g., scilliglaucogenin, or unsubstituted in position 14, e.g., the aglycones of neriumosides, the 14,16-dianhydrogitoxigenin from *Cryptostegia madagascariensis* Bojer ex. Decne, Madagascar rubbervine, and the 14-deoxybufadienolides from *Fusilium physodes* (Jacqu.) Raf. ex Speta (*Urginea physodes* (Jacq.) Baker) [163]. C-nor-D-homo homologs of cardenolides were isolated from *Cascabela thevetia* (L.) Lippold.

Over 40 different monosaccharides as components of cardioactive steroid glycosides are known. As sugar components serve, in addition to widespread monosaccharides, such as D-glucose, many unusual sugars, especially 6-deoxyhexoses, 2,6-dideoxyhexoses, and their 3-methyl ethers (Fig. 8.4).

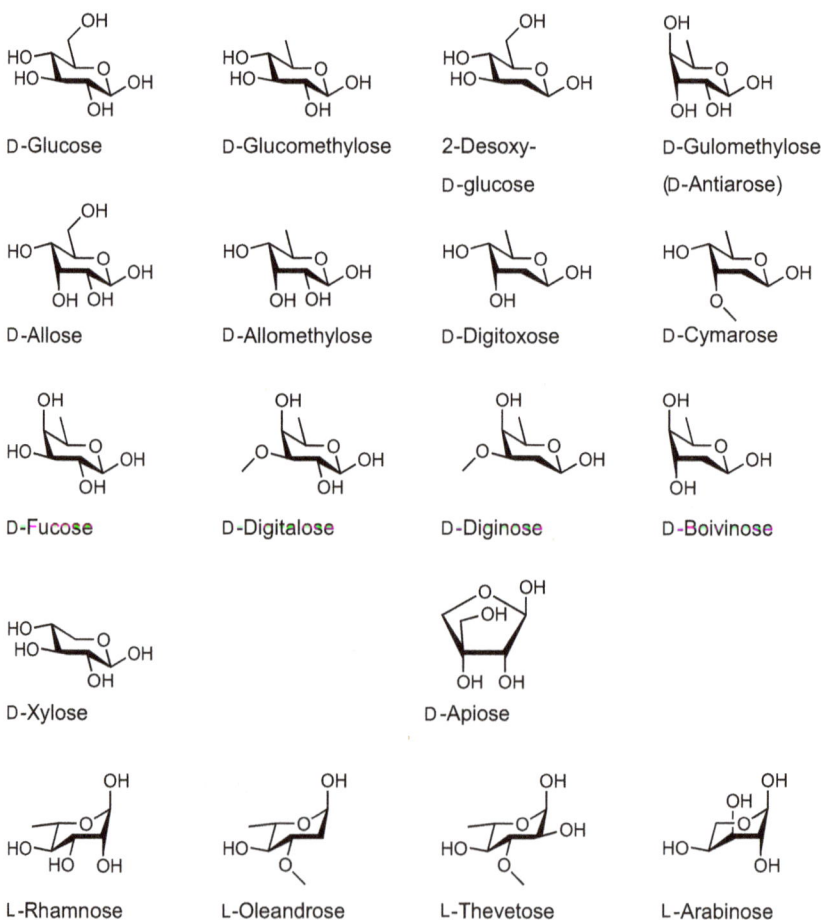

Fig. 8.4: Monosaccharide components of cardioactive steroid glycosides.

The monosaccharide or the oligosaccharide residues are almost bound at the OH group on C-3 of the aglycone. The oligosaccharide residues are always unbranched, usually 1→4-linked and consist of a maximum of five monosaccharide residues. D-Monosaccharides are almost β-glycosidically linked and L-monosaccharides are almost α-glycosidically linked (Klyne's rule). The β-glycosidically bound D-sugars are in $^4C_1$ conformation and the α-glycosidically bound L-sugars in $^1C_4$ conformation. As an example for a complete cardenolide, the structure of digitoxin is given (Fig. 8.5). Steroidal glycosides deviating from this rule, e.g., with α-glycosidic binding of a terminal D-deoxy sugar, are called neoglycosides. If other monosaccharides besides D-glucose are present in the sugar chain, the D-glucose is not directly linked to the aglycone but terminates the chain. Such glycosides are called primary glycosides. They are probably storage forms of steroid glycosides.

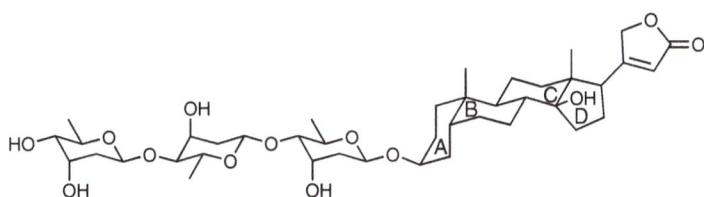

Digitoxin

**Fig. 8.5:** Digitoxin: perspective representation.

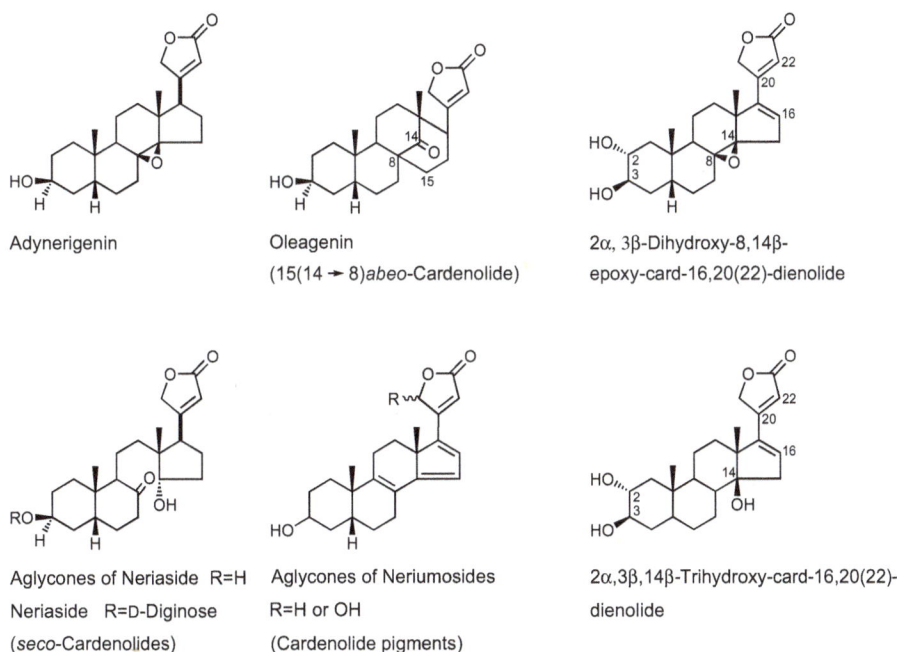

Adynerigenin

Oleagenin
(15(14 → 8)*abeo*-Cardenolide)

2α, 3β-Dihydroxy-8,14β-epoxy-card-16,20(22)-dienolide

Aglycones of Neriaside  R=H
Neriaside  R=D-Diginose
(*seco*-Cardenolides)

Aglycones of Neriumosides
R=H or OH
(Cardenolide pigments)

2α,3β,14β-Trihydroxy-card-16,20(22)-dienolide

**Fig. 8.6:** Cardenolide glycosides with unusual structure.

In the so-called affinosides from *Anodendron affine* (Hook. and Arnott) Druce, Apocynaceae, a 3-*O*-methyl-4,6-dideoxy-hexosulopyranose residue is attached in positions 2 and 3 by an acetal and a hemiketal bond [137]. Such compounds occur also in Asclepias species (Apocynaceae [81, **159**]). Sugar components with attached thiazoline rings have also been detected.

## 8.3.2 Biogenesis

The biogenesis of the aglycones of the cardioactive steroid glycosides starts from cholesterol by shortening the side chain. From the formed pregnan-3β,14β–triol-20-one-21-ol derivatives, the butenolide ring of the cardenolides is formed by reaction of the C-20 and the hydroxyl group at C-21 with malonyl coenzyme A by insertion of a $C_2$-residue under splitting from CoA and $CO_2$. The formation of the pentadienolide ring of the bufanolides takes place through insertion of a $C_3$-residue which probably originates from oxalo-acetate [6]. The binding of sugars presumably already occurs at the pregnane derivative stage. Further, hydroxylation, catalyzed by monooxygenases, and formation of the oxo-groups, catalyzed by dehydrogenases, occur possibly at the glycoside stage [240, 260].

The terminal D-glucose residue, on the other hand, is bound after the transition of the glucose-free glycoside into the vacuole [259]. The glucose residue is cleaved off again upon injury of the plant tissues by very effective β-glycosidases to form the so-called secondary glycosides.

Degradation of cardioactive steroid glycosides in the plant, especially at the end of the growing season, has been observed [229].

The cardioactive steroid glycosides are accompanied in many (all?) cases by pregnane glycosides, their biogenetic precursors. These compounds do not possess cardiac effects [409].

## 8.3.3 Distribution

Cardioactive steroid glycosides are found in about 20 plant families in over 60 genera. Among others, they occur in the families Apocynaceae (e.g., in the genera Acokanthera, Adenium, Anodendron, Apocynum, Asclepias, Beaumontia, Calotropis, Cascabela, Cerbera, Cryptolepis, Cryptostegia, Glossostelma, Gomphocarpus, Mandevilla, Marsdenia, Melodinus, Nerium, Oxystelma, Pachypodium, Pergularia, Periploca, Plumeria, Sarcostemma, Strophanthus, and Xysmalobium), Asparagaceae (Speirantha), Celastraceae (Elaeodendron, Euonymus, and Lophopetalum), Combretaceae (Terminalia), Asteraceae (Saussurea), Brassicaceae (Erysimum and Lepidium), Crassulaceae (Cotyledon, Kalanchoe, and Tylecodon), Euphorbiaceae (Mallotus), Fabaceae (Coronilla and Securigera), Iridaceae (Moraea), Liliaceae s.l. (Bowiea, Convallaria, Drimia, Fusillium, Melianthus,

Ornithogalum, Scilla, and Speirantha), Moraceae (Antiaris and Streblus), Plantaginaceae (Digitalis and Isoplexis), Ranunculaceae (Adonis and Helleborus), Santalaceae (Thesium), Solanaceae (Nierembergia), and Sterculiaceae (Mansonia).

Free aglycones of cardioactive steroids and their glycosides or esters are found in animals also. Toads, snakes, and some members of the leaf beetles produce cardioactive steroids themselves. The majority of insects in which the compounds occur have ingested them from their food plants.

### 8.3.4 Application

The first indication of the therapeutic use of plants with cardioactive steroid glycosides dates back to the sixteenth century B.C. It was mentioned in the Egyptian medical papyrus Papyrus Ebers (sea onion recommended for heart complaints), and to Greek antiquity (Helleborus as an emetic). Toad poisons have long been a part of the traditional Chinese medicine (TCM). The introduction of foxglove, *Digitalis purpurea*, into the therapy of heart diseases was made in 1785 by the English physician William Withering (1741–1799). In his famous book *An Account of the Foxglove, and Some of its Medical Uses* in 1785, he summarized his experience about the therapeutic effects of foxglove. In 1869, the French pharmacist Claude Adolphe Nativelle (1812–1889) created a digitalis preparation (Digitaline Nativelle), which formed the basis of digitalis therapy in Europe.

At present, because of the difficulty in standardizing plant extracts with cardioactive steroid glycosides and the narrow therapeutic indexes, mainly pure isolated glycosides are used. The positive inotropic effect occurs in both the healthy and the insufficient heart. However, the diseased heart reacts significantly more strongly. The use of cardioactive steroid glycosides in chronic heart failure is still relevant today, especially in high-risk patients. In atrial fibrillation associated with tachycardia, their use is recommended [46]. The therapy prevents the deterioration in the condition of patients with heart failure. A prolongation of life expectance can be accepted [**51, 149**].

Extracts of plants containing cardioactive steroid glycosides are used in complementary medicine, e.g., in TCM, also.

Extracts of plants of the genera Acokanthera, Adenium, Antiaris, Asclepias, Nerium, Periploca, and Strophanthus have long been used as arrow poisons by indigenous of East Asia, Africa, and South America, and in some cases as fish poisons [76, 312, **112**].

### 8.3.5 Cardiac Effects

Structural features essential for the action of cardioactive steroid glycosides are
- a carbonyl oxygen of the β-positioned, unsaturated lactone ring (active group), which is strongly electronegative and forms a receptor like hydrogen bond to the H-donor of the α-subunit of the $Na^+/K^+$-ATPase;

–  the steroid base body with *cis*-linkage of rings C and D (adhesive group), which in this conformation forms van der Waals bonds to the enzyme molecule; and
–  the β-position of the OH groups at C-3 and C-14.

Replacement of the butenolide ring by a pentadienolide ring increases the effect about 10-fold. The steroid glycosides with *trans–trans–cis* linkage of the rings of the steroid base body (5Hα, uzarigenin type) have only little cardiac effect.

The number of oxygen atoms on the sterane base body of the aglycone and the nature of the bound sugars mainly determine the pharmacokinetic behavior. Only the sugar closest to the aglycone influences the binding to the enzyme. An increasing number of oxygen functions in the steroid molecule increases its polarity. Thus, the rate of absorption from the intestine and the strength of plasma protein binding decrease, and the rate of elimination increases. Digitoxin with 5 OH groups will be absorbed at 90–100%, digoxin with 6 OH groups at 60–80%, lanatoside C with 9 free OH groups at 35–40%, convalloside with 8 OH groups at 10%, and k-strophanthoside with 12 OH groups at about 4%. The onset of action of strongly hydroxylated compounds is very rapid after parenteral (!) administration. The duration of action is short. Less polar compounds are absorbed from the intestine to a greater extent and are firmly bound to plasma proteins. Their decay rates are small (overview of structure–activity relationships and pharmacokinetics: [46, 151, 194, 301, 392]).

The positive inotropic effect of cardioactive steroid glycosides is based partially on inhibition of the Mg-dependent $Na^+/K^+$-ATPase, a P-type ATPase, which is localized in the cell membrane of the cardiac muscle cells and involved in cation transport.

Cardiac glycoside binding occurs to the α-subunits of the enzyme on the extraplasmic side of the membrane of the heart cells. The binding is reversible and meets all criteria for specific receptor binding [201, 206]. When therapeutic doses are administered, only 20–40% of the glycoside binding sites on the enzyme are occupied. $Na^+$ and $Mg^{2+}$ ions promote the glycoside binding at the $Na^+/K^+$-ATPase, whereas it is inhibited by $K^+$ ions [201].

The inhibition of $Na^+/K^+$-ATPase blocks the exchange of three $Na^+$ ions out of the cell for two $K^+$ ions into the cell, thereby increasing the intracellular $Na^+$ ion and decreasing the intracellular $K^+$ concentration. As a result, the exchange of intracellular $Ca^{2+}$ ions for extracellular $Na^+$ ions by the $Na^+/Ca^{2+}$ exchanger is restricted and less $Ca^{2+}$ ions are transported out of the cell. In addition, $Ca^{2+}$ ions, bound intracellularly, may also be released by exchange for the $Na^+$ ions. This increases the concentration of $Ca^{2+}$ ions in the sarcoplasm available for contraction processes in the cell. The result is an increase in the contractile force of the heart [51]. An increase in intracellular $Na^+$ concentration from 1 to 2 mM causes a doubling of contraction strength in vitro [133].

Besides this mode of action, cardioactive steroid glycosides activate the ryanodine receptor RyR2, a component of a $Ca^{2+}$ release channel [353]. By this way, the transport of $Ca^{2+}$ from intracellular stores, mainly from sarcoplasmic reticulum to the sarcoplasm, is promoted.

Which of the two mechanisms is mainly responsible for increasing the calcium concentration in the sarcoplasma is still unclear. It may be different for the different glycosides. It is suspected that lipophilic cardioactive steroid glycosides, like digitoxin, would gain better access to the intracellular site, which means to RyR2, than hydrophilic ones like ouabain. Thus, physicochemical differences determine not only the differences in the beginning and the duration of the effect between the different compounds but possibly also those in mode of action and effect strength [135, 168, 353].

An occasionally observed transient stimulation of $Na^+/K^+$-ATPase by cardioactive steroid glycosides is indirectly caused by the release of β-adrenergic transmitters [201].

It is becoming increasingly known that the cardioactive steroid glycosides, similar to other steroids from plants, e.g., saponins, also play a hormonal role in humans and animals. They can interact with nuclear receptors to activate cell signaling cascade that regulates cell growth and proliferation [21, 196].

The hormone-like endogenous compounds that occur in human blood, in adrenals, and in the hypothalamus and are called endogenous digitalis-like substances, digitalis-like factors, digitalis-like compounds, ouabain-like factors, or digitalis-like immunoreactive factors or substances (DLIF or DLIS). They act as transcription factors and affect hormonal management, immune processes, and cell proliferation and differentiation [196, 429].

Hence, the mechanism of action of the cardioactive steroid glycosides is much more complex than previously thought (reviews on action and mechanisms of action: [151, 164, 201, 263, 275, 346, 392, 428]).

Further investigations are necessary to clarify whether these activities, which are currently still being discussed in contradictory ways, will allow the development of new active substances [21, 196, 311].

## 8.3.6 Toxicity

At toxic doses, cardioactive steroid glycosides increase the $Ca^{2+}$ ion concentration in the cell to such an extent that the $Ca^{2+}$ storage capacity of the endoplasmic reticulum is exceeded in the relaxation phase and afterpotentials occur, leading to extrasystoles [171, 51]. Stimuli originating from the sympathetic and/or parasympathetic nervous system are thought to be involved in the toxic effect. According to a hypothesis by Heller [164], positive inotropic and toxic effects of the compounds are mediated by different mechanisms: toxic effects by attack at the $Na^+/K^+$-ATPase and positive inotropic effects by attack at the $Na^+$-$K^+$-Cl-cotransporter. Besides this it is assumed that cardioactive steroid glycosides increase intracellular reactive oxygen species which could contribute to arrhythmogenesis through redox modification of RyR2s [171].

The therapeutic breadth of the cardioactive steroid glycosides is small. In the case of digitalis, the toxic dose is only 50–60% higher than the therapeutic dose. However, intolerance symptoms may already occur at 30% of the optimal therapeutic dose, and they are frequent at 60% of the therapeutic dose (reviews on toxicity: [18, 106, 133, 151, 261, 318, 392, 5]).

Hypokalemia, e.g., caused by using diuretics or laxatives, hypercalcemia, and hypomagnesemia increase the risk of poisoning by cardioactive steroid glycosides. The strength of effects is also increased by various disease states (e.g., renal insufficiency, hypothyroidism, hypoproteinemia, dehydration, hypoxia, AV conduction disorders, and coronary sclerosis), by advanced age, and by numerous drug interactions, e.g., with verapamil, quinidine, amiodarone, clarithromycin, and cyclosporine. Patients with pacemakers require particularly intensive monitoring since the warning symptom of bradycardic arrhythmias does not occur here.

Animal studies show large differences in the effect and toxicity of the compounds depending on species, age, and sex [431], (reviews on the dependence of the effect on other factors: [106, 151, 263, 301, 392]).

Poisoning due to overdose of drugs containing cardioactive steroid glycosides by the patient or the physician constitutes a high proportion of poisonings treated in hospitals [419]. In addition, accidental poisonings, e.g., by 'snacking' on the preparations by children, suicide attempts, and homicides, are known.

🕱 Symptoms of poisoning by cardioactive steroid glycosides are alternating bradycardic and tachycardic arrhythmias (sinus bradycardia, atrioventricular (AV) block, death by asystole) in cardiac healthy people and ventricular arrhythmia (extrasystoles, atrial tachycardia, death by ventricular fibrillation) in cardiac patients. Extracardiac toxic effects extend to the CNS, especially the trigger zone in the medulla oblongata, the kidneys, and the vascular system. Nausea and vomiting almost always occur. Other symptoms include drowsiness, color vision (yellow sight), hallucinations, respiratory disturbances, and cyanosis. Recognition of intoxication is particularly difficult in multimorbid elderly patients and in children. Differential diagnosis by ECG (first-degree AV block) and determination of serum concentrations of cardioactive steroid glycosides and/or potassium ions are necessary [169]. Electrocardiography indicates an elevated PR interval, a reduced QRS-T interval, and T-wave inversion [125]. For the determination of serum concentration, fluorescence polarization immunoassay, digoxin immunoassay, and LC-MS/MS are recommended [93, 332].

⚕ Mild symptoms of intoxication during therapy with cardioactive steroid glycosides can be remedied by dose reduction or short-term interruption of therapy. After ingestion of toxic doses, administration of activated charcoal and gastric lavage with charcoal is useful for toxin elimination. After 1 h or later, charcoal shall be given again (enterohepatic circulation! [322]). Interruption of the enterohepatic circulation of digitoxin is accomplished by administration of colestyramine (initially 4–8 g, continuing for 3–5 days at 4- to 6-hourly 4 g), colestipol, or activated charcoal. Hemoperfusion and hemodialysis are ineffective. Hemofiltration has been successful in digoxin overdose [249].

Arrhythmias are treated with atropine (for extreme sinus bradycardia, for AV block of II and III degree, 0.5–1 mg), with lidocaine (for ventricular tachycardia, initially 50–100 mg, i.v., then infusion 2–4 mg/min), or by electrode fibrillation, preferably with phenytoin protection (for ventricular flutter and fibrillation, 5 mg/kg BW). In hypokalemia, cautious administration of K+ ions (maximum 10 mmol/h) could be done under ECG monitoring. The use of a temporary transvenous pacemaker may also be required. In cases of central signs of poisoning, phenothiazines or barbiturates may be applied [132, 1, 169].

A major improvement in treatment options for live threatening (!) suicidal or accidental poisoning by cardioactive steroid glycosides, less so for digitalis extracts, is the use of Fab fragments of antibodies directed against digoxin from the blood serum of immunized sheep (digitalis antidote, dry substance in injection vials, digitalis antitoxin, Digibind from the Wellcome Company, UK, or Digifab from Protherics Inc., USA). There is cross-reactivity with digitoxin and some other cardioactive steroid glycosides, e.g., oleandrin [93, 114, 130, 378]. Allergy test before administration of the antibodies is required.

Application of Fab fragments is indicated whenever high serum cardioactive steroid glycoside concentrations (>100 ng/mL digitoxin or >3 ng/mL digoxin), potassium levels >7 mmol/L, and threatening arrhythmias occur. The glycosides are complexed by the Fab fragments and excreted predominantly renally in the form of inactive glycoside–antibody complexes. Often, there is a decrease in arrhythmias during Fab infusion or shortly thereafter. If the glycoside content is known, the amount of antibody should be such that about half of the glycosides are bound in the blood (80 mg of Fab fragments should bind 1 mg of digitalis glycosides). If the glycoside content is unknown, it is recommended that about 80–160 mg should be given first in adults as a short infusion (or 40–80 mg in children) and then another 160 mg (or 80 mg) as an infusion over 6–8 h [130, **161**]. Further application may also be necessary later since release of glycosides from stores may occur when the antibody (plasma half-life 12–20 h) has already been excreted. In any case, the determination of cardioactive steroid glycoside and potassium and calcium levels in the blood is necessary; in case of arrhythmia, ECG monitoring is also determined (reviews on the therapy of intoxications: [18, 37, 106, 132, 159, **161**]).

⊗ Consequences of chronic use of Digitalis preparations, especially at high doses, may be headache, insomnia, depression, hallucinations, nausea with frequent vomiting, diarrhea, cardiac arrhythmias, blurred vision, and color vision. In men, gynecomastia may occur after prolonged use due to the estrogen-like effect [1, 263].

## 8.3.7 Noncardiac Effects

An increasing number of data show that the effects of the cardioactive steroid glycosides are not limited to their activity on the heart.

They act as inhibitors of cancer cell proliferation and as inducer of apoptosis in several solid tumors. The in vitro data are supported by epidemiological studies stating that patients treated with cardioactive steroid glycosides were protected from various types of cancer. Some compounds, e.g., digoxin, are candidates for the development of anticancer drugs [344, 345].

Oleandrin is known to target signaling cascades of cancer and inflammation, such as transcription factor NF-κB, mitogen-activated protein kinases, and phosphoinosite 3 kinase [195]. It prevents the activation of the activator protein-1 (AP-1) induced by tumor necrosis factors, phorbol esters, or lipopolysaccharides. These factors are responsible for the activation of genes expressing pro-inflammatory cytokines. Therefore, inflammatory responses are suppressed [274].

Ouabain is a senolytic agent with broad activity. Senescent cells are particularly sensitized to ouabain-induced apoptosis. Experimentally aged cells are killed selectively. The use against age-related diseases warrants further exploration [152].

Digitoxin, digoxin, convallatoxin, ouabain, and some other cardioactive steroid glycosides inhibit a variety of DNA and RNA viruses, e.g., herpes simplex virus, the chikungunya virus, corona virus, Ebola virus, influenza virus, rhino virus, human immune deficiency virus, human cytomegalovirus, and tick born encephalitis virus [387]. The possible use of cardioactive steroid glycosides to treat COVID-19 is under investigation [11, 385]. Glycosides from Periploca species are successfully used for the treatment of condylomata acuminata caused by viruses [24].

Antileishmanial effects of the compounds were demonstrated in animal experiments [134].

### 8.3.8 Plants with Cardenolides

#### 8.3.8.1 Foxglove Species (Digitalis Species)

The genus Digitalis, foxglove, Plantaginaceae (formerly assigned to Scrophulariaceae) comprises about 20 species, distributed throughout most parts of temperate Europe to Central Asia, naturalized in northern parts of North America and other parts of the world [97, 263]. Three shrubby species formerly assigned to the genus Digitalis, endemic to Madeira and the Canary Islands, are now assigned to the genus Isoplexis.

Native to Central Europe are *Digitalis purpurea* L., red foxglove (occurring in numerous varieties), *D. lutea* L., small yellow foxglove, *D. grandiflora* Mill., large yellow foxglove, and *D. nervosa* Steud. and Hochst. ex Benth. (*D. laevigata* Waldst. and Kit.), smooth foxglove. Widespread in southeastern Europe is *D. lanata* Ehrh., woolly foxglove. *Digitalis purpurea, D. grandiflora, D. lutea*, and *D. ferruginea* L., rusty foxglove, originate from southeastern Europe and are cultivated as ornamental plants. The plants were introduced to gardens in many other parts of the world including North America and escaped sometimes into the wild.

Digitalis species are perennials, 0.6–2 m tall, usually producing only a basal rosette of leaves in the first year and a tall, unbranched racemose inflorescence in the second year.

*Digitalis purpurea* (Ph. 8.1), biennial, has red flowers with numerous large purple spots outlined in white, rarely white flowers also. It occurs in sparse forests and on clear cuts of central and western Europe from Great Britain and southern Sweden to Austria, to the Pyrenean Peninsula, and to Morocco. It is naturalized throughout the Americas and is also cultivated as a garden plant. *D. grandiflora* has flowers over 3 cm long, pale ocher yellow, veined or spotted brown inside, and a glandular hairy stem. Locations are sparse forests from Belgium to the Pyrenean Peninsula and east to western Siberia. *D. lutea* has light lemon-yellow flowers with a distinct five-digit flower tube and a glabrous stem. The plant thrives on bushy slopes and in deciduous forests from Belgium to northern Spain and northwest Africa, in the east to Austria. *D. nervosa* has pale ocher-yellow, rusty brown veined flowers with a long protruding middle lobe of the lower lip of the corolla tube and is found only in Austria and the Balkan Peninsula. *D. lanata* (Ph. 8.2), biennial or perennial, has white or pale ochre-yellow, brown-veined flowers with a long lower lip and a woolly haired inflorescence. Native to the Balkan Peninsula and southwest Asia. *D. ferruginea* grows to 1.50 m height and has a long, loose inflorescence with yellowish gray bells that are rusty red inside.

Ph. 8.1: *Digitalis purpurea* (source: Eberhard Teuscher).

Ph. 8.2: *Digitalis lanata* (source: Wolfgang Kiefer).

More than 100 different cardenolide glycosides have been identified in Digitalis species to date. Aglycones of the species mentioned are digitoxigenin, gitoxigenin, digoxigenin, diginatigenin, and gitaloxigenin (Fig. 8.2).

According to the letter designation of the lanatosides, the glycosides are divided into A-series glycosides (aglycone digitoxigenin), B-series glycosides (aglycone gitoxigenin), C-series glycosides (aglycone digoxigenin), D-series glycosides (aglycone diginatigenin), and E-series glycosides (aglycone gitaloxigenin). The monosaccharide components detected were D-allomethylose, D-boivinose, D-digitoxose, D-digitoxose, 3-acetyl-D-digitoxose, L-fucose, D-glucomethylose, D-glucose, and D-xylose (Fig. 8.4).

The cardioactive steroid glycosides with D-glucose terminated genuine di-, tri-, and tetraglycoside sugar chains, the so-called primary glycosides (e.g., purpurea glycoside A), are, catalyzed by β-glucosidases contained in the plants, postmortem converted to the glucose-free secondary glycosides (e.g., digitoxin [111, 263]).

The content of cardioactive steroid glycosides in the plants is about 0.2–3% (up to 6% [83]). Information on major and important minor glycosides present in the leaves of the abovementioned Digitalis species, their structure and their toxicity is given in Tabs. 8.1 and 8.2 [177, 263, **93**].

The glycoside spectra of the seeds differ from that of the leaves. Thus, digitalinum verum, glucoverodoxin, and glucodigifucoside dominate in the seeds of *D. purpurea*. In seeds of *D. lanata*, desacetyllanatoside C, glucoevatromonoside, and glucogitoroside are present in greater quantity. The glycoside composition is highly dependent on the species' membership in each subspecies, varieties as well as chemotypes, on the age of the plant, and on environmental conditions.

Accompanying substances include pregnan glycosides (digitanol glycosides, e.g., diginine, digipurpurine, digitalonin, approximately 1%), anthraquinone derivatives, flavonoids, and steroid saponins, e.g., the furostane-type bisdesmosides purpureagitoside and lanagitoside, and monodesmosides of spirostan type, e.g., gitonin, tigonin, and digitonin, in the seeds up to 6%.

The Digitalis glycosides that are most frequently used in therapy and consequently give rise to the greatest number of poisonings are digitoxin and digoxin. When administered perorally, digitoxin is almost 100% absorbed and digoxin up to about 80%. Digitoxin is metabolized very slowly in the human organism. After successive split off the deoxysugars and epimerization of the OH group at C-3, conjugation with glucuronic or sulfuric acid occurs. Detoxification by hydrogenation of the double bond of the lactone ring can also occur. The greatest quantity of the glycosides is excreted unchanged renally. The decay rate for digitoxin is only 9% per day. This is responsible for its accumulation. The decay rate of digoxin is between 15% and 20% per day [301, 328].

For digitoxin, the therapeutic plasma concentration begins at 10 ng/mL, and arrhythmias can already be expected at 29 ng/mL. The therapeutic plasma level of digoxin is higher than 0.8 ng/mL, but already a concentration above 1.7 ng/mL leads to arrhythmias in 10% of patients [111, **97**].

Tab. 8.1: Cardioactive steroid glycosides of leaves of Digitalis species.

| Series | Glycoside | mg/100 g DW | | | | | |
|---|---|---|---|---|---|---|---|
| | | *D. purpurea* | *D. lutea* | *D. grandiflora* | *D. laevigata* | *D. lanata* | *D. ferruginea* |
| A | Purpurea glycoside A | 20–120 | + | | | + | + |
| | Lanatoside A | | | 80 | 50 | 80–240 | 100–400 |
| | Digitoxin | 2–40 | 10–90 | + | | 5–20 | |
| | Acetyl digitoxin | | 25 | | | | 60–150 |
| | Neoodorobioside | | 60–80 | 45–75 | | | |
| | Glucoevatromonoside | 2–25 | 10–35 | 2–120 | + | 20–160 | 80–170 |
| B | Purpurea glycoside B | 20–80 | | + | | + | |
| | Lanatoside B | | 8 | + | | 10–150 | |
| | Glucogitoroside | 2–62 | | 15 | 15 | 20–120 | 50–80 |
| | Digitalinum verum | 10–40 | | 15 | | 20–120 | |
| | Neodigitalinum verum | | 10–15 | 60–80 | 20 | | |
| C | Lanatoside C | | + | + | | 80–240 | 30 |
| | Desacetyl-lanatoside C | | | | | 5–20 | |
| D | Lanatoside D | | | | | + | |
| E | Lanatoside E | | + | | | 5–20 | |
| | Glucogitaloxin | 10–190 | | | | + | |
| | Glucoverodoxin | 10–40 | | + | | 20–100 | 30 |
| | Neoglucoverodoxin | | | 25–60 | | | |
| | Glucolanodoxin | 2–43 | | | | 20–120 | 40 |

Tab. 8.2: Structure and toxicity of selected Digitalis glycosides.

| Series | Glycoside | Structure | LD$_{50}$ mg/kg BW, i.v., cat |
|---|---|---|---|
| A | Purpurea glycoside A | DTX-Dx-Dx-Dx-Glc | 0.33–0.51 |
| | Lanatoside A | DTX-Dx-Dx-Dx(Ac)-Glc | 0.36–0.38 |
| | Digitoxin | DTX-Dx-Dx-Dx | 0.24–0.62 |
| | Acetyldigitoxin | DTX-Dx-Dx-Dx(Ac)- | 9.45–0.51 |
| | Neoodorobioside | DTX-Dtl-Glc(1→2) | 1.03 ± 0.13 |
| | Glucoevatromonoside | DTX-Dx-Glc | 6.63 ± 0.23 |
| | Evatromonoside | DTX-Dx | |
| B | Purpurea glycoside B | GTX-Dx-Dx-Dx-Glc | 0.37–0.55 |
| | Lanatoside B | GTX-Dx-Dx-Dx(Ac)-Glc | 0.39–0.40 |
| | Gitoxin | GTX-Dx-Dx-Dx | 0.50–0.81 |
| | Glucogitoroside | GTX-Dx-Glc | 4.92 ± 0.28 |
| | Gitoroside | GTX-Dx | 2.38 ± 0.17 |
| | Digitalinum verum | GTX-Dtl-Glc | 0.97 ± 1.33 |
| | Neodigitalinum verum | GTX-Dtl-Glc(1→2) | Inactive |
| C | Lanatoside C | DGG-Dx-Dx-Dx(Ac)-Glc | 0.23–0.28 |
| | Desacetyllanatoside C | DGG-Dx-Dx-Dx-Glc | 0.23 |
| | Digoxin | DGG-Dx-Dx-Dx | 0.23–0.28 |
| D | Lanatoside D | DNG-Dx-Dx-Dx(Ac)-Glc | 0.41 |
| | Diginatin | DNG-Dx-Dx-Dx | |
| E | Glucogitaloxin | GLG-Dx-Dx-Dx-Glc | |
| | Lanatoside E | GLG-Dx-Dx-Dx(Ac)-Glc | 0.60 |
| | Gitaloxin | GLG-Dx-Dx-Dx | 0.71–1.01 |
| | Glucoverodoxin | GLG-Dtl-Glc | |
| | Neoglucoverodoxin | GLG-Dtl-Glc(1→2) | |
| | Verodoxin | GLG-Dtl | 0.26 |
| | Glucolanodoxin | GLG-Dx-Glc | 5.74–0.27 |
| | Lanodoxin | GLG-Dx | |

Abbreviations: DTX, digitoxigenin; GTX, gitoxigenin; DGG, digoxigenin;
DNG, diginatigenin; GLG, gitaloxigenin; Dx, D-digitoxose; Dx(Ac), 3-acetyl-D-digitoxose;
Dtl, D-digitalose; Glc, glucose; Glc(1→2), D-glucose, 1→2-linked (for structures, see Figs. 8.2, 8.4 and 8.5)

Intoxications during Digitalis therapy are the most common type of poisonings caused by Digitalis glycosides. The US poison centers reported 2,632 cases of digitoxin toxicity and 17 cases of digitoxin-related deaths in 2008. Women are more commonly affected than men [64]. These medically induced intoxications are caused by dosage errors, accumulation (renal insufficiency!), or altered glycoside tolerance as a result of a preexisting medical condition (fever), hormonal influences, or drug interactions.

Accidental poisoning in children from Digitalis preparations available in households is not uncommon [244, 393]. There are also reports of poisoning by deliberate ingestion of Digitalis preparations, digitoxin, or digoxin with suicidal intent [28, 60, 62, 66, 248, 349, 391, 394, 407].

The lethality is reported to be 20% or more for accidental and suicidal poisoning with digitalis [159].

† A 57-year-old woman who had never had heart disease was murdered with the aid of a Digitalis preparation (8 mg digitoxin) placed in a glass of fruit juice [129].

† A 34-year-old woman with recurrent depressive disorders used the dried leaves of *D. purpurea* to make a tea that she drank. About 70 h later, she developed abdominal pain, emesis, and bradycardia. The plasma concentration of digoxin measured 21 h after ingestion of the tea was 3.53 nmol/L (therapeutic range 0.77–1.50 nmol/L). The patient was treated with antiemetics and simple analgesia only [253].

† A 27-year-old woman, suffering from severe depression, ingested an unknown quantity of leaves of foxglove in the garden of her parents with suicidal intend. After 4–6 h, the parents found her vomiting. In the emergency room nausea and a marked reduction of color vision was observed. Blood samples showed high level of digitoxin. Due to recurrent bradycardia (30/min) a temporary pacemaker was implanted. Activated charcoal and colestyramin were administered. Eleven days after intoxication, a subtherapeutic level of digitoxin was measured [187].

Severe poisoning by Digitalis plants in humans is rare due to the low risk of confusion, the strongly bitter taste, and the vomiting that sets in soon after ingestion. Poisonings by tea drinks made from foxglove leaves or by those contaminated with the leaves (confusion of leaves of *D. lanata* with those of *Plantago lanceolata* L.), sometimes with fatal outcome, have been reported [32, 88, 102, 256, 388]. Poisonings due to confusion of Digitalis leaves with leaves of *Borago officinalis* L. in the preparation of food also occurred, in some cases ending fatally [73, 308, **49**]. About 2–3 g of the dried leaves was reported to cause death in humans. The flowers, which have sometimes been tested for edibility by children, are hardly toxic because of their lower content of cardioactive steroid glycosides. Only nausea and vomiting have been observed as consequences [**161**].

☠ Acute poisoning by foxglove plant parts or extracts first manifests itself by nausea, vomiting, and diarrhea. As a resorptive effect, pulse slowing occurs initially after about 1 h. The pulse may drop to less than 40 beats/min and may be barely palpable. The most important cardioactive symptoms of poisoning are various types of cardiac arrhythmias, e.g., ventricular and superventricular arrhythmias, AV block, and AV nodal rhythm. The pulse becomes very rapid (up to 140 beats/min) and progressively weaker ('delirium cordis'). Extracardiac symptoms include centrally induced vomiting, weakness, anxiety, color vision, and other visual disturbances (not specific to digitalis poisoning as often stated [132]), headache, possibly psychosis, hallucinations, and delirium. Death occurs by cardiac arrest or asphyxia after a few hours or days, in case of i.v. application after minutes.

⚕ After ingestion of leaves or stems, primary poison removal (induced vomiting, activated charcoal) and inpatient monitoring should be performed, if necessary symptomatic treatment (for intoxications by isolated cardioactive steroid glycosides, see Section 8.2.6).

🐎 Animals react differently to Digitalis ingestion. Horses are the most sensitive. The lethal dose for a horse is about 25 g of dried leaves or 100–200 g of fresh leaves of *D. purpurea* [61]. A pig (50 kg BW) died after administration of 4–5 g of dried leaves of *D. lanata*. Poisoning also occurred in ducks, turkeys, cats, and dogs. Ruminants are less at risk since the glycosides are largely destroyed by the rumen flora. The lethal dose in cattle is more than 150 g of the dried leaves per animal [263, **60, 61**]. Red deer is known to be fond of foxglove leaves. When they graze on areas with abundant Digitalis plants, they show anorexia, vomiting, and arrhythmias as signs of intoxication [34, 89, 263].

*Isoplexis chalcantha* Svent. and O'Shan. and *I. canariensis* (L.) Lindl. ex G. Don, which have so far played no role in the toxicological literature, contain mainly uzarigenin glycosides (5α!), besides glycosides of pachygenin, xysmalogenin, digitoxigenin, and canarigenin. They are said to act similarly to *Digitalis purpurea* [46, 363, 395].

### 8.3.8.2 Lily of the Valley (*Convallaria majalis*)

*Convallaria majalis* L., lily of the valley (Ph. 8.3), Asparagaceae, is the only species of this very species-poor genus (three species) native to Central Europe.

It is a perennial plant that grows to a height of about 10–20 cm with a stoloniferous creeping rhizome. The leaves are entire-margined, long elliptic, and acuminate. The flower stalk is leafless and bears at the top a 5- to 13-flowered raceme of white, bell-shaped, fragrant flowers at the tip. The fruits are red berries with two to six whitish or blue seeds. Lily of the valley occurs in sparse deciduous and mixed forests throughout most of Europe and temperate Asia and is cultivated as an ornamental plant in gardens. It has been naturalized in many temperate zones including North America.

Ph. 8.3: *Convallaria majalis* (source: left: Isolde Kühn, right: Sabine Witt).

The content of cardioactive steroid glycosides in the underground organs of *C. majalis* is relatively constant at 0.19–0.24%. In the aerial organs, it decreases from 0.45% in the very young leaves to about 0.01% in the leaves at the end of the growing season [229]. The content in the flowers is about the same as in the young leaves. There are no or only trace amounts of cardenolide glycosides in the pulp of the ripe red berries, and about 0.45% in the seeds [369, 371].

To date, about 40 cardioactive steroid glycosides have been isolated from *C. majalis*. Table 8.3 provides information on its major cardenolide glycosides [221, 228, 365].

Tab. 8.3: Cardioactive steroid glycosides of *Convallaria majalis*.

| Glycoside | Content (%) | Structure | LD$_{50}$ mg/kg BW, i.v., cat |
|---|---|---|---|
| Convallatoxin | 4–40 | SDN-Rha | 0.07–0.08 |
| Convalloside | 4–24 | SDN-Rha-Glc | 0.22 |
| Desglucocheirotoxin | 3–15 | SDN-Gulom | 0.10 |
| Glycoside F | 3–8 | SDN-Allom-Rha | |
| Glycoside U | 2–5 | SDN-Allom-Ara | |
| Convallatoxol | 10–20 | SDO-Rha | 0.09 |
| Convallatoxoloside | 1–3 | SDO-Rha-Glc | |
| Desglucocheirotoxol | 2–15 | SDO-Gulom | |
| Lokundjoside | 1–25 | Bip.-Rha | 0.11 |
| Cannogeninrhamnoside | 5 | CON-Rha | |

Abbreviations: SDN, strophanthidin; SDO, strophanthidol; Bip., bipindogenin; CON, cannogenol; Gulom, D-gulomethylose; Rha, L-rhamnose; Glc, D-glucose; Allom, D-allomethylose; Ara, L-arabinose (for structures, see Figs. 8.2 and 8.4)

The glycoside spectrum is highly dependent on plant age, growth conditions, growth location, and chemotype. In the plants of western and northwestern Europe dominates convallatoxin, in plants of northern and eastern Europe convalloside. Central European provenances contain convallatoxin, convallatoxol, and convalloside in balanced proportions [49]. Concomitant compounds worth mentioning are steroid saponins, including the trisdesmosidic convallamaroside, and azetidine-2-carboxylic acid.

Poisonings can occur as side effects during therapeutic use of convallatoxin or convallatoxol (due to the low tendency to accumulate, however, much less frequently than during therapy with Digitalis glycosides), during folk-medicinal use of the herb against 'dropsy' and when the leaves were mistaken for wild garlic leaves. The red berries entice children to eat them and give rise to toxicological consultation cases. Because of the extremely low content of cardioactive steroid glycosides in the pulp (the seeds are usually not bitten) and the low rate of absorption of the glycosides, few sequelae are expected after ingestion of the berries. It is unlikely that the glycosides were responsible for the death of a 3-year-old girl after drinking water from a vase in which lily of the valley had stood [16]. Saponins are also involved in the effects of poisoning (promoting absorption of cardenolide glycosides?).

> 🐾 Poisoning from lily of the valley leaves or flowers may manifest as vomiting, diarrhea, dizziness, increased diuresis, weakness, and slow and irregular pulse.
>
> 📧 For treatment, see Digitalis intoxications.

🐕 Deaths have been observed in animals, especially geese and ducks, after eating lily of the valley leaves [**15, 138**]. An intoxication of a dog has been described [25].

### 8.3.8.3 Milk Star Species (Ornithogalum Species) and False Sea Onion Species (Albuca Species)

The genus Ornithogalum, milk star, Asparagaceae, includes about 100 species, native to Europe, Africa, and temperate Asia. For Central Europe, seven species are reported, of which *Ornithogalum umbellatum* L. (aggregate of several clades), star of Bethlehem, Chincherinchee, nap at noon, is the most widespread and best studied with respect to its cardioactive steroid glycosides. In addition to *O. umbellatum*, e.g., *O. arabicum* L., Arab's eye, *O. pyramidale* L., pyramidal milk star, *O. nutans* L, nodding milk star, and *O. narbonense* L., Pyrenean milk star, are used as garden plants.

Of interest is also the species *Albuca bracteata* (Thunb.) J.C. Manning and Goldblatt (*Ornithogalum longibracteatum* Jacq.), false sea onion, which is cultivated as an indoor plant.

Ornithogalum species are bulbous plants with basal linear long-sheath leaves. The flowers are in racemes or umbels; they have six white petals with a green central stripe.

*O. umbellatum* (Ph. 8.4) is common in South and Central Europe, the Near East, and North Africa. Preferred habitats are rich meadows, roadsides, and shrubberies. *O. collinum* Guss., crested milkstar, *O. gussonei* Ten., narrow-leaved milk star, *O. boucheanum* (Kunth) Asch., garden milk star, and *O. pyramidale*, are found rarely in gardens and fields. *O. nutans* is found as relict of former cultures in monastery gardens, especially in parks and gardens. *O. pyramidale* has its distribution center in Southwestern Europe.

In Ornithogalum and Albuca species, the aglycones digitoxigenin, gitoxigenin, oleandrigenin, sarmentogenin, strophanthidin, uzarigenin, syriogenin (Fig. 8.2) and as sugar components 6-deoxy-D-allose, D-apiose, L-arabinose and D-digitoxose (Fig. 8.4) were found. Convallatoxin (40 mg/100 g DW) and convalloside (10 mg/100 g DW), rhodexin A (0.1 mg/100 g DW), and rhodexoside (both with sarmentogenin as aglykon) are remarkable glycosides of bulbs of *O. umbellatum* [67, 142, 220, 305, 460]. For steroids of South African Ornithogalum species, see Section 8.5.

☠ Ingestion of Ornithogalum species, e.g., *O. umbellatum*, especially of the bulbs and flowers, by humans causes nausea, salvation, burning, and swelling of the lips, tongue and throat and gastroenteritis, abnormalities in the heart rate and heart rhythm. Fatal arrhythmias may occur.

🚑 For treatment, see Digitalis intoxications.

🐃 Poisonings of animals, e.g., sheep and dogs, by Ornithogalum species have been observed.

### 8.3.8.4 Spindle Tree Species (Euonymus Species)

The genus Euonymus (genus name masculine, by some authors also considered feminine), spindle tree, spindle bush, Celastraceae, includes about 180 species, whose main distribution is in East Asia. In Central Europe, three species are native: *Eu. europaeus* L., common spindle, *Eu. verrucosus* Scop., warty bark spindle, and *Eu. latifolius* (L.) Mill., broad-leaved spindle. In addition, a high number of other species in many varieties are cultivated in gardens and parks, e.g., *Eu. japonicus* Thun., Japanese spindle bush, *Eu. atropupureus* Roxb., purple spindle bush, and *Eu. nanus* M. Bieb., dwarf spindle bush. The hart wood of some species was used in the past for making hand spindles (name!).

Euonymus species are mostly deciduous, rarely evergreen shrubs or trees. *Eu. europaeus* (Ph. 8.5) is a deciduous shrub or tree. The inconspicuous, greenish-white flowers form leaf-axillary, three- to nine-flowered umbels. The four- or five-edged, rose- or carmine-red fruit capsules with whitish seeds with a bright orange-red seed coat are striking. It is common in deciduous forests, along shores and walls in most of Europe, and is also cultivated as an ornamental tree. *Eu. verrucosus* is distinguished by branches densely covered with dark corky warts. The seeds are black. It occurs in the Eastern Europe. *Eu. latifolius* has winged fruits on the edges. It inhabits the Alpine foothills and Alps.

Ph. 8.5: *Euonymus europaeus* (source: Wolfgang Kiefer).

Information about the cardioactive steroid glycosides of the seeds of *Euonymus europaeus* is given in Tab. 8.4 [212, 213, 404].

Tab. 8.4: Cardioactive steroid glycosides of seeds of *Euonymus europaeus*.

| Glycoside | Structure | Content mg/100 g DW | $LD_{50}$ mg/kg BW, i.v., cat |
|---|---|---|---|
| Evonoside | DTX-Rha-Glc-Glc | 10 | 0.84 |
| Evobioside | DTX-Rha-Glc | | |
| Evomonoside | DTX-Rha | | 0.28 |
| Evonoloside | CON-Rha | | |
| Glucoevonoloside | CON-Rha-Glc | | |
| Glucoevonogenin | EVO-Glc | | |

Abbreviations: DTX, digitoxigenin; CON, cannogenol; EVO, evonogenin; Rha, L-rhamnose; Glc, D-glucose (for structures, see Figs. 8.2 and 8.4)

In the root bark of *Eu. atropurpureus*, which is used as a medicinal drug, seven cardenolide glycosides were detected, among them euatroside (digitoxigenin (gluco)arabinoside), euatromonoside (digitoxigenin arabinoside), and evatromonoside (digitoxigenin digi-toxoside [410]). The leaves of Euonymus species can also contain cardenolide glycosides [138, 215].

The corky winged stem of *Eu. alatus* (Thunb.) Siebold, winged euonymus, that has been used thousands of years in East Asian countries for treatment of urticaria, arthritis, and other diseases contains the glycosides acovenosin, euonymoside, and euconymoside A. They possess potential for treatment of tumors and diabetes [124].

As accompanying substances, approximately 0.1% of alkaloid polyesters are present in the seeds of *Eu. europaeus*, e.g., evonin, evorin, and evonymin. They are polyesters of a sesquiterpene polyol with the basic structure of 5,11-epoxy-5β,10α-eudesm-4(14)-an (dihydro-β-agarofuran) esterified with pyridine carboxylic acids (e.g., evonic acid), acetic acid, and others. They irritate the gastrointestinal system [98]. Furthermore, the structure of armepavine, a 1-benzyl-tetrahydro-isoquinoline alkaloid, and peptide alkaloids, e.g., frangulamine, franganine, and frangufoline, is known.

Some Chinese species, e.g., *Eu. verrucoides* Peck, *Eu. fortunei* (Turcz.) Hand–Mazz., and *Eu. phellomanus* Loes., contain insecticidal sesquiterpene pyridine alkaloids with β-dihydroagarofuran skeletons (euverrine A and B, euphellin, eujaponin [185]) in the bark. Caffeine and theobromine were found in the seeds, pericarp, and leaves of *Eu. europaeus*, *Eu. japonicus*, and *Eu. latifolii* [48, 65].

The red, conspicuously shaped fruits of *Eu. europaeus* attract especially children to eat them and for this reason often give rise to cases of consultation or mild poisonings.

Poisonings by the root bark of *Eu. atropurpureus* has occurred sporadically with medicinal use of the drug, e.g., as a cardiac agent and are characterized by vomiting and abdominal pain.

♣ Symptoms of poisoning after ingestion of spindle tree fruits after a latency period of several hours are usually only gastrointestinal irritation, and in severe cases colic, circulatory disorders, collapse symptoms, and possibly convulsions. Kidney and liver damage may remain after the poisoning has passed. Thirty-six fruits reported to be fatal to humans [166, 182, 412, **53**].

⚕ In case of ingestion of more than three fruits, gastric emptying and administration of activated charcoal should be performed [**161**].

🐎 Poisoning by parts of the plant has also been observed in animals, such as horses, sheep, and goats.

Evonin is probably responsible for the insecticidal effect of the powdered seeds used in the past for scabies, mites, and other vermin. Decoctions of the fruits served as a diuretic.

### 8.3.8.5 Adonis Rose Species (Adonis Species)

The genus Adonis, Adonis rose, Ranunculaceae, includes about 32 annual or perennial species native to temperate Europe and Asia. In Central Europe occur *A. vernalis* L., spring Adonis, yellow pheasant's eye, *A. flammea* Jacq., flame Adonis, *A. aestivalis* L., summer Adonis, and *A. annua* L., pheasant's eye, the latter probably introduced.

In addition to the mentioned species, *A. aleppica* Bois., native to the Near East and conspicuous for its dark red flowers, and the yellow flowered *A. amurensis* Regel and Radde, native to Manchuria, are grown as ornamentals.

Adonis species are annual or perennial herbs with two- to five-pinnate leaves with narrow-linear segments. The mostly solitary flowers are radial in structure and have 5 sepals, up to 20 vividly colored petals and numerous stamens and carpels.

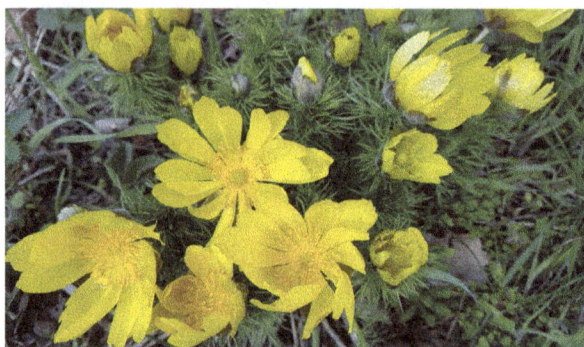

Ph. 8.6: *Adonis vernalis* (source: Margit Bukowski).

*Adonis vernalis* (Ph. 8.6) has flowers 3–8 cm in diameter with 10–20 yellow petals. The fruits are hairy. The plant is common in eastern, central, and southern Europe on dry grasslands and in loose pine forests. In *A. flammea*, the flower diameter is 0.5–4 cm. The flower has three to eight red or rarely yellow petals. The fruits are glabrous. It occurs from Morocco to Central Europe in fields rich in nutrients and lime. *A. aestivalis* has flowers up to 3 cm in size with mostly six red, rarely yellow, oblong-oval petals. The fruits are glabrous. It is found from North Africa to Central Europe, east to Afghanistan, especially in grain fields. *A. annua*, native to southern Europe, has 1.5–3 cm flowers with 6–10 broad-oval, dark blood-red petals.

The following data are given on the content of cardioactive steroid glycosides in the herb: for *A. vernalis* 0.24–0.8%, *A. flammea* 0.2–0.47%, *A. aestivalis* 0.20–0.39%, and *A. annua* 0.10–0.66%. The highest concentration is reached during flowering. The content of seeds is higher than that of the herb [239, 280, 287, 422, 430, 438, 453].

Regarding the structure of cardioactive steroid glycosides, *A. vernalis* is the best studied (Tab. 8.5). Until now, 32 cardenolides have been detected. Information about the main representatives is given in Tab. 8.5 [97, 189, 192, 239, 337, 439, 453].

Tab. 8.5: Structure and toxicity of cardioactive steroid glycosides of *Adonis vernalis*.

| Glycoside | Structure | LD$_{50}$ mg/kg BW, i.v., cat |
|---|---|---|
| Adonitoxin | ADI-Rha | 0.19 |
| 3-Acetyladonitoxin | ADI-3-Acetyl-Rha | |
| k-Strophanthoside | SDN-Cym-Glc-Glc | 0.13–0.19 |
| k-Strophanthoside-β | SDN-Cym-Glc | 0.13 |
| Cymarin | SDN-Cym | 0.11–0.13 |
| Strophanthidinfucoside | SDN-Fuc | |
| Strophanthidindigitaloside | SDN-Dtl | |
| Vernadigin | SDG-Dgn | |
| Adonitoxol | ADO-Rha | |

Abbreviations: ADI, adonitoxigenin; SDN, strophanthidin; SDG, strophadogenin; ADO, adonitoxilogenin (adonitoxigenol); Glc, D-glucose; Rha, L-rhamnose; Cym, D-cymarose; Fuc, L-fucose; Dtl, D-digitalose; Dgn, D-diginose (for structures, see Figs. 8.2 and 8.4)

The glycoside spectrum of other species is less well known. In *A. flammea*, 10 cardenolide glycosides were detected, including k-strophanthin-β and cymarin [239, 453]. More than 20 cardenolide glycosides occur in *A. aleppica*, of which periplorhamnoside (periplogenin-α-L-rhamnoside), strophanthidin-diginoside, and the free 3-epi-periplogenin were identified [191]. Cymarin, convallatoxin, and chorchoroside A (strophanthidin-boivinoside), among others, were found in *A. amurensis* [374].

(🐾) The cardioactive steroid glycosides of the Adonis species are only absorbed to a small extent after p.o. application and hardly accumulate. Intoxications are known from overdoses of drug preparations during therapeutic use. After ingestion, gastrointestinal irritation symptoms are in the foreground. After i.v. application, fatal outcome of poisoning is possible.

🐎 In animals, poisoning by Adonis weed rarely occurs. Young animals showed colicky diarrhea and eventual cardiac arrest after eating *A. vernalis* leaves in pasture [**138**]. It is reported from the USA that three horses that had eaten hay containing *A. aestivalis* developed distension of the gastrointestinal tract, endocardial hemorrhage, cardiac necrosis, and died [441].

### 8.3.8.6 Crown Vetch Species (Coronilla, Securigera, and Hippocrepis Species)

The species of the genus Coronilla, crown vetch, Fabaceae, were partially assigned to the genera Securigera and Hippocrepis by Lassen [251] in 1989. This assignment is controversial. Most species of these three genera occur in the Mediterranean region. Those found in Central Europe are *Securigera varia* (L.) Lassen (*Coronilla varia* L.), trailing crown vetch (Ph. 8.7), *Hippocrepis emerus* (L.) Lassen (*C. emerus* L.), shrub crown vetch, *C. coronata* L., scorpion vetch, *C. minima* L., small crown vetch, *C. vaginalis* Lam., vaginal crown vetch, and very rarely *C. scorpioides* (L.) W.D.J. Koch, scorpion senna, introduced from southern Europe. With the exception of *H. emerus,* all species contain cardioactive steroid glycosides [335, 401]. However, only *S. varia* and *C. scorpioides* have been well studied phytochemically.

The species are annual or perennial herbs, semishrubs, or small shrubs. The leaves are pinnate. The inflorescences are umbels. The flowers have yellow, more rarely pink, crimson, or white petals. The wings of the typical flowers are obovate to oblong in shape, the scape is curved and beaked.

Ph. 8.7: *Securigera varia* (source: Orest Lyzhechka/iStock/Getty Images Plus).

*Securigera varia* has a 5- to 20-flowered umbel with pink, rarely white flowers with a light red plume. It occurs in Central Europe, especially on semiarid grasslands. *Coronilla scorpioides* has yellow flowers. It is occasionally introduced with seed from southern and western Europe. *Hippocrepis emerus, Coronilla coronata, C. minima,* and *C. vaginalis* are perennials, semishrubs, or shrubs with yellow flowers. They are found only in the southwest and south of Europe.

*S. varia* contains cardioactive steroid glycosides and aliphatic nitro compounds as toxicants. Hyrcanoside (hyrcanogenin-3β-(4′-β-D-glucopyranosido)-β-D-xylopyranoside) and desglucohyrcanoside have been isolated and identified [165, 222]. In *C. scorpioides,* among others, glucocorotoxigenin (corotoxigenin-3β-D-glucopyranoside), frugoside (coroglaucigenin-3β-D-allomethylopyranoside), and coronillobioside (corotoxigenin-3β-(4′-β-D-glucopyranosido)-D-glucopyranoside [219, 221, 223, 224]) were found.

The cardioactive steroid glycosides from crown vetch species are only absorbed to a small extent when ingested and hardly accumulated. Hyrcanoside and desglucohyrcanoside act cytostatic in addition to their cardiac effects [165, 434].

&#x26D3; Poisonings by crown vetch herb are rare in humans. They occasionally occur after overdosage of diuretic crown vetch tea and manifest as nausea, vomiting, diarrhea, and, rarely, convulsions (**138**).

Coronilla species are used due to their diuretic effect.

### 8.3.8.7 Wallflower Species (Erysimum Species) and Other Brassicaceae

About 100 species are included in the genus Erysimum, wallflowers, Brassicaceae. They are distributed in Europe, North Africa, Western and Central Asia, and North America.

For the European area are specified:

- species belonging to the Pontic flora elements: *E. repandum* L., bushy wallflower, *E. crepidifolium* Rchb., pale wallflower, *E. durum* J. and C. Preisl., hard wallflower, and *E. diffusum* Ehrh., diffuse wallflower;
- Mediterranean-Pontic species: *E. odoratum* Ehrh., smelly wallflower, and *E.* × *cheiri* (L.) Crantz (*Cheiranthus cheiri* L.), common wallflower, grown in various cultivated forms as garden plants;
- species belonging to the Central European-alpine element: *E. hieraciifolium* L. f., alpine wallflower, *E. silvestre* Scop., lacquer knapweed, and *E. rhaeticum* (Schleich ex Hornem.) DC., Swiss wallflower;
- western alpine species: *E. ochroleucum* (Haller f. ex Schleich.) DC., pale yellow wallflower; and
- archaeophyte *E. cheiranthoides* L., wormseed mustard (Ph. 8.8).

*E. perofskianum* C.A. Fisch. and C. A. Mey., *E. pulchellum* (Willd.) J. Gray, *E. allonii* hort (*Cheiranthus allonii* hort.), *E. arkansanum* Nutt., *E. suffruticosum* Spreng., and *E.* × *cheiri* are used as ornamental plants. All species contain cardioactive steroid glycosides.

Erysimum species are annual or perennial herbs or semishrubs with undivided, entire to deeply toothed, hairy leaves. They have the typical flower structure of crucifers (Brassicaceae, four sepals, four corollas, mostly yellow, four long and two short stamens, fruit a pod). *E. cheiranthoides* is the most common, to find in gardens, fields, and along paths in temperate Europe and Asia. For *E. crepidifolium, E. hieraciifolium, E. durum,* and *E. odoratum,* the northern limit of distribution is approximate in the Harz Mountains (Germany). *E. repandum* and *E. diffusum* occur in the south and southeast of Europe. *E. ochroleucum, E. rhaeticum,* and *E. silvestre* are perennial mountain plants. *E.* × *cheiri* is native to the Mediterranean area and naturalized as a cultivated escape. The petals of the wild forms are golden yellow, those of the cultivated forms also dark brown, later mottled yellow, purple-violet, or scarlet; forms with double flowers also occur.

So far, more than 50 cardioactive steroid glycosides have been isolated from Erysimum species. Information on the structure, distribution, and toxicity of Erysimum glycosides is summarized in Tab. 8.6 [269, 271, 272, 373].

Ph. 8.8: *Erysimum cheiranthoides* (source: Ulrike Lindequist).

Erysimoside and helveticoside (identical to erysimine, erysimotoxin, desglucoery-simoside, and alloside A) occur in almost all Erysimum species studied [272]. Main components in *E. × cheiri* are probably cheirotoxin and cheiroside A [45, 181, 270, 303, 348, 373]. From seeds of *E. cheiranthoides* of Chinese origin, 13 cardioactive glycosides were isolated, among them olitoriside (strophanthidin–boivinose–glucose), erysimo-side, glucoerysimoside, and cheiranthosides I to XI [254, 255].

Tab. 8.6: Cardioactive steroid glycosides of Erysimum species.

| Glycoside | Structure | Detected in | LD$_{50}$, mg/kg BW, i.v. cat |
|---|---|---|---|
| Helveticoside | SDN-Dx | 1,2,3,4,5,6,7,8 | 0.05–0.08 |
| Erysimoside | SDN-Dx-Glc | 1,2,3,4,5,6,7,8,9 | 0.16 |
| Glucoerysimoside | SDN-Dx-Glc-Glc | 5,6,9 | |
| Canescein | SDN-Gulom | 4, 8 | 0.096 |
| Erychroside | SDN-Dx-Xyl | 2,7 | |
| Corchoroside A | SDN-Boi | 1,2,5,6,7,8 | |
| Erycanoside | SDN-3AcDx-Glc | 4 | |
| Eryscenoside | SDN-2DGal-Glc | 4 | |
| Strophalloside | SDN-Allom | 3 | |
| Cheirotoxin | SDN-Gulom-Glc | 3,4,6,9 | 0.12 |
| Erycorchoside | SDN-Boi-Glc | 2,3,4,5,7 | 0.3 |
| Erysimol | SDO-Dx-Glc | 7,8 | |

Tab. 8.6 (continued)

| Glycoside | Structure | Detected in | LD$_{50}$, mg/kg BW, i.v. cat |
|---|---|---|---|
| Erychrosol | SDO-Dx-Xyl | 7 | |
| Erycordin | CON-Glom-Glc | 1,7 | |
| Glucoerycordin | CON-Glom-Glc-Glc | 6 | |
| Alliside | BIP-Fuc | 6,9 | |
| Glucodigifucoside | DTX-Fuc-Glc | 7 | |
| Cheiroside A | UZA-Fuc-Glc | 6,9 | 0.68–0.79 |
| Glucobipindogeningulomethyloside | BIP-Gulom-Glc | 9 | |

Abbreviations: BIP, bipindogenin; CON, cannogenol; DTX, digitoxigenin; SDN, strophanthidin; SDO, strophanthidol; UZA, uzarigenin; 3AcDx, 3-acetyl-D-digitoxose; Allom, D-allomethylose; Boi, D-boivinose; 2DGal, 2-desoxy-D-galactose; Dx, D-digitoxose; Fuc, fucose; Glc, D-glucose; Glom, D-glucomethylose; Gulom, D-gulomethylose; Rha, L-rhamnose; Xyl, D-xylose (for structures, see Figs. 8.2 and 8.4).
1, *E. crepidifolium*; 2, *E. hieraciifolium*; 3, *E. repandum*; 4, *E. diffusum*; 5, *E. perovskianum*; 6, *E.* × *allionii hort.*; 7, *E. cheiranthoides*; 8, *E. suffruticosum*; 9, *E.* ×*cheiri*

The highest content is found in the seeds, decreasing sharply toward the root (e.g., in *E. crepidifolium* in seeds about 3.5%, leaves 2.9%, stem 0.59%, and root 0.51% [233, 348]).

Accompanying glucosinolates are found with methylthio-, methylsulfinyl-, or methylsulfonyl groups, yielding the corresponding isothiocyanates (e.g., 3-methylthiopropyl isothiocyanate).

Some other Brassicaceae also contain cardioactive steroid glycosides. For example, the herb of *Sisymbrium officinale* (L.) Scop., hedge mustard, a very widespread ruderal plant in Central Europe, contains about 50 mg/100 g of the compounds, especially helveticoside and corchoroside A [233], in addition to 0.6–0.9% glucosinolates [75]. Cardioactive steroid glycosides have also been detected in the seeds of *Alliaria petiolata* (M. Bieb.) Cavara and Grande, hedge garlic, *Hesperis matronalis* L., dame's violet, *Lepidium perfoliatum* L., clasping pepper weed, and *Lunaria rediviva* L., perennial honesty (Ph. 8.9 [232]). In other genera, e.g., Matthiola species, gillyflowera, their occurrence is suspected.

☠ Poisonings occurred only when the plants were used in folk medicine as a low-cumulative remedy for mild heart failure and as abortifacient. In toxic doses, they cause arrhythmia, tachycardia, hypotension, and in the lethal phase, ventricular fibrillation [15, 27, 425].

🐾 Among animals, geese are particularly threatened by Erysimum species, especially by *E. crepidifolium*. Cats and dogs can also be poisoned.

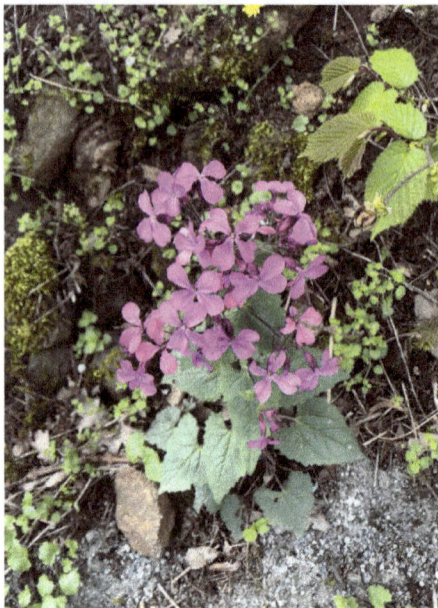

Ph. 8.9: *Lunaria rediviva* (source: Ulrike Lindequist).

### 8.3.8.8 Oleander Species (Nerium Species)

The genus Nerium, Oleander, Apocynaceae, includes only *Nerium oleander* L., olean-der, rosebay, and *N. indicum* Mill., Indian oleander, which is often assigned to the spe-cies *N. oleander.*

*Nerium oleander* occurs from the Mediterranean region through the East Indies to Japan and in Northwest and South Africa (Cape Province) and is also cultivated as a pot plant. *N. indicum* is distributed from Iran to central India, in the Himalayas up to altitudes of 2,000 m, and is cultivated in India.

Oleander (Ph. 8.10) is an evergreen, mostly shrubby plant, rarely also a tree up to 6 m high, with leathery, lanceolate leaves, which are mostly arranged in the threes in whorls, more rarely opposite or in fours on the stem. The handsome white, yellowish, pink to purple flowers have five petals. Forms with double flowers are also cultivated.

There are very different data on the content of cardioactive steroid glycosides. While Karawya et al. [197] determined amounts between 0.35% and 0.52% in the leaves (based on DW), Tittel and Wagner [405] found over 1% in all samples exam-ined. The content in leaves is highest during flowering. Reddish flowering varieties should have a slightly higher content than white-flowering ones [197].

About 40 cardenolide glycosides have been isolated from the leaves so far. The most relevant toxin is oleandrin (folinerin, oleandrogein-3α-L-oleandroside) (Tab. 8.7 and Fig. 8.6 [1, 71, 175, 379, 380, 449, 451, 455]).

Ph. 8.10: *Nerium oleander* (source: Wolfgang Kiefer).

Tab. 8.7: Cardioactive steroid glycosides of leaves of *Nerium oleander*.

| Glycoside | Structure | Content mg/100 g DW | LD$_{50}$ mg/kg BW, i.v., cat |
|---|---|---|---|
| Gentiobiosyloleandrin | OLE-Ole-Glc(1→6β)Glc | 315–580 | |
| Glucosyloleandrin | OLE-Ole-Glc | trace–511 | |
| Oleandrin | OLE-Ole | 12–117 | 0.20 |
| Nerigoside | OLE-Dgn | 0–58 | |
| Odoroside H | DTX-Dtl | trace–73 | 0.20 |
| Odoroside A | DTX-Dgn | 17–154 | 0.19 |
| Adynerin | ADY-Dgn | trace–13 | inactive |
| Oleaside A | OLA-Dgn | 89 | |
| Kaneroside | HEC-Dgn | | |
| Neriumoside | DHC-Dgn | | |
| Strospeside | DTX-DTl | | |

Abbreviations: ADY, adynerigenin; DTX, digitoxigenin; OLA, oleagenin; OLE, oleandrin; HEC, 2α-hydroxy-8,14β-epoxy-5β-card-16,20(22)-dienolide; DHC, 2α,14β-dihydroxy-5β-card-16,20(22)-enolide; Glc, D-glucose; Ole, L-oleandrose; Dgn, D-diginose; Dtl, D-digitalose (for structures, see Figs. 8.2, 8.4 and 8.6)

In fresh leaves of oleander, triosides with terminal gentiobiose occur almost exclusively. They are apparently primary glycosides and enzymatically cleaved to the corresponding biosides or monosides [38].

In addition to the glycosides listed in Tab. 8.7, oleasides B, C, D, and E, among others, are known, but, like oleaside A, they have no cardiotonic effect [1].

The glycoside spectrum of *N. indicum* is similar to that of *N. oleander* [451] in its major glycosides, except for quantitative differences. Among others, the *seco*-cardenolide neriaside [448] and the yellow-colored neriumosides A-1, A-2, B-1, B-2, and C-1 were detected (Fig. 8.6 [43, 379, 450]).

Pregnane derivatives occur as accompanying substances, including pregnendiones, so-called neridienones, in the root bark of *N. indicum*. Of note is the occurrence of ursane-type triterpenes esterified with phenylacrylic acids, e.g., neriucumaric acid (28-carboxy-2-*cis*-p-coumaroyloxy-3β-hydroxy-urs-12-ene), with depressor effects on the CNS [247]. 7,16-Dihydroxy- and 8,16-dihydroxy *trans*-palmitic acid (neriumol and nerifol) have also been isolated [382].

The cardiac effect of the glycosides of oleander in humans is somewhat weaker than that of the Digitalis glycosides. Their diuretic effect is particularly strong. Oxytocin-like effects on the uterus explain abortions after administration of oleander extracts [306].

Preparations from oleander leaves are only rarely used in treatment of heart failure and for skin rashes. Current trends show possible use of oleandrin or oleander preparations in the treatment of proliferative diseases including cancer [195]. In clinical trials with patients with advanced cancer, administration of oleander preparations alone or in combination with chemotherapeutic agents achieved tumor regression and prolonged survival with acceptable side effects [50, 216, 242].

In other studies, antiviral activity of oleandrin against enveloped viruses has been studied [242, 313]. It inhibited the production of infectious virus particles when applied prior as well as after infection of cells by SARS-CoV-2. But the authors warn about the toxicity [104, 333].

The poisonous effect of oleander was already known in ancient times. Even today, poisonings are frequently reported (overview: [250], report of some cases: [74]). Poisonings were caused by mistaking the leaves for Eucalyptus leaves when making tea (fatal outcome [162]), by eating the flowers (vomiting, heart trouble, and favorable outcome [332]), during self-medication in the form of infusions, by sucking the nectar from the flowers [161], by using the wood as meat skewers or by burning it in a barbecue fire [159], by folk medicinal use of the plants (e.g., for leprosy, malaria, fever, alcoholism, and psoriasis [31, 131, 443]), or for abortion, murder, and suicide [41, 54, 74, 77, 107, 146, 252, 320, 351]). In Germany, after ingestion of oleander leaves by children, there were frequent cases of counseling, but hardly any serious poisonings [49, 82]. In Switzerland, poisoning of children has not been reported (lack of attractive fruit, bitter taste [183]).

The cumulative lethal dose of dried leaves is 30–60 mg/kg BW for capuchin monkeys and likely in a similar range for humans [372]. For horses, 15–20 g, for cattle 10–20 g, and for sheep 1–5 g of the fresh leaves are lethal [61]. A daily dose of 60 mg dried leaves/kg BW caused death in sheep after 3–14 days [5]. Of six sheep, five died after a single dose of 110 mg dried oleander leaves/kg BW within 4–12 h [23]. Consumption of five leaves is said to cause lethal poisoning of humans. It was reported that only one leaf had severe toxic effect to children. Controversially, ingestion of three leaves by a 7-year-old child caused moderate symptoms only [125].

† A 71-year-old person committed suicide by ingestion of a self-prepared oleander leaf infusion. Toxic levels of oleandrin were detected (blood 37.5 ng/mL, vitreous humor 12.6 ng/mL, urine 83 ng/mL liver 205 ng/mg, gastric content 31.2 ng/mL). Death was attributed to fatal arrhythmias [74].

† A 47-year-old female, known to be depressive, ingested a bowl of oleander blooms for the purpose of suicide. She vomited some hours after ingestion. Heart rate was 40 beats/min. The patient was treated by oral administration of charcoal combined with sodium sulfate and with electrolyte solutions for correction of the potassium level. A temporary external pacemaker was attached because of an intermittent AV block III. The patient could be discharged on day 6 [325].

⚕ Symptoms of poisoning by oleander leaves or infusions from the leaves are gastrointestinal, neurological, and cardiovascular problems characterized by nausea, vomiting, abdominal pain, diarrhea, headache, mydriasis, visual disturbances, tremors, cyanosis, dyspnea, hyperkalemia, sinus bradycardia, ventricular arrhythmia, and increasing cardiac insufficiency up to cardiac paralysis. Death may occur in less than 1–2 h. Local application of oleander sap may cause skin irritation or eye inflammation [74].

🛏 For treatment, see Digitalis intoxications.

🐎 Animal poisonings, e.g., of cattle, horses, geese, guinea pigs, dogs, cats, and monkeys, have been described but are rare, probably because of the strong bitter taste of the plant parts. Poisonings occur in hungry animals or when the food is contaminated with the plants. In horses, the prognosis is particularly poor because of their inability to vomit. Symptoms in 50 lactating cows, fed with pruning wastes containing oleander leaves, were depression, anorexia, ruminal atony, diarrhea, serous nasal discharge, tachycardia, and irregular heartbeat. Thirteen cows from 50 died in 4 days. Transfer into milk and dairy products occurred [9, 22, 77, 82, 173, 183, 199, 211, 266, 291, 352, 389, **60**]. Pathological investigations showed damages of the lung, liver, and heart and changes in biochemical serum parameter [125].

In addition to its promising potential as a cytostatic and antiviral drug, *Nerium oleander* may act as an insecticide, pesticide, rodenticide, and antimicrobial agent [33].

### 8.3.8.9 Further Plants of Dogbane Family (Apocynaceae) Containing Cardenolides
In addition to the plants mentioned earlier, some further plant species from the family Apocynaceae deserve our interest because they are important poisonous plants in their native countries and are used there, e.g., as arrow poisons. Besides they are important as industrial drugs to obtain therapeutically used cardioactive steroid glycosides. Thus, they can be the cause of iatrogenic poisoning. Among them are Strophanthus species, *Apocynum cannabinum,* Cascabela species, Cerbera species, and Periploca species.

 **Strophanthus gratus** (Wall. and Hook.) Baill., **S. kombe** Oliv. and **S. hispidus** DC. are used for the extraction of cardioactive steroid glycosides.

*S. gratus* (Ph. 8.11) is a liana common in the coastal forests of tropical West Africa and is shrubby in culture. *S. kombe* is a shrub found in the southeastern African Lake district in Malawi, Zambia, and Mozambique. *S. hispidus* occurs from tropical West Africa to Tanzania. All parts of the Strophanthus species contain cardioactive steroid glycosides. They are particularly abundant in the seeds (3–8%).

Ph. 8.11: *Strophanthus gratus*, Semen (source: Nadin Schultze).

In the mixture of about 30 cardenolide glycosides in the seeds of *S. gratus* ouabain (g-strophanthin, ouabagenin-3α-L-rhamnoside) dominates with a proportion of over 80%. Ouabain, formerly known as g-strophanthin, was first obtained by Arnaud in 1882 from the root of the tree *Acokanthera schimperi* (A. DC.) Schweinf. (*A. ouabaio* Cathelineau ex Lewin), Bushman's fair poison, used by the natives of Somalia for arrow poisons.

'k-Strophanthin', isolated from the seeds of *S. kombe*, is a mixture consisting mainly of k-strophanthoside (strophanthidin-3-[(α-D-gluco)-β-D-gluco]-β-D-cymaroside [343, 406]).

Strophanthus glycosides are characterized by the rapid onset of action when administered parenterally and by low accumulation. They are hardly absorbed when given orally. For use in acute heart failure, ouabain must be administered i.v. Today, Strophanthus glycosides are only seldom used in the therapy of heart insufficiency [135]. However, the use of ouabain as senolyticum to treat age-related diseases seems to be conceivable [152].

☠ Unlike Digitalis glycosides, ouabain is only rarely the cause of intoxications because of the low resorption rate. Severe poisoning is possible only after overdose given by i.v. application or by rectal administration. Poisoning by rectal application of 100 mg ouabain is described in the literature [**169**].

Because of the low absorbability of Strophanthus glycosides, extracts from the plants containing them are suitable arrow poisons for hunting. The hunted game is edible without danger after cutting out the wound sites.

*Apocynum cannabinum* L., American hemp (Ph. 8.12), is a perennial that grows in North America on wasteland to 1 m height. It contains, especially in the root, cymarin (strophanthidin-3β-cymaroside, 0.2–0.4%) and apocannoside (cannogenin-3β-cymaroside) as main glycosides. Standardized extracts of the drug Apocyni cannabini radix are contained in finished medicinal products and used for the gentle flushing out of cardiac-induced edema.

Ph. 8.12: *Apocynum cannabinum* (source: Wolfgang Kiefer).

Ph. 8.13: *Cascabela thevetia* (source: Horst Pilgrim).

(☠) Poisoning of humans by *Apocynum cannabinum* is not known because of the low absorbability of its cardioactive steroid glycosides.

*Cascabela thevetia* (L.) Lippold (*Thevetia peruviana* (Pers.) K. Schum., Ph. 8.13), yellow oleander, is a shrub or small tree with milky latex, originally from American tropics, found in the southern USA (Florida), Central and South America, and the West Indies. It is cultivated as an ornamental tree in many tropical countries because of its orange yellow, pleasantly fragrant flowers. The seeds are called cabalonga blanca or codo de fraile.

The content of cardioactive glycosides in the seeds may be as high as 5% [317]. Aglycones of the 16 glycosides found are digitoxigenin, cannogenin, 19-carboxydigitoxigenin, and uzarigenin (Fig. 8.2) as well as the C-nor-D-homo homologues of cannogenin, thevetiogenin, and 19-formylthevetiogenin [2, 42]. The main active ingredients are thevetin B (cerberoside, 1.1% in seeds, digitoxigenin-3-(3-gentiobiosyl)-α-L-thevetoside), peruvoside (0.4%, cannogenin-α-L-thevetoside), and thevetin A (0.2%, cannogenin-3-(3-gentiobiosyl)-α-L-thevetoside [3]).

A similar glycoside spectrum is present in the seeds of *Cascabela ovata* (Cav.) Lippold (*Thevetia ovata* DC.), jade plant, money plant, a shrub native to Mexico and Central America [246]. All other parts of the two plants mentioned contain less than 0.1% of cardioactive steroid glycosides [317].

The cardioactive steroid glycosides of the Cascabela species are well absorbed after peroral ingestion and accumulate little [341]. Peruvoside is therapeutically used [153].

The seeds of both Cascabela species are frequently the cause of accidental poisonings [17, 63, 147, 193, 358, 361]. Every year, thousand cases of poisoning are reported worldwide. Some of which are fatal. One seed can seriously poison a child; 8–10 seeds are fatal to an adult [193]. In South Asia, the fruits and leaves of yellow oleander are frequently abused as suicide poisons, especially among younger people [112, 131, 147, 324, 341, 360]. In Sri Lanka, 415 cases of suicide attempts using Cascabela seeds, including many by children, often with fatal outcomes, had to be treated in 11 months at one hospital in 1995 [112]. Severe symptoms of poisoning have been observed in a patient after ingesting as few as two seeds of *C. thevetia* [276].

Yellow oleander herb, e.g., in capsules or seeds pulverized, is sold in some markets as well as via the internet as a 'safe' and 'natural' treatment for obesity. In Mexico, a herbal vendor recommended his clients the seeds to get rid of internal parasites. Four members of a family were seriously intoxicated, and two died [147, 289].

❖ A 37-year-old man with no known comorbidities consumed a single crushed seed of yellow oleander. After 4 h, he developed vomiting and dizziness. Upon arrival at the hospital, he had normal sinus rhythm. During the following hours, he developed acute ST-segment myocardial infarction [12].

🜊 Symptoms of poisoning by yellow oleander beginning 0.5–3 h after ingestion of seeds are headaches, nausea, vomiting, diarrhea, intense abdominal cramps, numbness of the extremities, restlessness, drowsiness, hypotension, arrhythmias, and metabolic abnormalities (hyperchloremia and metabolic acidosis [112, 113, 198, 273, 276, 324]).

▣ For treatment, see Digitalis intoxications.

🐎 Animal poisoning by Cascabela species has been described.

Extracts of *Cascabela peruviana* can be used as piscicides to eliminate fish in shrimp farms [359].

**Periploca** species, formerly assigned to Asclepiadaceae, are perennial, deciduous climbing plants, hedges, or trees that occur in southern Europe, Asia Minor, and North Africa. They are cultivated rarely in Central Europe as ornamental woody plants. A total of 46 cardenolide glycosides and 92 pregnane derivatives were isolated so far from them.

Ph. 8.14: *Periploca graeca* (source: Nahhan/iStock/Getty Images Plus).

Periplocin (periplogenin-3-(β-D-gluco)-β-D-cymaroside), isolated from the bark of *P. graeca* L., silk vine (Ph. 8.14), found in the southwestern Caucasus, was first reported by Lehmann 1897. Periplocymarin (periplogenin-3β-D-cymaroside) was produced by enzymatic hydrolysis with strophantobiase from periplocin. Both compounds occur besides others also in other Periploca species. The aglycones of the cardenolide glycosides found in the barks of Periploca species are periplogenin, its C-7 and/or its C-8 hydroxy-, C-7–C-8, or C-8–C-14-epoxy derivative [172, 258].

Periploca species show digitalis like cardiotonic capacity and furthermore antiinflammatory, immunosuppressive, antitumor, antibiotic, and insecticidal effects. The pregnane glycosides show no cardiotonic effect, but they are involved in the other mentioned effects [172, 258].

*P. sepium* Bunge, Chinese silk vine, and *P. forrestii* Schltr. are used in Chinese folk medicine since historical times. Their bark is recorded in the Chinese Pharmacopoeia. It is used for the treatment of autoimmune diseases, especially rheumatoid arthritis, and strengthens bones and tendons [258]. Periplocin serves for the treatment of heart circulatory insufficiency and also for other diseases [172, 258].

☠ The plants offer no incentives for human consumption. Poisonings by using Cortex Periplocae have been recorded. Symptoms of poisoning are cardiotoxic ones. The $LD_{50}$ of periplocin for mice is 15 mg/kg BW i.p. [258].

🐾 Poisoning of animals, e.g., sheep, horses, and elephants, by Periploca species is recorded [59, 87, **159**].

*Cerbera odollam* Gaertn., suicide tree, pong-pong tree, mintola (Ph. 8.15), is a tree with a maximum height of 15 m, which is also grown as a hedge. It is native to South India, Madagascar, and other parts of southern Asia and grows preferentially in wet areas, especially in costal salt swamps. The flowers are large with white petals with a yellowish center. The spherical fruit turns from green to red as it matures. A fibrous shell encloses ovoid kernel (2 × 1.5 cm). The whole plant contains a milky white latex. The seeds were sometimes used as weight loss supplement [315]. The cardioactive steroid glycosides of the seeds are, e.g., cerberin, cerebroside, neriifolin, tanghinin, and tanghinoside [289].

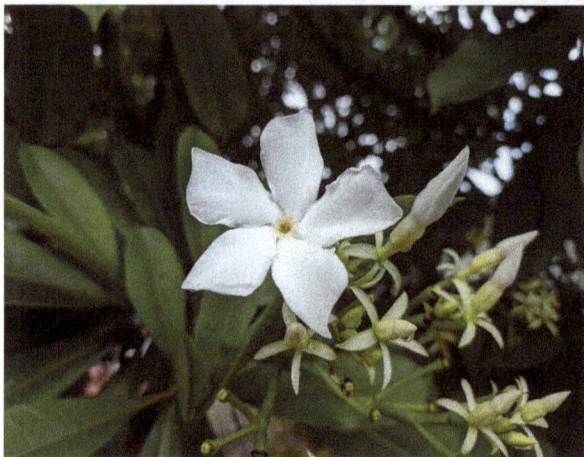

Ph. 8.15: *Cerbera odollam* (source: Sarno Markosasi/iStock/Getty Images Plus).

The seeds of the suicide tree are a frequently used agent of homicide and suicide especially in India. More than 500 cases of fatal Cerbera poisoning took place in Indian state Kerala between 1989 and 1999. Poisonings are also known from other countries, e.g., the USA, and caused there by the fruits grown from imported seeds [290, 140].

⚕ A 33-year-old woman, who has used pong-pong kernels as a weight loss supplement suffered from vomiting, bradycardia, severe hyperkalemia, and slow atrial fibrillation followed by cardiovascular collapse. Despite the administration of nine vials of digitoxin-specific Fab, the patient could not be resuscitated [315].

🕯 Symptoms of poisoning after ingestion of the seeds of *Cerbera odollam* are nausea, vomiting, hyperkalemia, thrombocytopenia, and EEG abnormalities. The risk of death is high [289].

💊 Initial management includes supportive therapy and administration of atropine, followed by temporary pacemaker insertion. In severe cases, the administration of digoxin immune Fab may be helpful [289].

*Cerbera manghas* L., sea mamba, naturally distributed from the Seychelles in Indian Ocean to French Polynesia, resembles in appearance and toxicity *C. odollam*. It can be found as an ornamental plant in other tropical area, e.g., on Hawaii [265].

### 8.3.9 Plants with Bufadienolides

#### 8.3.9.1 Hellebore Species (Helleborus Species)

The genus Helleborus, hellebore, Ranunculaceae, includes about 20 species, found especially in southern and southeastern Europe and in Asia. Two species occur in Western China. *Helleborus niger* L., Christmas rose (Ph. 8.16), *H. foetidus* L., bear's root, stinking hellebore, *H. viridis* L., green hellebore, *H. dumetorum* Waldst. and Kit. ex Willd, hedge hellebore, *H. odorus* Waldst. and Kit. ex Willd., fragrant hellebore, and *H. multifidus* Vis. are grown in Europe. Ornamentals include *H. orientalis* ssp. *guttatus* (A. Braun and Sauer) B. Mathew, Lenten rose, and *H. purpurascens* Waldst. and Kit., purple hellebore. Many hybrids are grown.

Helleborus species are perennial, lime-loving herbs with well-developed rhizomes. The stem is mostly herbaceous, and rarely woody in the lower part. The basal leaves are partly wintergreen. They are foot-shaped, divided, and toothed, and those of the stems are often entire. The handsome flowers have a simple perianth with 5-

Ph. 8.16: *Helleborus niger* (source: Isolde Kühn).

segmented perianth (green, white, reddish, purple), 5–15 short cone-shaped honey leaves, numerous stamens, and 3–5 free carpels.

*Helleborus niger*, a plant with white or reddish flowers, blooms as early as February to April. It occurs in Central and South Europe and is frequently cultivated in various varieties as an ornamental plant. *H. foetidus* has greenish, red-edged perianth leaves and a richly leafy stem. It is native to southern and western Europe and grows in oak and beech forests and on hills covered with shrubs. In *H. viridis*, the stem is without leaves up to the branching point. The perianth is pure green. It occurs scattered in sparse forests and shrubberies in France, and rarely in northern Europe. *H. dumetorum* and *H. odorus* have yellow-green to green flowers.

Mainly the underground organs of the abovementioned species have been studied for their cardioactive steroid glycosides so far. For *H. multifidus* and *H. odorus*, results are also available on the bufadienolide glycoside content of the seeds [214].

Bufadienolide derivatives of the roots and seeds of Helleborus species have been found to be hellebrigenin (bufotalidin), helleborogenon, 11α-hydroxy-hellebrigenin, and 14β-hydroxy-3-oxo-20,22-bufadienolide, the latter two so far only in *H. odorus*. Additionally, telocinobufagin was found in *H. niger* [145]. Most glycosides, among them hellebrin (hellebrigenin-3β-(4'-β-D-glucopyranosyl) α-L-rhamnopyranoside) possesses the aglycon hellebrigenin [329, 440]. The highest content of hellebrin has been found in *H. purpurescens* (up to 1.50% DW). No cardioactive steroid glycosides were found in the rhizome and roots of *H. foetidus*.

In addition to a rapid onset and long-lasting cardiac effect, hellebrin exhibits analgesic and spasmolytic effects at therapeutic doses [108]. The lethal dose in the cat ($LD_{50}$ of 0.1 mg/kg BW, i.v.) is less than that of Digitalis glycosides. The aglycone hellebrigenin is more toxic (LD 0.077 mg/kg BW, cat) than its glycosides [262].

Steroid saponins may be considered as secondary agents, which may be the main agents in species poor in cardioactive steroid glycosides. Aglycones identified were 3β-hydroxy-22α-methoxy-furost-5-ene (in *H. viridis*), spirosta-5,25(27)-diene-1β,3β,11α-triol (in *H. dumetorum*, *H. multifidus*, and *H. odorus*), and macranthogenin (25(27)-dehydrosarsapogenin) in *H. niger* ssp. *macranthus* (Freyn) Schiffn., wild Christmas rose, possibly also in *H. niger*, *H. foetidus*, and *H. purpurascens*. The steroid saponins exert strong local mucosal irritation, cause vomiting, and have laxative effects. After resorption they initially cause CNS excitation and later paralysis.

The aboveground plant parts of Helleborus species are capable to produce protoanemonin, which is highly irritating to the skin and mucous membranes.

Ecdysterones have been detected in a variety of Helleborus species. These are steroids with cholestane bodies that are known as molting hormones of insects and crustaceans and are thought to function as allomones in plants.

In addition, α- and β-thionine derivatives, the hellethionines A – D, have been isolated from *H. purpurascens*. They are short-chain cystine-rich peptides that are thought to serve as defenses against pathogenic microorganisms [293].

Cardiotoxic alkaloids (named celliamine, sprintillamine, and sprintilline, $C_{28}H_{45}NO_2$ and $C_{25}H_{41}NO_2$, respectively, structure unknown) have been isolated from *H. viridis.*

Already in ancient times Helleborus species were used for healing purposes, e.g., for dropsy, kidney diseases, as anthelminthics, as purgatives, in epilepsy, apoplexy, and as 'training resources' for the immune system and for mental illnesses [127]. Because of its irritating effect on the nasal mucosa, powdered Helleborus roots were part of the 'Schneeberger Schnupftabak' ('Schneeberg snuff'). 'Hellebore sticking' was formerly used in treatment of animals. In this process, rhizome pieces of Helleborus species were implanted with a thick needle in chronically ill animals as part of a nonspecific stimulation therapy to activate healing processes. The pieces were placed in the earlobes of sheep and pigs, and in the chest area of cattle and horses. After about 24 h, the sticks were removed. This resulted in an increase in the number of leukocytes and neutrophil granulocytes in the blood and in an increase in phagocytosis activity [51]. In humans, the rhizome has been used to intentional delay wound healing processes by insertion into wounds [350].

Today, the use of extracts or fermented extracts (cleavage of the glycosides!) of *Helleborus niger* in tumor therapy is of interest [367]. The extracts inhibit the proliferation of tumor cells, including such of primary tumors of children, by induction of apoptosis via the intrinsic mitochondria signaling pathway with consecutive processing of caspase 3 [184]. A case description reports minor regression and long-time survival in a patient with malignant pleural mesothelioma under *Helleborus niger* extract [325]. The effects are attributed to the content of protoanemonin, saponins [367], and deglucohellebrin [415].

Only rarely *H. purpurascens* is still used for rheumatic diseases in Hungary [243].

☠ Symptoms of poisoning after ingestion of plant parts of hellebore species are vomiting, diarrhea, cardiac arrhythmias, and cardiac paralysis. Poisonings by *H. niger* are known but occur rarely. They are due to confusion of Christmas rose with other plants, consumption of the seeds, or overdose during therapeutic use. Symptoms of saponins and protoanemonin poisoning predominate, e.g., scratching in the mouth and throat, salivation, nausea, vomiting, diarrhea, dizziness, shortness of breath, possibly convulsions, and central respiratory paralysis.

🚑 Treatment is symptomatic after primary toxin removal by activated charcoal and laxatives.

🐎 Fatal poisonings of grazing animals by the herb of *H. viridis* [186] and *H. foetidus* [169] have been described.

## 8.3.9.2 Winter Aconite (*Eranthis hyemalis*)
*Eranthis hyemalis* (L.) Salisb., winter aconite, Ranunculaceae, seems to contain bufadienolide glycosides. The information on their occurrence is contradictory [190, 364]. γ-Chromone derivatives, e.g., partly cardioactive eranthins A and B, were isolated [190].

☙ Ingestion of the bulbous rootstock can lead to nausea, vomiting, visual disturbances, bradycardia, arrhythmia, and respiratory distress [127].

⚑ Treatment is symptomatic after primary toxin removal by activated charcoal and laxatives.

### 8.3.9.3 Cape Tulip Species (Moraea Species)

The genus Moraea, Cape tulip, Iridaceae, includes 230 perennial herbal species. Of special toxicological importance are *M. flaccida* Steud. (*Homeria flaccida* Sweet), one-leaf Cape tulip, and *M. miniata* Andrews (*Homeria miniata* Sweet) two-leaf Cape tulip, native to southern Africa, introduced to Australia as garden plants because of their attractive flowers. Major constituents are the bufadienolides 1α,2α-epoxyscillirosidine and 16β-formyloxybovogenin A.

🐎 The fresh plants or hay containing dried plants are often the cause of animal poisoning in Central Africa and Australia. Particularly affected are cattle, sheep, goats, and donkeys. About 700 g of the plant material kills a sheep in 2 h ('tulp poisoning' [134, 159]).

### 8.3.9.4 Sea Onion Species (Drimia Species)

Drimia species, Asparagaceae, sea onion, are perennial bulbous plants. Their erected flowering stem appears after growth of the lanceolate leaves. Flowers are arranged in dense racemes. The bulbs are pear shape with a diameter of up to 30 cm. The plants are widespread in Mediterranean area, Iran, Africa, and India.

*D. maritima* (L.) Stearn (*Urginea maritima* (L.) Baker, *Scilla maritima* L., Ph. 8.17), true sea onion, sea squill, is a collective species with six subspecies. All minor species occur in the Mediterranean region and the Canary Islands. The plant is cultivated, although rarely today, as a houseplant.

The true sea onion *D. maritima* s.l. is distinguished by a largely aboveground, white, yellow, reddish, or brown bulb that can weigh up to 3 kg in its natural habitat. The flowering stem, up to 1.5 m long, bears a raceme of up to 60 flowers with 6 white petals. The broad-lanceolate leaves are gray-green. In the false sea onion, *Albuca bracteata* (see section 8.3.8.3), the petals have a green central stripe. Its bulb is green.

The bulbs of all subspecies contain cardioactive steroid glycosides in concentrations of 0.2–4%. Approximately equal concentrations are also present in the aboveground organs [227]. There is no correlation between onion color (red or white) and the spectrum and content of active ingredients [58, 237]. The first cardioactive glycoside isolated from *D. maritima* by Stoll 1933 was scillaren A [400].

Table 8.8 summarizes the most active constituents of *D. maritima* [209, 210, 227, 231, 238].

Ph. 8.17: *Drimia maritima* (source: Peter Schönfelder).

Tab. 8.8: Cardioactive steroid glycosides of bulbs of *Drimia maritima*.

| Glycoside | Structure | LD$_{50}$ mg/kg BW, i.v., cat |
|---|---|---|
| Scillaren A | SCI-Rha-Glc | 0.14–0.17 |
| Scillaren A diacetate | SCI-Rha(diAc)-Glc | |
| Proscillaridin A | SCI-Rha | |
| Glucoscillaren A | SCI-Rha-Glc-Glc | 0.17 |
| Glucoscillaren A diacetate | SCI-Rha(diAc)-Glc-Glc | |
| Glucoscillaren(1-3) | SCI-Rha-Glc-Glc(1-3) | |
| Scillarenin-β-D-glucoside | SCI-Glc | |
| Gamabufotalin-rhamnoside | GBU-Rha | |
| Scilliphaeoside | SPA-Rha | 0.09 |
| Scilliphaeoside-β-D-glucoside | SPA-Glc | |
| Glucoscilliphaeoside | SPA-Rha-Glc | |
| Diglucophaeoside | SPA-Rha-Glc-Glc | |
| Scilliroside | SRO-Glc | 0.12–0.13 |
| Desacetylscilliroside | DA-SRO-Glc | |
| Scillirubroside | SRU-Glc | 6.13–0.46 |
| Scillirubroside-α-L-rhamnoside | SRU-Rha | |
| Scilliglaucoside | SCA-5β-Glc | 0.07 |
| Scilliglaucoside-α-L-rhamnoside | SCA-Rha | |
| Scillicyanoside | SCY-Glc | |

Abbreviations: SCI, scillarenin; GBU, gamabufotalin; SPA, scilliphaeoside; SRO, scilliroside; DA-SRO, desacetyl-scilliroside; SRU, scillirubroside; SCA, scilliglaucoside; SCY, scillicyanoside; Rha, L-rhamnose; Rha(diAC), 2,3-diacetyl-L-rhamnose; Glc, D-glucose; Glc(1, 2,3), D-glucose, 1–3 linked

Bufadienolides have also been identified in *D. altissima* (Lf.) Ker Gawl., African squill [334], *D. aphylla* (Forssk.) Spera, sea squill [235], *D. hesperia* (Webb and Bertel.) C.J. Manning and Goldblatt [234], *D. indica* (Roxb.) Jessop (Ayurvedic medicine for cardiac diseases [24]), *D. numidica* (Jord. and Fourr) J.C. Manning and Goldblatt [235], and *D. pancration* (Steinh.) J.C. Manning and Goldblatt [226, 230, 236].

True sea onion is one of the longest known medicinal plants. It was also cultically revered to avert disaster [52]. Today preparations from the bulbs are traditionally used for respiratory disorders like asthma, and medical problems affecting the joint, bones, and skin [58, 356]. Scillaren A, scillaren B, and proscillaridin A are rarely used in therapy for congestive heart failures and rheumatic heart diseases.

After p.o. uptake, 20–35% of the glycosides were absorbed. Elimination is predominantly renal with a decay rate of 50% per day. The toxic plasma level for proscillaridin is approximately 0.4 ng/mL. Scilliroside induces lethal convulsions in rats (LD 0.1–0.2 mg/animal [**149, 169**]).

Poisonings are predominantly medicinal. A case of poisoning by a cough preparation containing true sea onion extract was described [278]. The fatal poisoning of a 55-year-old woman who had eaten two cooked sea onions to relieve arthritic symptoms is reported from Turkey [411]. The death of a man was caused by eating pancakes made with red sea onions [**169**].

Intoxications by ingestion of Drimia preparations, especially of the red type, to be used as rat or mice poison occurred in the past.

♦ A 36-year-old man ingested the contents of three bottles of a raticide with suicidal intent and was admitted to a hospital with an extremely rapid pulse (140 beats/minute), a sharp drop in blood pressure, and gastrointestinal irritation. Unaware of the poison, ouabain was initially given. Thereafter, severe arrhythmias, massive vomiting diarrhea, and central symptoms developed immediately. Treatment with atropine saved the patient [128].

In South Africa, poisoning by *D. sanguinea* (Schinz) Jessop, which is used in folk medicine there, occurs frequently. In 1 year, 44 fatal poisonings were recorded [**159**].

☠ Symptoms of poisoning by sea onions are nausea, vomiting, dizziness, hyperkalemia, disturbances of the cardiovascular system, and kidney irritation with tenesmus and possibly hematuria [58, **97, 107, 169**].

🚑 Treatment is symptomatic after primary toxin removal by activated charcoal and laxatives.

🐐 In sheep, severe poisoning by *D. sanguinea* ('slangkop poisoning') [188, **159**] and *Fusilium physodes* [309] is observed in northern South Africa. Numerous cases of animal poisoning by Drimia species have also been reported from North Africa. Animals exhibit anorexia, apathy, movement disorders, and arrhythmias. The poisonings are often fatal. Scillaren A and 5α-4,5-dihydroscillaren have been isolated as responsible compounds from *D. sanguinea* [268] and physodin A–D (14-deoxybufadienolides) from *F. physodes*. The $LD_{50}$ of physodin A in the guinea pig is 0.22 mg/kg BW, s.c. [309].

Apart from internal use as a cardioactive agent, slices of the fresh bulbs of true sea onion are used externally in folk medicine for burns and erysipelas (cooling effect of the cut fresh bulb, antibiotic, and local anesthetic effect). Because of the reflex bronchial irritation, it is also a component of expectorants.

### 8.3.9.5 Kalanchoe and Bryophyllum Species

The genus Kalanchoe, Crassulaceae, is divided into the sections Kalanchoe and Bryophyllum, including about 125 species. The plants are perennials, semishrubs, small trees, and lianas occurring in the tropics from South Africa to Madagascar, India to Taiwan and Java, even sporadically growing in tropical America. They usually have fleshy leaves and four-digit flowers. In the notches of the leaves, small daughter plants are formed, which serve for vegetative propagation.

The Bryophyllum and Kalanchoe species are kept as houseplants because of the interesting brooding plantlets under the name brooding leaf. They are also named Goethe plants. These include *B. daigremontianum* (Raym.-Hamet and H. Perr.) A. Berger (*K. daigremontiana* Raym.-Hamet and H. Perr., Ph. 8.18), mother of thousands, devil's backbone, *B. delagoense* (Eckl. and Zeyh.) Bruce (*B. tubiflorum* (Harv.) Raym.-Hamet, *K. tubiflora* Harv.), mother of millions, chandelier plant, *K. blossfeldiana* Poelln., flaming Katy, and *B. pinnatum* (Lam.) Pers. (*K. pinnata* Lam.), life plant, cathedral bell. The former three species are native to Madagascar, and the latter's native country is unknown.

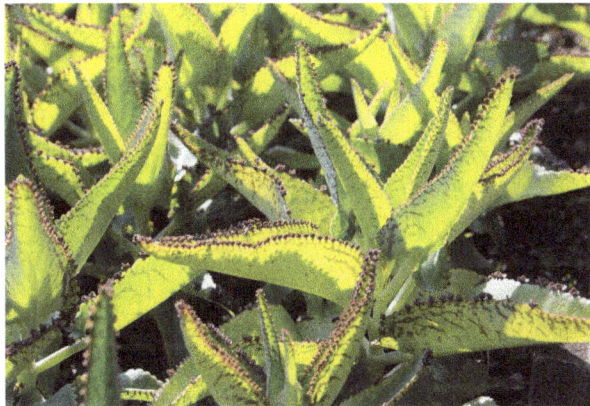

Ph. 8.18: *Bryophyllum daigremontianum* (source: WALA Heilmittel GmbH).

The abovementioned species and *B. proliferum* (Bowie) Raym.-Hamet (*K. prolifera* Bowie), blooming boxes, are of special toxicological interest. They are naturalized garden escapees in the Australian state of Queensland and are responsible for numerous poisonings of grazing livestock.

From the large variety of bufadienolide glycosides occurring in both genera, only examples can be given. With respect to the spectrum of bufadienolides, Bryophyllum and Kalanchoe do not differ.

From *B. daigremontianum*, daigremontianin, bersaldegenin-1,3,5-orthoacetate, dai-gredorigenin-3-acetate, bryotoxins A, B, and C (bryophyllin A, Fig. 8.7) were isolated [72, 446]. From the root, eight new bufadienolide glycosides were isolated, named as kala-daigremoside A, GB, C, D, E, F, G, and H [218, 300]. In *B. tubiflorum*, bersaldegenin-1,3,5-orthoacetate [427] and, as in *B. proliferum* and *B. pinnatum*, bryotoxins A to C were found [72, 283, 447]. The aerial parts of *K. ceratophylla* Haw. (*K. gracilis* Hance) contain kalanchosides A, B, C, bryophyllins A and G, thesiuside, hellebrigenin, and hellebrigenin acetate [218].

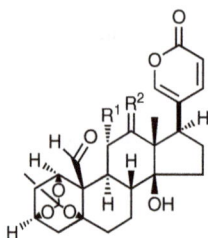

Daigremontianin  R¹=OH
                 R²=O

Bersaldegenin-1,3,5-orthoacetate
                 R¹=H,   R²=H₂

Bryotoxin C      R¹=OH, R²=H₂     Fig. 8.7: Bufadienolides of *Bryophyllum daigremontianum*.

The biosynthesis of bufadienolide glycosides is increased in response to biotic stresses [141, 218].

Reports about the pharmacological action of the isolated bufadienolide glycosides are limited. Besides their cardiotonic effect, cytostatic, antitumor, antibacterial, viro-static, and sedative actions were proven. Bryophyllins A and B are highly cytotoxic to KB cells with an $ED_{50}$ of less than 80 ng/mL [446, 447]. Bryophyllin A shows a strong antiviral effect on Epstein–Barr viruses ($IC_{50}$ = 0.4 µM) [403].

Bersaldegenine-1,3,5-orthoacetate has a similar pharmacological profile as the Digi-talis glycosides in terms of positive inotropic effect [369]. In addition, like daigremontia-nin, it has a strong sedative effect (0.1–0.5 mg/kg BW, mouse) but a centrally excitatory effect at higher doses (convulsions, paralysis [426]). In guinea pigs, i.v. administration of 760 µg/kg BW of bersaldegenin-1,3,5-orthoacetate resulted in death [369].

(☠) Although there are no reports of serious poisoning by Bryophyllum and Kalanchoe species in humans, ingestion of plant parts should be avoided, especially by children. Poisoning of humans or animals by ingestion of meat or organs from animals being poisoned by Kalanchoe species is conceivable [218, 403, **49**].

🐄 Animal poisoning by Bryophyllum and Kalanchoe species occurs frequently. Between 1960 and 1984, poisonings were recorded in 379 cattle in Queensland. Symptoms of poisoning included salivation, anorexia, rumen atonia, diarrhea, cardiac arrhythmia, and dyspnea. Death of the animals often occurred. The lethal dose of *B. tubiflorum* inflorescences in cattle is about 0.7 g DW (approximately = 7 g FW)/kg BW [174]. Similar poisoning symptoms have been reported in sheep by *K. lanceolata* Forssk. [8] and *K. crenata* (Andrews) Haw. (*K. integra* (Medik.) Kuntze [279, 340, 417]), by *B. pinnatum*, *B. blossfeldiana*, and *B. tubiflorum* in cattle [284, 288, 347], and by various Kalanchoe species in chickens [433]. *K. rotundifolia* (Haw.) Haw. and *K. lanceolata* Forsk., native to southern Africa, can be the cause of 'crimpsiekte' disease, as can Cotyledon or Tylecodon species (see Section 8.3.9.6). Main constituents are the bufadienolide glycosides lanceotoxins A and B (aglycone 5-acetylhellebrigenin, $LD_{50}$ 0.2 and 0.1 mg/kg BW, respectively, s.c., guinea pig [13, **159**]).

*Bryophyllum pinnatum* and *B. daigremontianum* are widely used in traditional medicine of Africa, South America, and Asia for common diseases such as burns, cough, wounds, insect bites, skin diseases, fever, kidney stones, gastric ulcer, rheumatoid arthritis, bacterial and viral infections, gynecological disorder, and chronic diseases like diabetes, and neurological and neoplastic diseases. The active constituents are today of interest as potential cancerostatic, cardiotonic, or antiviral drugs. The therapeutic use is, however, limited by the lack of clinical evidence and possible toxicity [19, 141, 218].

## 8.3.9.6 Cotyledon Species

Cotyledon, Crassulaceae, is a genus of succulent plants, mostly from southern Africa, but they grow also in other dry parts of Africa and Asia, e.g., on the Arabic peninsula. They are shrublets with fleshly woody stems and persistent succulent leaves. Best known is *Cotyledon orbiculata* L., pig's ears (Ph. 8.19). It includes a high number of varieties and cultivated forms. The shrub with orange-red or yellow flowers is native to South Africa and popular in gardens in many countries.

Tylecodon (Crassulaceae), also native to South Africa, is 1970 split off from the genus Cotyledon. Its leaves are deciduous. *Tylecodon wallichii* (Harv.) Toelken, pegleg butterbush, is a representative of this genus. The shrub with a thick tuberous stem and a height of 50–100 cm is also cultivated as a garden plant [**159**].

Bufadienolide glycosides were detected in *C. orbiculata* L. var. *orbiculata* [392, **159**], *T. grandiflorus* (Burm. f.) Toelken [14], *T. ventricosus* (Burm.) Toelken, and *T. wallichii* [57, 426]. From *C. orbiculata*, the structurally remarkable orbicusides A to C, tylecoside C, and cotyledoside (Fig. 8.8) were isolated, and from Tylecodon species, the tylecosides A, B, C, D, F, and G [55, **159**].

Ph. 8.19: *Cotyledon orbiculata* (source: dan_ray/iStock/Getty Images Plus).

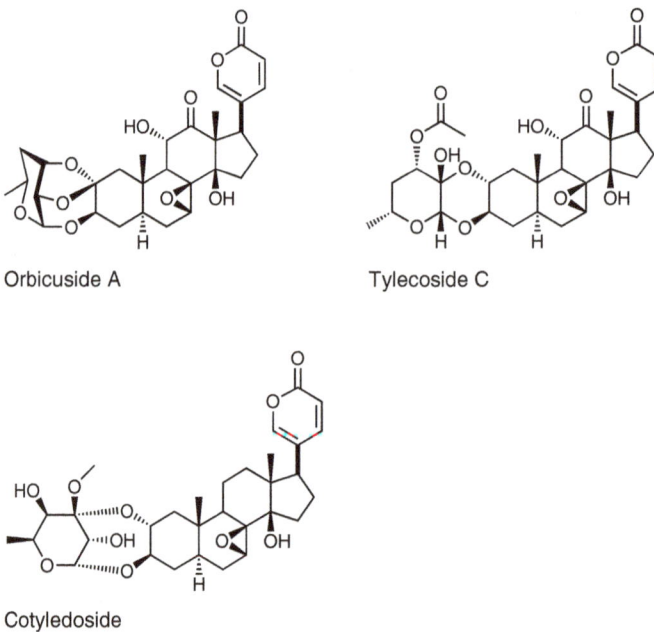

Orbicuside A

Tylecoside C

Cotyledoside

Fig. 8.8: Bufadienolides of Cotyledon and Tylecodon species.

Humans are at risk by ingestion of meat of poisoned animals. There are reports about intoxications of indigenous people in South Africa following ingestion of such 'diseased meat' [55].

🐈 Chronic poisoning of animals by Cotyledon, Tylecodon, and Kalanchoe species, termed 'cotyledosis' or 'krimpsiekte', is very common in South Africa in the Karoo semidesert. Ingestion by grazing animals leads either to immediate death ('opblaas krimpsiekte') or, after long-term ingestion of small amounts, often after weeks or months, to neurological symptoms, especially convulsions, respiratory paralysis, and paralysis of the head and neck. The animals often lie paralyzed on their sides ('dun krimpsiekte') for weeks while fully conscious. Presumably, these symptoms result from brain damage such as vacuolization of the thalamic nucleus and cerebral edema [56]. Goats and sheep are particularly affected [55, 57, 159].

Also dangerous to humans and animals are the bufadienolide glycosides, e.g., bovoside A, of the bulbs of *Bowiea volubilis* Harv. ex Hook. (Hyacynthaceae, native to South Africa [**159**]).

# 8.4 Withanolides

## 8.4.1 Chemistry and Distribution

Withanolides (withasteroids), about 400 representatives known, are $C_{28}$ derivatives of ergostane (24-methyl-cholestane), oxidized at C-22 and C-26, forming a six-membered lactone ring (e.g., withaferin A and withanolide E, Fig. 8.9), more rarely a five-membered lactone ring (ixocarpalactone B, Fig. 8.10), or connected by a methylene bridge of a bicycled seven-membered lactone ring (physalins, Fig. 8.11), usually unsaturated. Other typical structural features include an oxo group at C-1, which can sometimes be reduced to a hydroxy group, a double bond between C-2 and C-3 or C-3 and C-4, rarely absent, a hydroxy group at C-3, and a high degree of oxidation of the molecule (hydroxy or epoxy groups, cyclic ether grouping). The lactone rings may be connected to ring D or to C-21 or may be present in reduced form as a lactol. A second lactone ring may also be present in the molecule. Similarly, 13,14-*seco* derivatives and spiroforms between ring D and the side chain are known (physalins).

Withanolides occur freely, rarely as glycosides and glycoside esters. A 3-*O*-sulfonyl compound has been detected with cilistol Y. Reduction products in which the O atom of the carbonyl group of the lactone ring is reduced to an OH group are derived from the withanolides [79, 84, 156, 296, 342, 383, 416].

The withanolides were usually named after the genus or species names of the plants in which they were first detected, e.g., acnistine, chantriolide, cilistol, coaguline, daturalactone, daturacin, exodeconolide, iochromolide, ixocarpalactone, jaborosalactol, jabortone, icandrine, nicanlode, nic-lactole, nic-lactone, nicalbine, perulactone, physaline, physachenolide, physaguline, physalolactone, pubescenolide, sitoindoside, trechenolide, withacnistine, withanone, withanoside, withaperuvine, and withaphysaline [342].

Withanolides have been found mainly in Solanaceae, namely in the genera Acnistus, Datura, Deprea, Discopodium, Dunalia, Exodeconus, Iochroma, Jaborosa, Lycium, Nicandra, Physalis, Salpichroa, Solanum, Tubocapsicum, Withania, and Witheringia.

Sporadically, they have also been found in plants of other families, e.g., the highly oxidized plantagiolides A to E and chantrolide A, which, in addition to saponins, are found in *Tacca plantaginea* (Hance) Drenth (Taccaceae). The rhizome of this plant is used in TCM as an analgesic, antipyretic, and antimalarial [342]. Occurrence in Ajuga species (Lamiaceae) is also postulated [179, 383].

From **Withania somnifera** (L.) Dunal (Ph. 8.20), Indian ginseng, winter cherry, Solanaceae, the first representatives of withanolides (withanol, somnirol, and withanon) were obtained. The perennial, which grows to a height of 2 m, is native to India but also occurs in subtropical dry areas around the Mediterranean, throughout Africa, and in China.

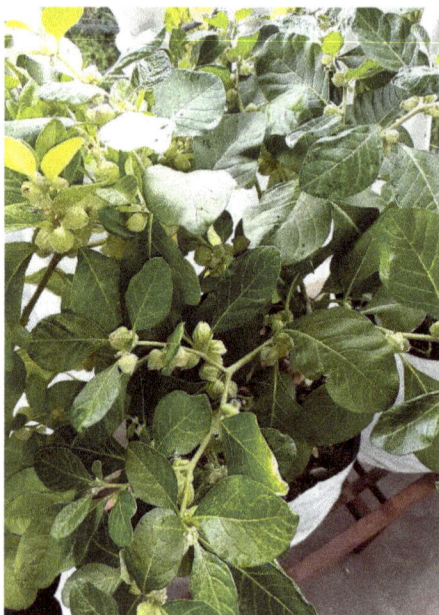

Ph. 8.20: *Withania somnifera* (source: Meena Rajbhandari).

*W. somnifera* contains about 40 different withanolides (Fig. 8.9) in varying amounts in different parts of the plant. The main component is withaferin A. Other withanolides are, e.g., withaferin D, withasomniferin-A, the withanolides D, E, G, J, the sitoindosides VII to X, and the withanosides I to VII [4, 26, 68, 207, 281, 295, 319, 420]. In addition, calystegins, nortropane derivatives, effective as glycosidase inhibitors, were found [44].

The genus **Physalis (Alkekengi)**, Japanese lantern, lampion flower, bladder cherry, Solanaceae, includes about 120 species native mainly to tropical and subtropical America. In Central Europe, especially in the south of the area, and in Western Asia, *Alkekengi officinarum* Moench (*Physalis alkekengi* L.), strawberry tomato (Ph. 8.21), thrives. It is very similar to *Ph. alkekengi* L. var. *franchetii* (Mast.) Makino, Chinese lantern

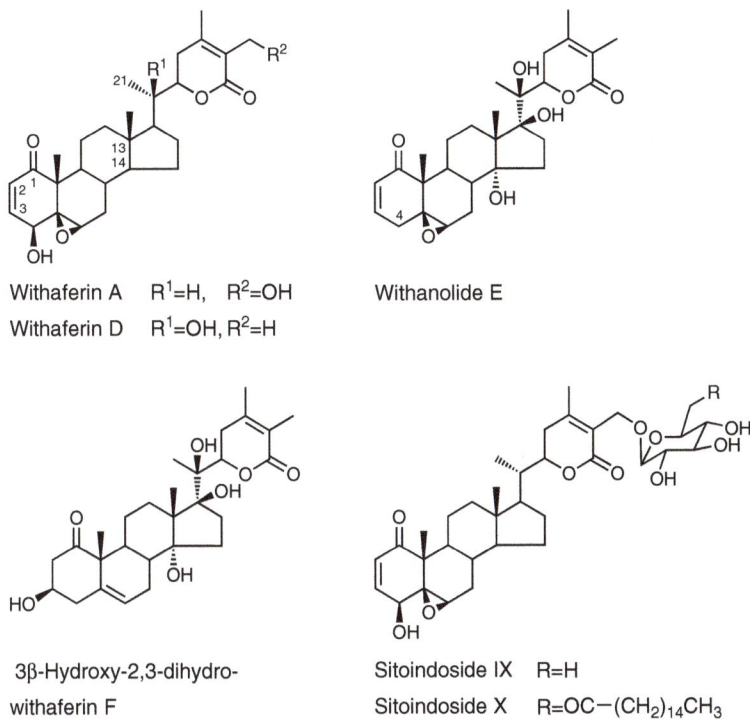

Fig. 8.9: Withanolides of *Withania somnifera*.

(often separated as a separate species *Ph. franchetii* Mast.), which is common in Japan, Korea, and northern China.

Some Physalis species, e.g., *Ph. angulata* L., balloon cherry, *Ph. peruviana* L., Cape gooseberry (Ph. 8.22), *Ph. philadelphica* Lam., tomatillo, and *Ph. pruinosa* L., strawberry tomato, yield fruits edible raw or cooked. *Ph. virginiana* Mill., Virginia groundcherry, is edible only cooked. In Central Europe, these species are rarely cultivated.

The bitter compound of *Alkekengi officinarum* was first isolated in 1852 by Dessaigne and Chantard [99]. Structural elucidation of physalins A, B, and C was performed by Matsura and coworkers in 1969 [282].

To date, about 350 withanolides and their derivatives have been identified in Physalis species [84], including, in addition to already known compounds such as withaferin A, the so-called physalins, withaphysalins, ixocarpalactones, physalolactones, and withaperuvins (Figs. 8.10 and 8.11 [1, 10, 85]).

From *A. officinarum*, the physalins A to C, E, F, L to OF, were obtained [84, 202, 203, 204, 205]. Approximately 40 withanolides were isolated from *Physalis peruviana*, including physalin A, withanolide E, and the withaperuvins B to H [7, 29, 84, 115, 241, 310, 319, 156]. *Ph. philadelphica* contains ixocarpolide, withaphysacarpin, and ixocarpalactone, among others [84].

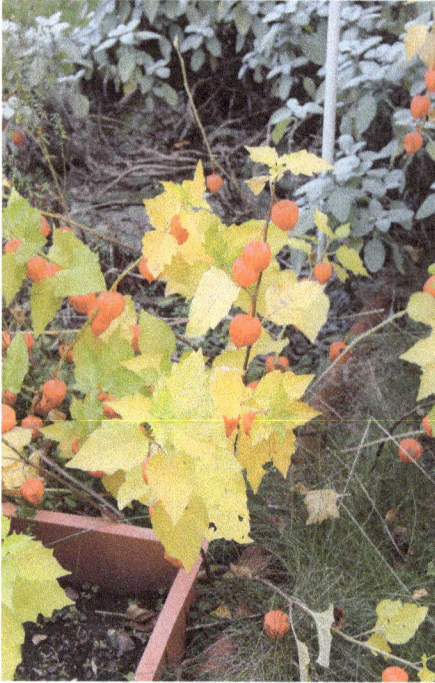

Ph. 8.21: *Alkekengi officinarum* (source: Eberhard Teuscher).

Ph. 8.22: *Physalis peruviana* (source: Peter Schönfelder).

In Physalis species, the withanolides occur in the green parts of the plants and in the roots. Constituents of the unripe fruit worth mentioning are physanols A and B (3β-benzoyloxy-6-oxo-stigmasta-7,20-diene and -stigmast-7-ene)-11α-ol, respectively

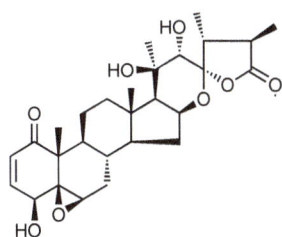

Ixocarpalactone B

Fig. 8.10: Ixocarpalactone B, a withanolide of Physalis species.

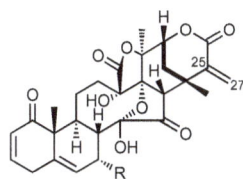

Physalin A  R=OH

Physalin C  R=H

Physalin O  R=OH

25,27 Dihydro

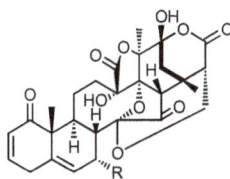

Physalin B  R=H

Physalin E  R=OH

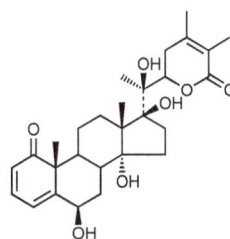

Withaperuvin C

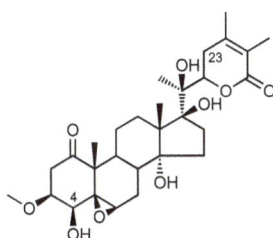

Physalactone

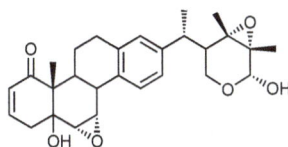

Nicandrenone

Fig. 8.11: Withanolides of Physalis and Nicandra species.

[375]). The withanolides are degraded when the berries ripen, but the insecticidal
4β-hydroxywithanolide E remains in the calyx to protect the fruit [39].

A variety of tropane alkaloids and their precursors have also been detected in the rhizome of Physalis species, among them tigloidine (3β-tigloyloxytropane), 3β-acetoxytropane,
physoperuvin, tropin, hygrin, and cuskhygrin [35, 355].

*Nicandra physalodes* (L.) Gaertn., erroneously also *N. physaloides*, apple of Peru
(Solanaceae, Ph. 8.23), is native to Peru but occurs worldwide. It contains about 20
withanolides (nicphysatones, nicandrenones (with aromatic ring D, Fig. 8.11), nicaphy-
sialines [444, 454, 456]) and the tropane derivatives tropinone and hygrin. This is used

Ph. 8.23: *Nicandra physalodes* (source: Tom Meaker/iStock/Getty Images Plus).

Ph. 8.24: *Lycium barbarum* (source: VladimirFLoyd/iStock/Getty Images Plus).

as an ornamental plant. Cultivated in vegetable beds, it reduces the infestation of vegetables with insect pests.

**Lycium barbarum** L. (*L. halimifolium* Mill.), boxthorn, matrimony vine, Solanaceae (Ph. 8.24), is native probably in China. The shrub is now widespread in all southern countries through overgrowth from cultivation. The fruits can be eaten raw or cooked. No symptoms of poisoning occurred after ingestion of 15–30 raw berries of *L. barbarum*, *L. europaeum* L., European wolfberry, or *L. chinense* Mill., Chinese wolfberry.

In the older literature, the occurrence of withanolides in Lycium species has been described [157, 158]. Nothing was known about their structure. In later investigations, neither withanolides nor steroid glycoalkaloids were found [217]. The occurrence of toxic tropane alkaloids in Lycium species, which is frequently postulated in the literature, does not correspond to the facts [49].

The minabeolides found in the soft coral Minabea species also belong to the withanolides.

## 8.4.2 Pharmacology, Current and Possible Therapeutic Use, and Toxicology

Withanolides have numerous pharmacological effects. So far, representatives with adaptogenic, antidiabetic, antineoplastic, anti-inflammatory, hepatoprotective, cardio-protective, immunomodulatory, antiviral, antibacterial, antiprotozoal, and insecti-cidal activities have been found [91, 100, 110, 156, 342, 354, 383, 430].

For more than 3,000 years, the roots, more rarely the leaves and fruits, of *Withania somnifera* are used medicinally under the name Ashwagandha (Sanskrit, means smell of the horse), English Indian ginseng, in Asia. Ashwagandha is a component of about 200 medicines of Ayurveda and mainly used as a geriatric and aphrodisiac, but also to treat hypertension, stress, diabetes, asthma, and cancer [326]. The bitter leaves serve as anthelmintic and antipyretic. Extracts of leaves and fruits are used externally for the treatment of skin tumors, boils, and ulcers. Furthermore, *W. somnifera* serves as an 'all-rounder', similar to ginseng, and is used worldwide in folk medicine and as a dietary supplement, as an adaptogen, tonic, roborans, for muscle building, as an agent to enhance sexual potency, as an antirheumatic, antitussive, sedative, hypnotic, antianxiety, antihyperglycemic, diuretic, or antiseptic agent and for the treatment of hypothyroidism [295].

In rats, extracts of *W. somnifera* protected against myonecrosis of the heart induced by an overdose of isoprenaline. The maximal cardioprotective effect was achieved by administration of 50 mg/kg BW, p.o., for 4 weeks [299]. Antitumor effects were demonstrated and are thought to be due to colchicine-like inhibition of mitosis at the metaphase stage or to induction of detoxifying phase II enzymes, e.g., quinone oxidoreductase [296].

Aqueous extracts from the berries of *Physalis alkekengi* are used as contraceptives in Iran. In China, Mexico, Indonesia, Peru, and Brazil, *Ph. angulata* is used for malaria, dermatitis, tracheitis, rheumatism, and hepatitis [156, 357]. The fruits and lampion-like sheath goblets of *Ph. alkekengi* var. *franchetii* are used in China to treat respiratory mucus, cough, pharyngitis, and pemphigus. The calyxes of *Ph. philadelphica* are used for diabetes. In Unani medicine, Physalis species are considered liver cleansing, diuretic, antiseptic, and sedative [156].

Goji berries, the fruits from Lycium species, are considered adaptogens, tonics, geriatrics, and immunostimulants.

Withaferin A and withanolide E are the best studied withanolides from a pharmacological point of view. Withaferin A seems to be mainly responsible for the observed antitumor effects. It inhibits the growth of various experimentally induced tumors in animals, e.g., in mice sarcoma 180 and Ehrlich ascites carcinoma. In estrogen-responsive and nonresponsive breast cancer cells, it led to apoptosis. Resistance to subsequently reintroduced tumor cells was developed via the formation of cytotoxic antibodies [110, 156, 354, 386, 396, 423, 445]. Development of potential antitumor drugs containing with-aferin A is in progress [154].

The immunosuppressive and anti-inflammatory effects of withaferin A include inhibition of mitogen-stimulated lymphocyte proliferation, graft-versus-host reaction (immune response directed by the graft against the host) in chicken, and adjuvant arthritis in rats [30, 136, 377]. They may be due to the structural similarity of withanolides to glucocorticoids [156].

Recent investigations indicate a potential benefit of withanolides in the treatment of COVID-19 infections. Withanolides bind to the spike protein (S-) of SARS-CoV-2, thereby hindering the access of the viruses into the host cells. They impede the activities of SARS-CoV-2 protease (Mpro) and alleviate the cytokine storm caused by COVID-19 infections by reducing the excretion of various procytokines [100, 101, 402].

Noteworthy is also the insecticidal action of withanolides caused by antagonism against the molting or pupal hormone ecdysterone. This gives the plants a selection advantage [103].

Withanolides from *W. coagulans* (Stock) Dunal have antifungal, antidiabetic, hypolipidemic, neuroprotective, anti-inflammatory, and wound healing activities [383].

Physalines have a spectrum of activities similar to that of the withanolides [202]. Nicandrenones from *Nicandra physalodes* show cytotoxic activity in vitro [241]. Sitoindosides have immunostimulatory and central stimulatory activity (antistress effect and improvement of learning and memory performance) [143].

The toxicity of the withanolides is low. The $LD_{50}$ of extracts from the roots of *Withania somnifera* was not reached in rats at 2 g/kg BW [323]. The $LD_{50}$ of withaferin A is 400 mg/kg BW, i.p., in mice [136]. The $LD_{50}$ of physalin X is 2 g/kg BW, p.o., and 1 g/kg BW, i.p., rats. It has abortive effects [298]. I.v. injections of extracts of berries of Physalis species resulted in a reduction in litter size in rats despite unchanged implantation rates [418].

(☗) Herb and roots of the mentioned plants are considered slightly toxic. Despite the frequent use of the roots and leaves of *Withania somnifera* as a medicine, no cases of acute (!) poisoning have been described in the literature. Nausea and diarrhea occasionally occurred as mild symptoms of poisoning after medicinal use. Cases of hepatotoxicity have been reported with chronic use (more than 2–12 weeks). Symptoms include nausea, cholestasis, jaundice, pruritus, and lethargy.

Presumably, ripe berries of the plants, including those of *Nicandra physalodes*, are nontoxic, despite their name 'poison berries'. The ripe berries of Physalis and Lycium species were consumed without symptoms.

The berries of *Alkekengi officinarum* nevertheless frequently give rise to toxicological consultation cases. However, serious symptoms of poisoning, apart from diarrhea and colic in young children, have not been observed even after ingestion of the immature berries or the calyx sheaths [75].

🐐 Animal poisonings thought to be caused by the herb of *Nicandra physalodes* [86] could not be clearly attributed to the plant. Feeding trials with sheep and goats yielded negative results [123].

## 8.5 Toxic Steroid Glycosides of South African Milk Star Species (Ornithogalum Species)

The toxicity of some South African Ornithogalum species, Asparagaceae, which are also invaded in Western Australia, is probably not, or not solely, due to the cardenolide content as in other toxic Ornithogalum species (see Section 8.3.8.3) but rather to the content of other toxic steroid glycosides, e.g., prasinoside G (Fig. 8.12), and cytotoxic ornithosaponins A–D [245, 294, **62**].

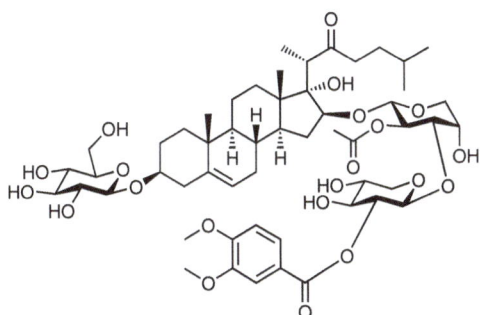

Prasinoside G

**Fig. 8.12:** Prasinoside G, a toxic steroid glycoside of South African Ornithogalum species.

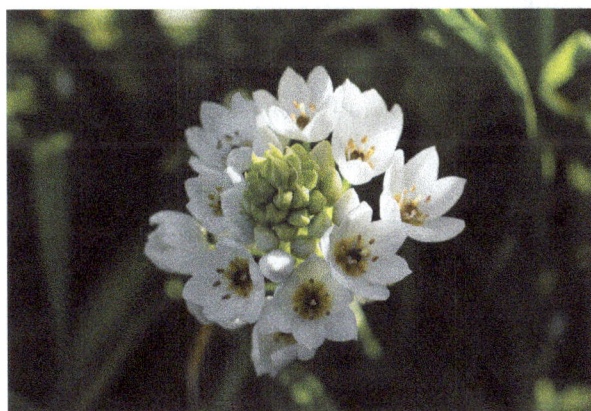

**Ph. 8.25:** *Ornithogalum thyrsoides* (source: ChrisDoAl/iStock/Getty Images Plus).

Of particular interest among the South African species is *Ornithogalum thyrsoides* Jaqu., Cape milk star, chincherinchee, wonder flower (Ph. 8.25), a bulbous plant, which also serves as an indoor plant. It is also cultivated, grown from imported bulbs, as an annual in gardens or as a cold-housing plant for obtaining cut flowers that keep for several weeks in the vase. Imports of the flower stems from South Africa take place.

🜪 Poisoning by South African Ornithogalum species results in violent diarrhea that can last for several weeks. In severe cases, death occurs within a few days with convulsions.

✻ External contact with *O. thyrsoides* causes contact dermatitis in sensitive individuals.

🐾 Severe poisoning by *O. thyrsoides* has been described in horses. Symptoms of poisonings were gastrointestinal irritation and convulsions. Temporary or permanent blindness also occurred in grazing cattle [144]. Poisonings of sheep and goats by *O. nanodes*, F. M. Leighton and *Albuca toxicaria* (C. Archer and R. H. Archer) J. C. Manning and Goldblatt (*O. toxicarium* (C. Archer et R. H. Archer)) are reported from South Africa. In one season, 3,000 animals were killed by *A. toxicaria*. The animals eat it when other feed is scarce. Death of sheep usually occurs several days after ingestion of the green plants [32, 302, **159**].

## 8.6 Petuniasterones and Petuniolides as Active Ingredients of Petunia Species

Petuniasterones and petuniolides are ingredients of the members of the genus Petunia, Solanaceae. The about 35 Petunia species are herbs with handsome flowers native to southern Brazil and Argentina. Crossing *Petunia axillaris* (Lam.) Britton, Sterns and Poggenb. and *P. integrifolia* (Hook.) Schinz and Thell., more than 100 years of breeding efforts have resulted in a high number of varieties of the garden Petunia, *P. hybrida* E. Vilm. (*P.* × *atkinsiana* D. Don, Ph. 8.26), which are among our most popular ornamental plants on the balcony.

The studies of *P. hybrida* were prompted by its resistance to insect feeding and the immunosuppressive effect of its extracts [116, 160]. They led to the discovery of the petuniasterones in 1988 by Elliger and coworkers.

Forty petuniasterones are currently known (Fig. 8.13). They are ergostane derivatives with a high degree of oxidation. In many of them, ring A has a dienone structure with a carbonyl group in position 3. On the side chain, many representatives have an acetate or a [(methylthio) -carbonyl]acetate residue, orthoester-like attached to the hydroxy groups at C-22, C-24, and C-25, or an acetate or *O*-methyl-malonyl residue esterlike attached. The orthoesters may arise from the 24,25-epoxy-22-acyloxy compounds, which are also found in Petunia species along with 22,25-epoxy compounds. Epoxy groups also occur on the sterane basic body. C-7 carries a permanent hydroxy group,

Ph. 8.26: *Petunia hybrida* (source: Ulrike Lindequist).

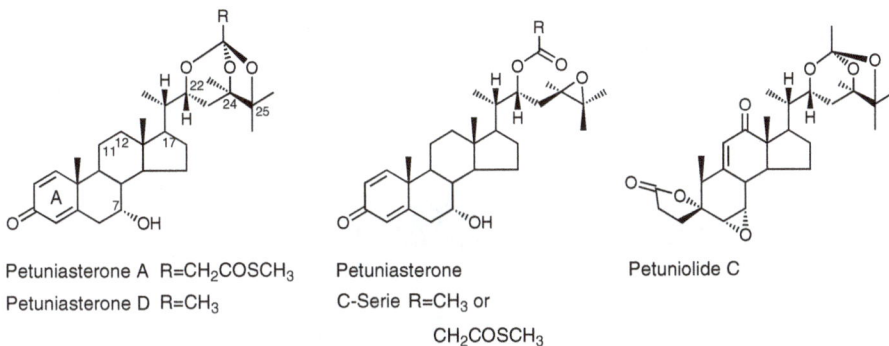

Petuniasterone A  R=CH$_2$COSCH$_3$
Petuniasterone D  R=CH$_3$

Petuniasterone
C-Serie  R=CH$_3$ or
         CH$_2$COSCH$_3$

Petuniolide C

Fig. 8.13: Steroids of Petunia species.

which may be esterified with acetic acid. Additional hydroxy groups may occur at positions 11, 12 and/or 17 [116, 120, 160]. Petuniolides arise from petuniasterones by rearrangement of ring A with loss of a C atom [118]. Petuniasterones and petuniolides have also been detected in other Petunia species [116, 118, 119, 120, 121].

Petuniasterones from the leaves are transported into their glandular hairs (trichomes) by the plant pleiotropic drug resistance (PDR)-type ABC transporter (PhPDR2), a transporter for secondary metabolites [79, 362].

The inhibitory effect of petuniasterones and petuniolides on insect growth and reproduction is more pronounced for petuniolides than for petuniasterones. Essential for the effect is the orthoester grouping [116, 118, 120]. In nymphs of the desert locust *Schistocerca gregaria*, petuniasterone C was shown to be a convulsant toxin that, like picrotoxinin (see Section 4.2.6), binds to proteins coupled with the chloride channel of the ionotropic GABA receptor and blocks the channel. This impairs presynaptic inhibitory mechanisms and induces convulsions [178].

In tests for antiviral activity, steroid mixtures from *P. hybrida* caused strong inhibition of the replication of influenza virus types A and B and of parainfluenza viruses. Some petuniasterones, e.g., petuniasterone D-diacetate and 30-hydroxypetuniasterone A, inhibit mitogen-stimulated proliferation of human lymphocytes at noncytotoxic concentrations [160].

(⚥?) Nothing is known about poisoning of humans or higher animals by Petunia species, probably because there is no incentive to eat the sticky, unpleasant-smelling leaves and stems of the plant.

## 8.7 Pregnane and *seco*-Pregnane Glycosides of Swallow Wort Species (Vincetoxicum and Cynanchum Species)

*Seco*-pregnane glycosides are the active ingredients of *Vincetoxicum hirundinaria* Medik., white swallow wort (Ph. 8.27), and other Vincetoxicum species or the related Cynanchum species (Ph. 8.28), Asclepiadaceae. The plants have articulated lactiferous tubes in all parts [155].

Ph. 8.27: *Vincetoxicum hirundinaria* (source: Wolfgang Kiefer).

Ph. 8.28: *Cynanchum acutum* (source: Peter Schönfelder).

*Vincetoxicum hirundinaria*, of which 11 subspecies exist, is a perennial, 30–150 cm tall plant, that, when crowded by other plants, continues to grow as a climbing plant. The fruit is a 5–7 cm long bellows capsule. The seeds have a white hairy head up to 18 mm long. The lime-loving plant occurs scattered in Europe in sunny, dry locations. Some species are cultivated as garden plants, e.g., *V. nigrum* (L.) Moench, black swallow wort, with red flowers.

The *seco*-pregnane derivatives of white swallow wort are 15-oxapregnane glycosides, which means C-15 is replaced by an O-atom (Fig. 8.14). It is likely that they are formed by oxidative opening of ring D between C-14 and C-15. They are abundant in Vincetoxicum and Cynanchum species [424].

The 15-oxapregnane glycosides occur in all parts of the white swallow wort, but they are preferentially present in the underground organs (up to 6% of DW). From the mixture of glycosides of the rhizome and roots known as vincetoxin, asclepiadin, cynanchin, or asclepion, the aglycones hirundigenin, anhydrohirundigenin, vincetoxigenin, and the glycoside hirundoside A (Fig. 8.14), presumably a secondary glycoside, were detected. Sugar components are, among others, D-cymarose, D-digitoxose, D-boivinose, and D-thevetose [126, 398, 399].

Compounds in which ring C (8,14-*seco*-pregnane derivatives) or rings C and D are open (13,14:14,15-di-*seco*-pregnane derivatives) and a nine-membered lactone ring is formed have also been isolated. To them belong the paniculatumosides A and B (Fig. 8.15) from *C. paniculatum* (Bunge) Kitag. ex H. Hara. The plant is native to China and is used in TCM for rheumatic pain and other diseases. 18-Nor-14,15-*seco*-pregnane derivatives also occur [225, 257, 307].

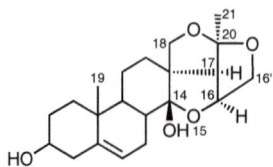

Hirundigenin

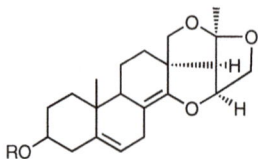

Anhydrohirundigenin  R=H

Hirundoside A(R=D-Oleandropyranosyl)

**Fig. 8.14:** Oxapregnane glycosides of *Vincetoxicum hirundinaria.*

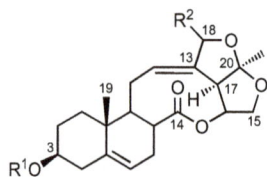

R¹= $\quad$ HO $\quad$ R²=β-OH

β-D-Oleandrose

R¹= $\quad$ HO $\quad$ R²=α-OH

β-D-Oleandrose

Paniculatumosides A and B

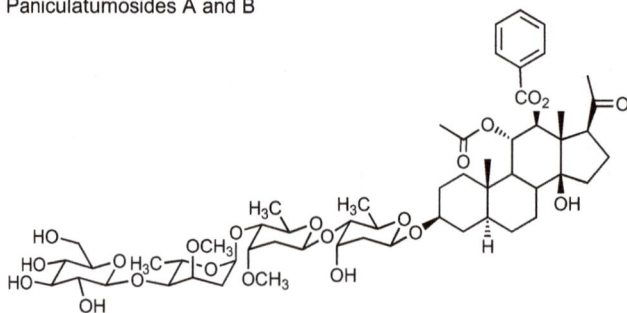

Cynafoside B

**Fig. 8.15:** Pregnane glycosides of Cynanchum species.

14,15-*seco*-Pregnane derivatives (the cynantrosides A–O) and 13,14, 14,15-di-*seco*-pregnane derivatives (the atroglaucosides A–C) were isolated from *C. atratum* Bunge. The plant is found in northern China and Japan. Its rhizome and roots are used in TCM against infections, and inflammatory skin diseases [457]. The cynafosides A and B (LD 10 mg/kg BW) and the pregnane glycosides B–H are components of the pregnane glycoside mixture of *C. africanum* (L.) Hoffmanns [397]. Toxic pregnane and *seco*-pregnane glycosides, e.g., sarcovimisides A–C, occur also in *C. viminale* (L.), caustic bush, sacred soma, a climbing plant, native to tropical Africa [421].

Other compounds are cytostatically active indolizidine alkaloids of the dibenzo[f, g]pyrrolo[1,2-b]isoquinoline type, e.g., tylophorine [432] or vincetene, a representative of the benzopyrroloisoquinoline type [69].

A variety of these pregnane and *seco*-pregnane glycosides have cytotoxic and immunosuppressive effects and promote apoptosis of tumor cells.

♟ According to older data, extracts from the seeds of white swallow-wort cause elevation of blood pressure, contraction of smooth muscles, ascending paralysis, and inhibition of diuresis [331]. Poisonings by ingestion of the rhizome are said to result in salivation, nausea, vomiting, diarrhea, brief agitation, and death by central paralysis [**53, 138**].

To our knowledge, cases of poisoning have not been described in recent years. Since the folk medicinal use, e.g., as emetic, diuretic, and diaphoretic, has fallen into oblivion, and the aerial parts of the plant offer no incentive for consumption, poisonings are not to be expected. The roots of *Vincetoxicum hirundinaria* occurred as impurities of the drug Primulae radix [278].

🐎 Cattle and sheep avoid the swallow wort. Horses eat it only after exposure to frost. Animal poisonings by these plants have been observed only rarely in Europe. The poisoning of a cat is described as an isolated case [167]. In Africa, poisonings of grazing animals by *Cynanchum africanum, C. obtusifolium* L. f., and *C. viminale* occur frequently. The poisonings are called cynanchosis ('klimop poisoning', also considered a form of 'krimpsiekte'). It is characterized by stilt-like walk, tremors, convulsions, and later partial paralysis that may last for days to weeks. If large amounts of the plants are ingested, death of the animals occurs. Similar poisonings occur in Australia and are caused there by *Cynanchum viminale* ssp. *australe* (R.Br.) (*Sarcostemma australe* R. Br.) [**159**].

## 8.8 1α,25-Dihydroxy-Cholecalciferol as Active Ingredient of Plants

1α,25-Dihydroxy-cholecalciferol (calcitriol, 1,25-$(OH)_2$-vitamin-$D_3$, Fig. 8.16) is the most active component of vitamin D group (a group of lipophilic secosteroids regulating the calcium balance of the body of animals and humans).

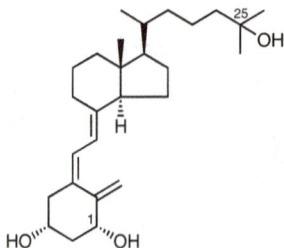

1α,25-Dihydroxycholecalciferol     Fig. 8.16: Active form of vitamin D₃.

In the first step of its biosynthesis in humans, 7-dehydrocholesterol, diverted from the cholesterol biosynthesis, goes to the skin, where its sterane ring is broken with the help of UVB radiation of the sunlight, leading to cholecalciferol (vitamin D₃). Then by hydroxylation in position 25 in the liver and subsequently in position 1 in the renal tubuli calcitriol arises. It was first identified in 1971 by M. F. Hollick. Calcitriol, transported to the cell nucleus, alters the transcription of some genes. That leads to enhancement of the intestinal and renal resorption of calcium and the mobilization of calcium from the bones.

Calcitriol is used therapeutically, e.g., in renal diseases associated with low blood calcium concentrations. The required daily dose is between 0.25 and 0.5 µg daily for an adult with a body weight of 50 kg.

> ☡ Calcitriol overdose in humans leads to hypercalcemia, characterized by states of weakness, head-ache, constipation, abdominal pain, vomiting, polyuria, and pruritus.

1α,25-Dihydroxycholecalciferol has been detected also in a number of plants, in addition to toxicologically safe levels of vitamin D₃ and 25-hydroxycholecalciferol. It is there glycosidically bound.

1α,25-Dihydroxycholecalciferol derivatives were found in relatively high concentration in young plants of some Poaceae species (sweet grasses), e.g., *Trisetum flavescens* (L.) P. Beauv., yellow oat grass, occurring in Central Europe in mountain fat meadows, and *Stenotaphrum secundatum* (Walter) Kuntze, buffalo grass, St. Augustine grass (Ph. 8.29), native to South Africa and cultivated in many tropical and subtropical areas.

Numerous Solanaceae also contain 1α,25-dihydroxycholecalciferol glycosides, e.g., the hammer shrub, *Cestrum diurnum* L. (Ph. 8.30), a tree, 1–10 m tall, thriving in the Antilles and in Florida, *C. laevigatum* Schltdl., widespread in Brazil, *C. parqui* L'Her., native to Chile, and *Nierembergia veitchii* Hook., white cupflower, a creeping plant, native to central and western parts of Argentina. The white cupflower is naturalized as a popular ornamental shrub in southern Europe. To the Solanum species that contain 1α,25-dihydroxycholecalciferol glycosides belong *S. glaucophyllum* Desf. (*S. malacoxylon* Sendtner), waxy leave nightshade, a shrub or little tree, and *S. erianthum*

Ph. 8.29: *Stenotaphrum secundatum* (source: Arousa/iStock/Getty Images Plus).

Ph. 8.30: *Cestrum diurnum* (source: Hemjaa/iStock/Getty Images Plus).

D. Don (*S. verbascifolium* L.), big eggplant, potato tree (Ph. 8.31), common in southeast Asia, northern Australia, and tropical America. Standardized extracts of *S. glaucophyllum* are used therapeutically.

To assess the content of 1α,25-dihydroxycholecalciferol, the vitamin D activity in hydrolyzed extracts of some plant species was determined. For *Trisetum flavescens*, it

Ph. 8.31: *Solanum erianthum* (source: kunphel/iStock/Getty Images Plus).

corresponded to 300 µg vitamin D/kg, for *Nierembergia veitchii* to 410 µg/kg, for *Cestrum diurnum* to 1,580 µg/kg, and for *Solanum glaucophyllum* to 2,070 µg/kg vitamin D [6, 20, 53, 70, 134, 161, 200, 275, 286, 308, 327, 338, 385].

   To determine the toxic dose, 1α,25-dihydroxycholecalciferol was added to the feed of piglets in increasing doses. Up to a dose of 10 µg/kg, the development of animals was normal. Doses of 20 µg/kg onward led to calcification of the soft tissues [6].

🐎 Animal poisonings, e.g., of cattle, horses, pigs, sheep, rabbits, and feathered cattle, by ingestion of the mentioned plants on pasture or in hay were frequently described. The animals show manifestations similar to that of vitamin $D_3$ hypervitaminosis. The calcium content of the blood is increased, and the magnesium and phosphate content are decreased. Symptoms of poisoning are disturbances in coordination of movements, inability to stand, weight loss, circulatory and respiratory disorders, skin changes (epidermis atrophy, regression of hair follicles, and fat and sweat glands), reduced milk production, and convulsions until death occurs. About 30 g/kg BW of the fresh herbage of *Cestrum diurnum* caused death of cattle within 48 h. Histopathological studies showed calcification of the heart, arteries, some tendons, ligaments, kidneys, and sometimes lung tissue, atrophy of parathyroid glands, necrosis in the liver, and hemorrhages in the intestinal tract. When 50 mg of dried leaves of *S. glaucophyllum* was fed for 180 days, hyperostosis occurred. The damage was irreversible [6, 36, 47, 61, 90, 94, 105, 109, 122, 144, 161, 200, 264, 285, 286, 304, 316, 327, 336, 339, 344, 368, 413, 442].

Because calcinosis also occurred in the offspring when *S. glaucophyllum* was fed to pregnant rats or rabbits, 1α,25-dihydroxycholecalciferol is thought to cross the placental barrier [6, 150].

In Argentina, poisoning of grazing animals by *Solanum glaucophyllum* is relatively common. Calcinosis caused by *Trisetum flavescens* has also been observed in the foothills of the Alps in Germany and Austria. *Cestrum laevigatum* was the cause of grazing animal poisoning in Brazil [47, 105, 327, 338]. Calcinosis was also caused by *Stenotaphrum secundatum* [20], and by *Nierembergia veitchii* [286, 388].

Feeding trials have shown that the addition of *Cestrum diurnum* leaves to poultry diets can significantly increase eggshell thickness and thus prevent losses due to egg breakage [80].

## References

(for numbers in bold, see cross-chapter literature)

[1]    Abe F, Yamauchi T (1979) Chem Pharm Bull 27: 1604
[2]    Abe F et al. (1992) Phytochemistry 31(1): 251
[3]    Abe F et al. (1994) Phytochemistry 37(5): 1429
[4]    Abou-Dough AM (2002) Arch Pharm (Weinheim) 335 (6): 267
[5]    Ada SE et al. (2001) Am J Clin Med 29(3/4): 525
[6]    Agraval AA et al. (2012) New Phytologist 194(1): 28
[7]    Akera T, Brody TM (1985) Trends Pharmacol Sci 6: 296
[8]    Akera T (1986) In: Cardiac Glycosides Eds Erdmann E, et al. Steinkopf-Verlag, Darmstadt, p. 19
[9]    Alfonso HA et al. (1994) Vet Hum Toxicol 36(1): 47
[10]   Ali A et al. (1984) J Nat Prod 47(4): 648
[11]   Amarelle L et al. (2019) Am J Physiol Lung Cell Mol Physiol 316(6): L1094
[12]   Anandhi D et al. (2018) J Postgrad Med 64(2): 123
[13]   Anderson LAP et al. (1983a) Onderstepoort J Vet Res 50: 295
[14]   Anderson LAP et al. (1983b) Onderstepoort J Vet Res 50: 301
[15]   Angarskaja MA et al. (1966) Farmakologija I Toksikologija 29: 39
[16]   Anonym (1954) Pharm Ztg: 510
[17]   Ansford AJ, Morris H (1981) Med J Austr 1 (7): 360
[18]   Antman EM, Smith TW (1985) Ann Rev Med 36: 357
[19]   ERD A et al. (2019) Biomed Pharmacother 113: 108721
[20]   Arnold RM, Fincham ICH (1997) Trop Anim Health Prod 29 (3): 174
[21]   Askari A (2019) Pharmacol Res Perspect 7 (4): e00505
[22]   Aslani MR, Rezakhani A (2000) Int J Trop Agr 18 (2): 18
[23]   Aslani MR et al. (2004) Vet Res Commun 28(7): 609
[24]   Aswal S et al. (2019) Medicina (Kaunas) 55(6): 255
[25]   Atkinson KJ et al. (2008) J Vet Emer Critic Care 18(4): 399
[26]   Jsa A-U-R et al. (1991) Phytochemistry 30: 3824
[27]   Babulova A et al. (1963) Arzneim Forsch 13: 412
[28]   Bachmann M (1960) Z Kreislaufforsch 49: 982
[29]   Bagchi A et al. (1984) Phytochemistry 23(4): 853
[30]   Bähr V, Hänsel R (1982) Planta Med 44 (1): 32
[31]   Bahri El L et al. (2000) Vet Hum Toxicol 42(2): 108, Comment in 42(3): 108
[32]   Bain RJI (1985) Br Med J 290 (6482): 1624
[33]   Bandara V et al. (2010) Toxicon 56(3): 273

[34]   Banikol H, Hofmann W (1973) Tierärztl Umsch 28: 612

[35]   Basey K, Wolley JG (1973) Phytochemistry 12 (10): 2557

[36]   Basudde CDK (1982) Poult Sci 61: 1001

[37]   Bateman DN (2004) Toxicol Rev 23 (3): 135

[38]   Bauer P, Franz G (1985) Planta Med 51 (3): 202

[39]   Baumann TW, Meier CM (1993) Phytochemistry 33 (2): 317

[40]   Baumgarten G (1969) Pharmazie 24: 778

[41]   Bavunoğlu I et al. (2016) Balkan Med J 33(5): 559

[42]   Begum S et al. (1993) J Nat Prod 56(4): 613

[43]   Begum S et al. (1999) Phytochemistry 50(3): 435

[44]   Bekkouche E et al. (2001) Phytochemistry 58(3): 455

[45]   Belokon VF, Makarevich IF (1980) Khim Prir Soedin: 424

[46]   Belz GG (2001) Eur J Clin Invest (Suppl 2): 10

[47]   Benesch C, Steng G (1999) Tierärztl Praxis Ausg G 272 83

[48]   Bishay DW et al. (1973) Phytochemistry 12: 693

[49]   Bleier W et al. (1967) Pharm Acta Helv 42: 423

[50]   Block KI et al. (2014) Cancer Res 74(19 Suppl): 4658

[51]   Bogdan L et al. (1990) Dtsch Tierärztl Wochenschr 97(12): 525

[52]   Bogs U (1947) Pharmazie 2: 408

[53]   Boland RL (1990) Biomed Environ Sci 1 (4): 414

[54]   Bors G et al. (1971) Pharmazie 26: 764

[55]   Botha C (2016) Molecules 21 (3): 348

[56]   Botha CJ et al. (1997) Onderststepoort J Vet Res 64(3): 189

[57]   Botha CJ et al. (1998) Onderstepoort J Vet Res 65(1): 17

[58]   Bozorgi M et al. (2017) J Trad Chin Med 37(1): 124

[59]   Brain C, Fox VEB (1994) J S Afr Vet Assoc 65 (4): 173

[60]   Brandes G, Suchowski G (1954) Ärztl Wochenschr 9: 134

[61]   Braun U et al. (2000) Vet Rec 146(6): 161

[62]   Braun V (1959) Münch Med Wochenschr 101: 1187

[63]   Brewster D (1986) Ann Trop Paediatr 6: 289

[64]   Bronstein AC et al. (2009) Clin Toxicol 47(10): 911

[65]   Brüning R, Wagner H (1978) Phytochemistry 17 (11): 1821

[66]   Brunner G et al. (2000) Lancet 356: 1406

[67]   Buchvarov Y et al. (1976) Farmatsia 26: 31

[68]   Budhiraja RD et al. (1986) Planta Med 52(1): 28

[69]   Budzikiewcz HP et al. (1979) Liebigs Ann Chem: 1212

[70]   Campero CM, Odriozola E (1990) Vet Hum Toxicol 32 (3): 238

[71]   Cao YL et al. (2018) Fitoterapia 127: 293

[72]   Capon RJ et al. (1986) Aust J Chem 39: 1711

[73]   Cardano S et al. (2002) Recenti Prog Med 93(4): 245

[74]   Carfora A et al. (2021) Forensic Sci Med Pathol 17(1): 120

[75]   Carnat A et al. (1998) Ann Pharm Fr 56(1): 36

[76]   Cassels BK (1985) J Etnopharmacol 14: 273

[77]   Ceci L et al. (2020) Toxins (Basel) 12(8): 471

[78]   Chapman JC, et al. (1977) Eur J Biochem 81(2): 293

[79]   Chen LX et al. (2011) Nat Prod Rep 28(4): 705

[80]   Chenianaiah S et al. (2004) J Steroid Biochem Mol Biol 89/90(1/5): 589

[81]   Cheung HTA et al. (1988) J Chem Soc, Perkin Trans 1: 1851

[82]   Chizzola R et al. (1998) Wiener Tierärztl Monatsschr 85(10): 359

[83]   Chkheidze NM et al. (1988) Izv Akad Nauk Gruz SSSR Ser Biol 14: 286
[84]   Christen P (1986) Pharm Acta Helv 61: 242
[85]   Christensen P (1989) Pharm Unserer Zeit 18 (5): 129
[86]   Cohen RDH (1970) Aust Vet J 46 (12): 599
[87]   Cook DR et al. (1990) Aust Vet J 67(9): 344
[88]   Cooper L et al. (1977) Morb Mortal Wkly Rep 26(32): 257
[89]   Corrigall W et al. (1978) Vet Rec 102: 119
[90]   Dämmrich K et al (1970) Dtsch Tierärztl Wochenschr 77: 342
[91]   Dar NJ et al. (2015) Cell Mol Life Sci 72(23): 4445
[92]   Dasgupta A, Datta P (2004) Ther Drug Monit 26 (6): 658
[93]   Dasgupta A, Emerson L (1998) Life Sci 63 (9): 781
[94]   De Barros S et al. (1981) Virchows Arch B Cell Pathol 35: 169
[95]   De Pinto F et al (1981) Clin Vet 104: 15
[96]   Dec GW (2003) Med Clin North Am 87 (2): 317
[97]   Deepak D et al. (1996) Fortschr Chemie Org Naturst 69: 71
[98]   Descoins C et al. (2002) Chem Pharm Bull 50(2): 199
[99]   Dessaigne V, Chantard J (1852) J Prakt Chem 55: 323
[100]  Dhanjal JK et al. (2021) Int J Biol Macromol 184: 297
[101]  Dhawan M, et.al (2021) J Appl Pharm Sci 11(4): 6
[102]  Dickstein ES, Kunkel FW (1980) Am J Med 69 (1): 167
[103]  Dinan L et al. (1996) Entomol Exp Appl 80: 415
[104]  DiPietro MA, Mondie CH (2021) J Am Coll Emerg Physicians Open 2(2): e12411
[105]  Dirksen G et al. (1974) Dtsch Tierärztl Wochenschr 81(1): 1
[106]  Dökert B (1979) Herzglykosidtherapie. Gustav Fischer Verlag, Jena
[107]  Driggers DA et al. (1989) West J Med 151(6): 660
[108]  Drogomir N et al. (1979) Timisoara Med 24: 17
[109]  Durand R et al. (1999) Vet Hum Toxicol 41(1): 26
[110]  Dutta R et al. (2019) Int J Mol Sci 20(21): 5310
[111]  Dwenger A (1973) Arzneim Forsch 23: 1439
[112]  Eddlestone M et al. (1999) Trop Med Int Health 4(4): 266
[113]  Eddlestone M et al. (2000) Heart (London) 83(3): 301
[114]  Eddlestone M et al. (2000) Lancet 355(9208): 967
[115]  Eguchi T et al. (1988) Chem Pharm Bull 36: 2897
[116]  Elliger CA et al. (1988) J Chem Soc, Perkin Trans 1: 711
[117]  Elliger CA et al. (1989) Phytochemistry 28(12): 3443
[118]  Elliger CA et al. (1990) J Chem Soc, Perkin Trans 1: 525
[119]  Elliger CA et al. (1992) J Chem Soc, Perkin Trans 1(1): 5
[120]  Elliger CA et al. (1992) J Nat Prod 55(1): 129
[121]  Elliger CA et al. (1993) Phytochemistry 33(2): 471
[122]  Esparza MS et al. (1983) Planta Med 47(1): 63
[123]  Everist SL (1981) Poisonous Plants of Australia, Angnus. Robertson Publishers, London
[124]  Fan L et al. (2020) J Ethnopharmacol 259: 112942
[125]  Farkhondeh T et al. (2020) Hum Exp Toxicol 39(6): 773
[126]  Fawcett JK et al. (1968) Tetrahedron Lett 9: 3799
[127]  Felenda JE et al. (2019) BMC Complement Altern Med 19: 105
[128]  Fiegel G (1954) Med Wochenschr 292
[129]  Fischer I (1962) Acta Med Leg Soc 4: 85
[130]  Flanagan RJ, Jones AL (2004) Drug Safety 27 (14): 1115
[131]  Fonseka MM et al. (2002) Hum Exp Toxicol 21(6): 293

[132] Forth W (1986) Klin Wochenschr 64: 96

[133] Fozzord HA, Sheets MF (1985) J Am Coll Cardiol 5: 10A

[134] Freitas CF et al. (2021) Parasitol Res 120(1): 321

[135] Fuerstenwerth H (2014) Am J Ther 21

[136] Fügner A (1973) Arzneim Forsch 23: 932

[137] Fukuyama Y et al. (1993) Phytochemistry 32(2): 297

[138] Fung SY (1986) Biochem Syst Ecol 14: 371

[139] Gabriel O, Leuten LV (1978) In: Biochemistry of Carbohydrates II Ed Manners DJ. Univ Park Press, Baltimore

[140] Gaillard Y et al. (2004) J Ethnopharmacol 95((2-3): 123

[141] García-Pérez P et al. (2020) Pharmaceuticals (Basel) !3(12): 444

[142] Ghannamy U et al. (1987) Planta Med 53(2): 172

[143] Ghosal S et al. (1989) Phytother Res 3: 201

[144] Gimeno EJ et al. (2000) J Vet Med A Physiol Pathol Clin Med 47(4): 201

[145] Glombitza KW et al. (1989) Planta Med 55(1): 107

[146] Goerre S, Fröhle P (1993) Schweiz Rundsch Med (Praxis) 82 (4): 121

[147] González-Stuart A, Rivera JO (2018) J Diet Suppl 15 (3): 352

[148] Goodwin TW (1980) The Biochemistry of Plants, vol. 4. Acad Press, New York, 485.

[149] Goodwin TW (1985) In: Steroles and Bile Acids, vol. 12. Elsevier, Amsterdam, 175.

[150] Gorniak SL et al. (2003) Reprod Toxicol 17(1): 67

[151] Greeff K (1981) Cardiac Steroid Glycosides. In: Handbuch der Experimentellen Pharmakologie vol. 56, I. II. Springer, Berlin

[152] Guerrero A et al. (2019) Nat Metab 1(11): 1074

[153] Gupta OP, Misra KC, Arora RB et al. (1974) Indian J Exp Biol 12: 399

[154] Hahm ER et al. (2020) Cancer Prev Res 13(9): 721

[155] Han L et al. (2018) Molecules 23(5): 1194

[156] Hang M et al. (2020) J Pharmacol 72(5): 649

[157] Hänsel R, Huang JT (1977) Arch Pharm (Weinheim) 310 (1): 35

[158] Hänsel R et al. (1975) Arch Pharm (Weinheim) 308(8): 653

[159] Harrlaß WD, Schöcke G (1987) Z Ges Inn Med 42: 282

[160] Hassan R et al. (1989) Pharmazie 44: 484

[161] Haussler MR et al. (1976) Life Sci 18: 1049

[162] Haynes BE et al. (1985) Ann Emerg Med 14: 350

[163] Heerden, van, FR et al. (1988) S Afr J Chem 41: 145

[164] Heller M (1990) Biochem Pharmacol 40: 919

[165] Hembree JA et al (1979) J Nat Prod 42(3): 293

[166] Hermkes L (1941) Münch Med Wochenschr 88: 1011

[167] Hess MO (2014) J Small Anim Pract 55 (7): 386

[168] Ho HT et al. (2011) J Physiol 589(19): 4697

[169] Holliman A, Milton D (1990) Vet Rec 127 (13): 339

[170] Holzinger F et al. (1992) FEBS Lett 314(3): 477

[171] Ho HT et al. (2011) J Physiol 589(19): 4697

[172] Huang M, et al. (2019) Molecules 24(15): 2749

[173] Hughes KJ et al. (2002) Aust Vet J 80(7): 412

[174] Hunziker H, Reichstein T (1945) Hel Chim Acta 28: 1472

[175] Huq MM et al. (1999) J Nat Prod 62(7): 1065

[176] Iizuka M et al. (2001) Chem Pharm Bull 49(3): 282

[177] Ikeda Y et al. (1995) J Nat Prod 58(6): 897

[178] Isman MB et al. (1997) Pestic Biochem Physiol 58(2): 103

[179]  Israil ZH et al. (2009) Pak J Pharm Sci 22(4): 425
[180]  Jacobs WA, Elderfield RC (1935) J Biol Chem 108: 497
[181]  Jaretzky R, Wilke M (1932) Arch Pharm 270: 81
[182]  Jaspersen-Schib R (1984) Schweiz Apoth Ztg 120: 1065
[183]  Jaspersen-Schib R (1987) Schweiz Apoth Ztg 125: 129
[184]  Jesse P et al. (2009) Pediatr Blood Cancer 52: 464
[185]  Jinbo Z et al. (2002) Phytochemistry 61(6): 699
[186]  Johnson CT, Routledge JK (1971) Vet Rec 89: 202
[187]  Joost A et al. (2010) www.kardio.hr/wp-content/uploads/2010/11/4-2010-77-81.pdf
[188]  Joubert JPJ, Schultz RA (1982) J S Afr Vet Assoc 53: 25
[189]  Junginger M, Wichtl M (1989) Planta Med 55 (1): 107
[190]  Junior P (1979) Phytochemistry 18 (12): 2053
[191]  Junior P, et al. (1985) Dtsch Apoth Ztg 125: 1945
[192]  Junior P, Wichtl M (1980) Phytochemistry 19 (10): 2193
[193]  Kakrani AL et al. (1981) Indian Heart J 33(1): 31
[194]  Kamernitsky AV, Reshetova IG (1977) Khim Prir Soedin: 156
[195]  Kanwal N et al. (2020) Food Chem Toxicol 143: 111570
[196]  Karás K et al. (2020) Biomed Pharmacother 127: 110106
[197]  Karawya MS et al. (1973) Planta Med 23(1): 70
[198]  Karthik G et al. (2020) J Family Med Prim Care 9(1): 136
[199]  Karunanidhi PS et al. (1997) Ind Vet J 74(11): 977
[200]  Kasali OB et al. (1977) Cornell Vet 67: 190
[201]  Katz AM (1985) J Am Coll Cardiol 5, Suppl A: 16 A
[202]  Kawai M et al. (1987) Phytochemistry 26(5): 3313
[203]  Kawai M et al. (1988) Bull Chem Soc Jpn 61: 2696
[204]  Kawai M et al. (1989) Chem Express 4: 97
[205]  Kawai M et al. (1992) Phytochemistry 31(12): 4299
[206]  Keenan SM et al. (2005) J Mol Graph Model 23(6): 465
[207]  Khajuria RK et al. (2004) J Sep Sci 27(7/8): 541
[208]  Kiliani H (1895) Arch Pharm 233: 311
[209]  Kirchner H (1978) Dissertation, Universität Wien
[210]  Kirchner H et al. (1981) Sci Pharm (Wien) 49: 281
[211]  Kirsch M (1990) Tierärztl Praxis 25 (4): 398
[212]  Kislichenko SG et al. (1967) Khim Prir Soedin: 193+241
[213]  Kislichenko SG et al. (1969) Khim Prir Soedin: 386
[214]  Kissmer B, Wichtl M (1986) Planta Med 52 (2): 152
[215]  Kitanaka S et al. (1996) Chem Pharm Bull (Tokyo) 44(3): 615
[216]  Ko SY et al. (2018) Int J Mol Sci 19(11): 3350
[217]  Kokotkiewicz K et al. (2017) Food Chem 221: 535
[218]  Kolodziejczyk-Czepas J, Stochmal A (2017) Phytochem Rev 16 (6): 1155
[219]  Komissarenko NF, Beleckij JN (1968) Khim Prir Soedin: 56
[220]  Komissarenko NF, Krivenchuk PE (1974) Khim Prir Soedin: 257
[221]  Komissarenko NF, Stupakova EP (1989) Rastit Resur 3: 453
[222]  Komissarenko NF (1969) Khim Prir Soedin: 141
[223]  Komissarenko NF et al. (1969) Khim Prir Soedin: 381
[224]  Komissarenko NF et al. (1969) Planta Med 17(2): 170
[225]  Konda Y et al. (1992) J Nat Prod 55(8): 1118
[226]  Kopp B (1983) Sci Pharm 51: 238
[227]  Kopp B, Dauner M (1983) Sci Pharm 51: 227

[228]  Kopp B, Kubelka W (1982) Planta Med 45 (8): 195

[229]  Kopp B et al. (1981) Sci Pharm 49: 265

[230]  Kopp B et al. (1984) Pharm Tijdschrift 3: 380

[231]  Kopp B et al. (1996) Phytochemistry 42(2): 513

[232]  Kowalewski Z (1963) Dissertat Pharm 15: 65

[233]  Kowalewski Z et al. (1978) Ann Pharm 13: 163

[234]  Krenn L, et al. (1988) Planta Med 54(3): 227

[235]  Krenn L et al. (1989) 37. Ann Congr Soc Med Plant Res, Braunschweig: P. 45

[236]  Krenn L et al. (1992) Planta Med 58(3): 284

[237]  Krenn L et al. (1994) Planta Med 60(1): 63

[238]  Krenn L et al. (2000) Fitoterapia 71(2): 126

[239]  Kretsu LG, Florya VN (1982) Iswesstija Akad Nauk Mold SSR Serija Biol Chim Nauk 2: 69

[240]  Kubelka W et al. (1977) Phytochemistry 16 (6):687

[241]  Kubwabo C et al. (1993) Planta Med 59(2): 161

[242]  Kumar A et al. (2013) Pharmacogn Rev 7(14): 131

[243]  Kumar VK, Lalitha KG (2017) Anc Sci Life 36 (3): 151

[244]  Kurlemann (1985) Monatsschr Kinderheilk 133: 495

[245]  Kuroda M, et al. (2006) Steroids 71(3): 199

[246]  Kyerematen G et al. (1985) Acta Pharm Suec 22: 37

[247]  Labeyrie E et al. (2003) J Chem Ecol 29(7): 1665

[248]  Lacassie E et al. (2000) J Forens Sci 45(5): 1154

[249]  Lai K et al. (1986) Arch Int Med 146: 1219

[250]  Langford SD, Boor PJ (1996) Toxicology 109 (1): 1

[251]  Lassen P (1989) Willdenowia 19: 49

[252]  Le Couteur DG, Fisher AA (2002) Clin Toxicol 40 (4): 523

[253]  Lehmann A et al. ((2021)) Glob Cardiol Sci Pract 2021: e202102

[254]  Lei ZH (2002) Chem Pharm Bull 50 (6): 861

[255]  Lei ZH et al. (2000) Chem Pharm Bull 48(2): 290

[256]  Lewis WH (1978) J Am Med Assoc 240 (2): 109

[257]  Li SL et al. (2004) J Nat Prod 67(1): 82

[258]  Li Y et al. (2016) Molecules 21(12): 1702

[259]  Löffelhardt W, Kopp B (1981) Phytochemistry 20 (6): 1219

[260]  Löffelhardt W et al. (1978) Phytochemistry 17(9): 1581

[261]  Longhurst JC, Jr RJ (1985) J Am Coll Cardiol 5: 99A

[262]  Luckner M, Diettrich B (1979) Pharmazie 34: 477

[263]  Luckner M, Wichtl M (2000) Digitalis – Geschichte, Biologie, Biochemie, Chemie, Physiologie, Molekularbiologie, Pharmakologie, Medizinische Anwendung, Wiss Verlagsges Stuttgart

[264]  Machado M et al. (2020) Toxicon 187: 1

[265]  Maharana PK (2021) J Biosci 46 (1):25.

[266]  Mahin L et al. (1984) Vet Hum Toxicol 26: 303

[267]  Maier MS et al. (1986) Phytochemistry 25(6): 1327

[268]  Majinda RRT et al. (1997) Planta Med 63(2): 188

[269]  Makarevich IF (1965) Khim Prir Soedin 160

[270]  Makarevich IF, Belokon VF (1975) Khim Prir Soedin 662

[271]  Makarevich IF et al. (1987) Khim Prir Soedin 119

[272]  Makarevich IF et al. (1989) Khim Prir Soedin 73

[273]  Mallick BK (1984) J Indian Med Assoc 82: 296

[274]  Manna SK et al. (2000) Cancer Res 60(14): 3838

[275]  Marck PV, Pierre SV (2018) Int J Mol Sci 19 (8): 2336

[276] Maringhini G et al. (2002) Ital Heart J 3(2): 137
[277] Maroyi A (2019) J Pharm Sci Res 11 (10): 3491
[278] Mason AB et al. (1987) Hum Toxicol 6: 251
[279] Masvingwe C, Masvenyengwa M (1997) J S Afr Vet Assoc 68 (1): 18
[280] Mathe A, Mathe I (1979) Herba Hungaria 18: 21
[281] Matsuda H et al. (2001) Bioorg Med Chem Lett 9(6): 1499
[282] Matsura T et al. (1969) Tetrahedron Lett 10: 1083
[283] McKenzie RA et al. (1987) Aust Vet J 64(10): 298
[284] McKenzie RA et al. (1998) Aust Vet J 66(11): 374
[285] McLennan MW, Kelly WR (1984) Aust Vet J 61 (9): 289
[286] Mello JRB, Habermehl GC (1998) Dtsch Tierärztl Wochenschr 105 (1): 25
[287] Melnikova TM, Voloshina DA (1981) Chim-Farm Shurnal 15: 69
[288] Mendonça FS, et al. (2018) Trop Anim Health Prod 50(3): 693
[289] Menezes RG et al. (2018) J Forens Legal Med 58: 113
[290] Menon MS et al. (2016) Heart Views 17(4): 136
[291] Meyer HP et al. (1993) Tijdschr Diergeneesk 118(13): 436
[292] Meyer K (1947) Helv Chim Acta 30: 1976
[293] Milbradt AG et al. (2003) Biochemistry 42 (17):2404
[294] Mimaki Y et al. (1996) Tetrahedron Lett 37: 1245
[295] Mirjalili MH et al. (2009) Molecules 14(7)
[296] Misico RI et al. (2002) J Nat Prod 65(5): 677
[297] Misico RI et al. (2011) Prog Chem Org Nat Prod 94: 127
[298] Mohana K et al. (1973) Indian J Exp Biol 17: 690
[299] Mohanty I et al. (2004) Basic Clin Pharmacol Toxicol 94: 184
[300] Moniuszko-Szajwaj B et al. (2016) Molecules 21(3): 243
[301] Mooradian AD (1988) Clin Pharmacokinet 15: 165
[302] Moore JH (2002) Thirteenth Australian Weed Conference. Perth
[303] Moore IA et al. (1954) Helv Chim Acta 37: 755
[304] Morris KML, Levack VM (1982) Life Sci 30: 1255
[305] Mrocik H (1959) Helv Chim Acta 42: 683
[306] Murard TMC et al. (1980) ref CA 93: 230921a
[307] Nakagawa T et al. (1983) Chem Pharm Bull 31(3): 870
[308] Negroni MS et al. (2019) Case Rep Cardiol 9707428
[309] Nell PW et al. (1987) Onderstepoort J Vet Res 54: 641
[310] Neogi P et al. (1986) Phytochemistry 26(1): 243
[311] Nesher M et al. (2007) Life Sci 80(23): 2093
[312] Neuwinger HD (2004) Toxicon 44 (4):417
[313] Newman RA (2020) Exp Pharmacol 12: 503
[314] Nogawa T et al. (2001) J Nat Prod 64(9): 1148
[315] Nordt SP et al. (2020) Am J Emer Med 38 (8): 1698. E5
[316] Odriozola ER et al. (2018) J Vet Diagn Invest 30(2): 286
[317] Oji O, Okafor QE (2000) Phytother Res 14 (2): 133
[318] Orrego F (1984) Gen Pharmacol 15: 273
[319] Oshima Y et al. (1989) J Chem Soc Chem Commun
[320] Osterloh J et al. (1982) J Am Med Assoc 247: 1596
[321] Paist WD et al. (1941) J Org Chem 6: 273
[322] Park GD et al. (1985) Drug Intell Clin Pharm 19: 931
[323] Patel SB (2016) J Ayurveda Integr Med 7 (1):30
[324] Pathare AV et al. (1987) J Postgrad Med 33: 216

[325] Paul G et al. (2017) J Thorac Dis 9(12): E1064

[326] Paul S et al. (2021) Biomed Pharmacother 143: 112175

[327] Peixoto PV et al. (2000) Vet Hum Toxicol 42(1): 13

[328] Peters U (1982) Eur Heart J 3 (Suppl D): 65

[329] Petricic J (1974) Acta Pharm Jugosl 24: 179

[330] Petricic J et al. (1977) Acta Pharm Jugosl 7: 127

[331] Pfeifer E (1954) Sci Pharm 22: 1

[332] Pietsch J et al. (2005) Int J Legal Med 119(4): 236

[333] Plante KS et al. (2021) Biomed Pharmacother !38: 111457

[334] Pohl T et al. (2001) Phytochemistry 58(4): 557

[335] Preira A (1949) Portugaliae Acta Biol 2: Ser

[336] Puche RC et al. (1980) Planta Med 40(12): 378

[337] Pusz A, Büchner S (1962/1963) Arzneim Forsch 12: 932 und 13: 409

[338] Rambeck WA, Zucker H (1982) Zentralbl Veterinärmed Reihe A 29: 289

[339] Rambeck WA et al. (1987) Z Naturforsch Sect C Biosci 42(4): 430

[340] Ramirez Ortega MDLC et al. (2002) Arch Cardiol Mex 72: Suppl, 1:S171

[341] Rao EV (1973) Indian J Pharm 35: 107

[342] Ray AB, Gupta M (1994) Prog Chem Org Nat 63: 1

[343] Reber H (1969). Dissertation, Universität München

[344] Reddy D et al. (2020) Molecules 25(16): 3596

[345] Ren Y et al. (2021) Molecules 26(12): 3672

[346] Repke KH, Schönfeld N (1984) Trends Pharm Sci 5: 393

[347] Reppas GP (1995) Aust Vet J 72 (11): 425

[348] Reuter G, Reichel R (1970) Zentralbl Pharm 109: 1241

[349] Rietbrock H et al. (1978) Dtsch Med Wochenschr 103: 1841

[350] Roberg M (1952) Pharmazie 7: 182

[351] Romano GA, Mombelli G (1990) Schweiz Med Wochenschr 120 (16): 596

[352] Rubini S et al. (2019) Toxins (Basel) 11(8): 442

[353] Sagawa T et al. (2002) Am J Physiol Heart Circ Physiol 282(3): 1118

[354] Saggam A et al. (2020) J Ethnopharmacol 255: 112759

[355] Sahai M, Ray AB (1980) J Org Chem 45 (16): 3265

[356] Saket K et al. (2020) Curr Drug Discov Technol 17(3): 318

[357] Saleem S et al. (2020) Iran J Basic Med Sci 23(12): 1501

[358] Samal KK et al. (1989) J Assoc Phys India 37(3): 232

[359] Sambasivam S et al. (2003) J Environ Biol 24(2): 201

[360] Saraswat DK, et al. (1992) J Assoc Phys India 40(9): 628

[361] Saravanapavananthan N, Ganeshamoorthy J (1988) Forensic Sci Int 36: 247

[362] Sasse J et al. (2016) Plant Cell Environ 39(12): 2725

[363] Schaller F, Kreis W (2006) Planta Med 72 (12): 1149

[364] Schaub C (1933) Dissertation, TH Braunschweig

[365] Schenk B et al. (1980) Planta Med 40(1): 1

[366] Schild CO et al. (2021) Toxicon 204: 21

[367] Schink M et al. (2015) J Ethnopharmacol 159: 129

[368] Schlegel P et al. (2017) Res Vet Sci 112: 119

[369] Scholtysik G et al. (1986) Cardiac Glycosides 1785–1985. Eds Erdmann E et al. Verlag Steinkopff, Darmstadt, p, p. 171

[370] Schoner W (2002) Eur J Biochem 269 (10): 2440

[371] Schrutka-Rechtenstamm R, et al. (1985) Planta Med 51(5): 387

[372] Schwartz WL, et al. (1974) Vet Pathol 11: 259

[373] Schwarz H et al. (1946) Pharm Acta Helv 21: 250
[374] Shang X et al. (2019) Front Pharmacol 10 Art 25. Doi: 10.3389/fphar.2019.00025
[375] Sharma NK et al. (1974) Phytochemistry 13(10): 2239
[376] Sheu JH et al. (1999) J Nat Prod 62(2): 224
[377] Shohat B et al. D (1978) Biomedicine 28: 18
[378] Shumiak GM et al. (1988) Ann Emerg Med 17: 732
[379] Siddiqui BS et al. (1997) J Nat Prod 60(6): 540
[380] Siddiqui S et al. (1986) Phytochemistry 26(1): 237
[381] Siddiqui S et al. (1987) Planta Med 53(1): 47
[382] Siddiqui S et al. (1987) Planta Med 53(5): 424
[383] Singh A et al. (2022) Molecules 27(3): 886
[384] Singh B, Rastogi RP (1970) Phytochemistry 9 (2): 315
[385] Siniorakis E et al. (2020) Ann Acad Bras Cienc 92(4)
[386] Sivasankarapillai VS et al. (2020) Environ Sci Pollut Res 27: 26025
[387] Škubnik J et al. (2021) Molecules 26(18): 5627
[388] Slifman NR et al. (1998) N Engl J Med 339(12): 806
[389] Smith PA et al. (2003) J Vet Int Med 17(1): 111
[390] Smith RA et al. (2000) Vet Hum Toxicol 42(6): 349
[391] Smith TW, Willerson JT (1971) Circulation 44: 29
[392] Smith TW et al. (1984) Prog Cardiovasc Dis 26: 413+495
[393] Smolarz A, Abshagen U (1985) Monatsschr Kinderheilkd 133: 682
[394] Sonne H, Haan D (1967) Arch Toxicol 22: 223
[395] Spengel S et al. (1967) Helv Chim Acta 50(7): 1893
[396] Stan SD et al. (2008) Cancer Res 68(18): 766
[397] Steyn P et al. (1989) South African J Chem 42(1): 29
[398] Stöckel K, Reichstein T (1969) Sci Pharm 37: 47
[399] Stöckel K et al. (1969) Helv Chim Acta 52(1175)
[400] Stoll A, Kreis W (1973) Helv Chim Acta 16: 1049
[401] Stoll A et al. (1949) Helv Chim Acta 32: 293
[402] Straughn AR, Kakar SS (2020) J Ovarian Res 13 (1): 79
[403] Supratman U et al. (2001) Biosci Biotech Biochem 65(4): 947
[404] Tamm C, Rosselet JP (1953) Helv Chim Acta 36: 1309
[405] Tittel G, Wagner H (1981) Planta Med 43 (11): 252
[406] Tittel G et al. (1982) Planta Med 45(8): 207
[407] Troster S et al. (1992) Dtsch Med Wochenschr 117: 1149
[408] Tschesche R (1934) Z Physiol Chem 229: 219
[409] Tschesche R, Buschauer G (1957) Liebigs Ann Chem 603: 59
[410] Tschesche R et al. (1955) Chem Ber 88: 1619
[411] Tuncok Y et al. (1995) Clin Toxicol 33(1): 83
[412] Urban G (1943/44) Samml von Vergiftungsfällen 13: A 935, 27
[413] Van der Lugt JJ et al. (1991) J Vet Res 58: 211
[414] Van Wyk AJ, Enslin PR (1969) J South Afr Chem Inst 22: 186
[415] Vartholomatos E et al. (2020) Anticancer Agents Med Chem 20(1): 103
[416] Vasina OE et al. (1986) Khim Prir Soedin: 263
[417] Verma RK et al. (1981) Indian J Anim Sci 51(15): 522
[418] Vessal M et al. (1991) J Ethnopharmacol 34(1): 69
[419] Viertel A et al. (2001) Dtsch Med Wochenschr 126(42): 1159
[420] Vitali G et al. (1996) Planta Med 62(3): 287
[421] Vleegaar R et al. (1993) J Chem Soc Perkin Trans 1: 4

[422] Voloshina DA, Melnikova TM (1978) Chim Farm Shurnal 12: 103
[423] Vyas AR, Singh SV (2014) AAPS J 16 (1): 1
[424] Wada K et al. (1982) Chem Pharm Bull 30: 3500
[425] Wagner H et al. (1970) Arzneim Forsch 20: 215
[426] Wagner H et al. (1985) Planta Med 51(2): 169
[427] Wagner H et al. (1985) Z Naturforsch B: Anorg Chem, Org Chem 40 B: 1226
[428] Wallick ET et al. (1986) Cardiac Glycosides 1785–1985. Verlag Steinkopff, Darmstadt, p. 27
[429] Walter U et al. (1993) Planta Med 59(7): A639
[430] Wang LQ et al. (2017) J Funct Foods 31: 229
[431] Weinhouse E et al. (1989) Life Science 44: 441
[432] Wiegrebe WE et al. (1969) Liebigs Ann Chem 721: 154
[433] William MC, Smith MC (1984) Am J Vet Res 45 (3): 543
[434] Williams M, Cassady JM (1976) J Pharm Sci 65: 912
[435] Windaus A, Freese C (1925) Ber Dtsch Chem Ges 58: 2503
[436] Windaus A, Stein G (1928) Ber Dtsch Chem Ges 61: 2436
[437] Winkler C (1986), Dissertation, Universität Marburg/Lahn
[438] Winkler C, Wichtl M (1985) Pharm Acta Helv 60: 243
[439] Winkler C, Wichtl M (1986) Planta Med 52 (1): 68
[440] Wissner W, Kating H (1974) Planta Med 26(7): 228, 26 (8):364
[441] Woods LW et al. (2004) Vet Pathol 41(3): 215
[442] Woodward JC et al. (1994) Bone 15(1): 1
[443] Wojtyna W, Enseleit F (2004) Pacing Clin Electrophysiol 27 (12): 1686
[444] Xiao Q et al. (2018) Steroids 131: 32
[445] Xu YM et al. (2017) J Nat Prod 80(7): 1981
[446] Yamagishi TE et al. (1989) J Nat Prod 52(5): 1071
[447] Yamagishi TE et al. (1988) Chem Pharm Bull 36: 1615
[448] Yamauchi TE, Abe F (1978) Tetrahedron Lett: 1825
[449] Yamauchi TE et al. (1975) Phytochemistry 14(5-6): 1379
[450] Yamauchi TE et al. (1976) Tetrahedron Lett 14: 1115
[451] Yamauchi TE et al. (1983) Phytochemistry 22(10): 2211
[452] Yamauchi TE et al. (1987) Chem Pharm Bull 35: 4993
[453] Yatsyuk VYE et al. (1983) Khim Prir Soedin 641
[454] Yu MY et al. (2017) Phytochemistry 137: 148
[455] Zhai J et al. (2022) Front Pharmacol 13: 822726
[456] Zhang L et al. (2019) Steroids 150: 108424
[457] Zhang Y et al. (2022) J Ethnopharmacol 284: 114748
[458] Zhao HM et al. (2019) Front Biosci 24(3): 576
[459] Zjawiong JK (2004) J Nat Prod 67 (2): 300
[460] Zoz IG, Chernykh NA (1969) Farm Shurnal 24: 77

# 9 Saponins

## 9.1 Chemistry and Biogenesis

Saponins are glycosides consisting of a lipophilic aglycone – a steroid or a polycyclic triterpene – linked to a hydrophilic residue – an oligosaccharide or an oligomer of monosaccharides and uronic acids.

Saponins are soluble in water, molecularly dispersed or colloidally. Due to their amphipathic nature, they exhibit surface activity. This means that they can reduce the surface tension of aqueous solutions. As a result, they have membrane activity either directly or after partial hydrolysis, i.e., they alter the properties of cell membranes, can form pores in the membranes, or can destroy them by attaching themselves to the membranes and intercalating or detaching lipophilic building blocks from the membrane.

The name saponin is based on the ability of their aqueous solution, like a solution of soaps, to form long-lasting foam when shaken (Latin sapo = soap).

According to the type of aglycones of the glycosides, called sapogenins, steroid saponins and triterpene saponins are distinguished. Steroid alkaloid glycosides and cucurbitaceous glycosides are also classified as saponins by some authors [13, 269].

Steroid sapogenins of plants are usually $C_{27}$ compounds with the spirostane ($16\beta,22:22,26$-diepoxy-cholestane) or furostane ($16\beta,22$-epoxy-cholestane-type base body, Fig. 9.1). ($25S$)-Spirostane-type saponins, so-called true saponins, ($25R$)-spirostane-type saponins, so-called isosaponins, ($25S$)-furostane-type saponins, and ($25R$)-furostane-type saponins occur. Rare sapogenins are derivatives of furanofurostane (spirofuran ($22R$)-$16\beta,22:22,25$-diepoxy-cholestane) and pyrostane (($22R$)-$22,26$-epoxy-cholestane). Cholestane derivatives without condensed $O$-heterocycles, e.g., cholestane-23-one derivatives, and pregnane derivatives, e.g., pregnane-20-one derivatives, occur as sapogenins too.

The spirostane- or furanofurostane derivatives are formally ketals of cholestane-22-one-16,26-diol or cholestane-22-one-16,25-diol resp. The furostane or pyrostane derivatives are half ketals of cholestane-22-one-16-ol or cholestane-26-one-22-ol resp. The rings A/B/C/D are linked *trans/trans/trans* or *cis/trans/trans* [97, 167, 180, 270, 279, 281].

Triterpene sapogenins (Fig. 9.2) are almost universally $C_{30}$ compounds. The most frequently occurring basic bodies of the triterpene saponins are the pentacyclic oleanane and the tetracyclic dammarane. In addition, derivatives of the likewise pentacyclic compounds ursane, lupane, and hopane frequently occur. Rarely occurring ring systems include cycloartenane, isolupane, lanostane, 27-nor-lanostane, 30-nor-lanostane, 30-nor-oleanane, 3,4-*seco*-lupane, 3,4-*seco*-oleanane, and 21,22-*seco*-onocerane. For oleanane and ursan derivatives, the rings A/B/C/D/E are *trans/trans/trans*, for lupan derivatives *trans/trans/trans/cis*, and for dammarane and holostane derivatives *trans/trans/trans* linked.

https://doi.org/10.1515/9783110724738-009

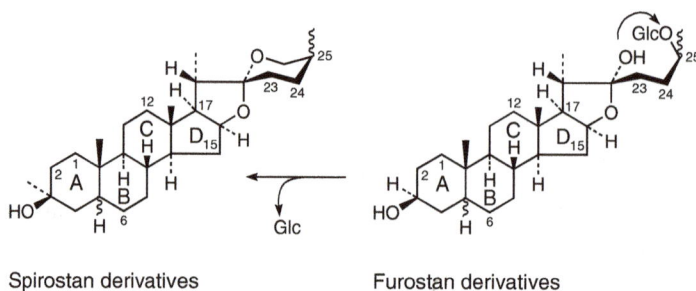

Spirostan derivatives    Furostan derivatives

| Name | Substituents at spirostan basic body | | | | | Double binding | Configuration at | |
|---|---|---|---|---|---|---|---|---|
| | 1β | 2α | 15β | 25 | Other | | C-5 | C-25 |
| Abamagenin | OH | | | CH$_3$ | 23 or 24CCl$_2$H | Δ$^5$ | – | R |
| Chlorogenin | | | | CH$_3$ | 6αOH | | α | R |
| Digitogenin | | OH | OH | CH$_3$ | | | α | R |
| Diosgenin | | | | CH$_3$ | | Δ$^5$ | – | R |
| Gitogenin | | OH | | CH$_3$ | | | α | R |
| Hecogenin | | | | CH$_3$ | 12=O | | α | R |
| Neoruscogenin | OH | | | =CH$_2$ | | Δ$^5$ | – | – |
| Pennogenin | | | | CH$_3$ | 17αOH | Δ$^5$ | | R |
| Sansevierigenin | OH | | | =CH$_2$ | 23βOH | Δ$^5$ | – | – |
| Sarsapogenin | | | | CH$_3$ | | | β | S |
| Smilagenin | | | | CH$_3$ | | | β | R |
| Tigogenin | | | | CH$_3$ | | | α | R |
| Yamogenin | | | | CH$_3$ | | Δ$^5$ | – | S |

Fig. 9.1: Steroid sapogenins.

The great diversity of triterpene sapogenins results from the hydroxylation pattern. An OH group is always found at C-3, OH groups frequently occur at C-16, C-21, and C-22 and less frequently at C-2 and C-15. The methyl groups C-23, C-24, C-28, C-29, and C-30 may be oxidized to CH$_2$OH or COOH groups, less frequently to CHO groups. Epoxy groupings, keto functions, and double bonds, the latter esp. between C-12 and C-13 occur. The hydroxy groups may be acylated (ester saponins). Acid components found include formic acid, acetic acid, n-butyric acid, iso-butyric acid, isovaleric acid, methylbutyric acid, angelic acid, tiglic acid, 3,3-dimethylacrylic acid, benzoic acid, cinnamic acid, ferulic acid, and rarely sulfuric acid [35, 100–102, 168, 180, 219, 270, 279, 293].

The oligosaccharide components of saponin molecules include: D-glucose, D-galactose, D-fructose, 3-methyl-D-glucose, D-xylose, L-arabinose (also present in furanoid form), L-rhamnose, L-fucose, D-glucomethylose (6-deoxy-D-glucose), D-apiose (present in furanoid form), and D-quinovose. D-Glucuronic acid and less commonly D-galacturonic acid also occur glycosidic bound in the oligosaccharide portion of the saponins.

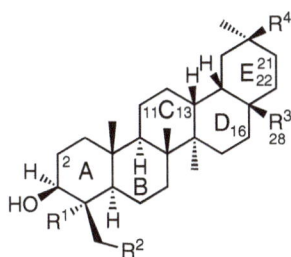

| Name | Substituents | | | | | | | | | | |
|------|:---:|:---:|:---:|:---:|:---:|:---:|:---:|:---:|:---:|:---:|:---:|
| | 2β | 11 | 13/28 | 16α | 21β | 22α | R¹ | R² | R³ | R⁴ | Δ¹² |
| **Neutral Triterpenes** | | | | | | | | | | | |
| Barringtogenol C | | | | OH | OH | OH | CH₃ | H | CH₂OH | CH₃ | + |
| Cyclamigenin A₂ | | | -O- | =O | | | CH₃ | H | CH₂- | CH₃ | |
| Cyclamigenin B | | | -O- | =O | | | CH₃ | H | CH₂- | CHO | |
| Cyclamiretin A | | | -O- | OH | | | CH₃ | H | CH₂- | CHO | |
| Protoaescigenin | | | | OH | OH | OH | CH₃ | OH | CH₂OH | CH₃ | + |
| **Triterpene monocarboxylic acids** | | | | | | | | | | | |
| Bayogenin | OH | | | | | | CH₂OH | H | COOH | CH₃ | + |
| Echinocystic acid | | | | OH | | | CH₃ | H | COOH | CH₃ | + |
| Glycyrrhetinic acid | | =O | | | | | CH₃ | H | CH₃ | COOH | + |
| Gypsogenin | | | | | | | CHO | H | COOH | CH₃ | + |
| Hederagenin | | | | | | | CH₂OH | H | COOH | CH₃ | + |
| Oleanolic acid | | | | | | | CH₃ | H | COOH | CH₃ | + |
| Quillajic acid | | | | OH | | | CHO | H | COOH | CH₃ | + |
| **Triterpene dicarboxylic acids** | | | | | | | | | | | |
| Acinosolic acid | OH | | | | | | CH₃ | H | COOH | CH₃ | + |
| Esculentic acid | | | | | | | CH₂OH | H | COOH | CH₃ | + |
| Jagilonic acid | OH | | | | | | CH₂OH | H | COOH | CH₃ | + |
| Spergulagenic acid | | | | | | | CH₃ | H | COOH | CH₃ | + |

Fig. 9.2: Triterpene sapogenins.

Some of the saponins isolated may be artifacts formed from the genuine saponins by hydrolytic cleavage of monosaccharide residues or acyl groups, by migration of acyl groups, or by splitting off water to form lactone or epoxy groups during the course of isolation.

To exert membrane activity, acidic saponins must bear at least three monosaccharide residues, neutral saponins at least two. To date, saponins with a maximum of 12 monosaccharide residues per molecule have been detected.

In contrast to the cardioactive steroid glycosides, saponins mostly carry branched oligosaccharide chains that are often terminated by pentoses. According to the number of sugar chains directly attached to the sapogenin, a distinction is made between monodesmosides (with a single chain), bisdesmosides (with two chains), and trisdesmosides (with three chains).

A sugar chain is almost always attached to position C-3. In the furostan-type steroid saponins, the hydroxyl group at C-26 also carries a glucose residue. Only in a few cases another sugar residue or a methyl group was found there. After splitting off this glucose residue, the furostane derivative (hemiketal) spontaneously changes into a spirostane derivative (ketal) (Fig. 9.1). In some cases, this ring closure fails to occur because of the absence or substitution of the hydroxyl group required for ring closure at position C-22. In the triterpene saponins, a second sugar chain is often acyl-glycosidic attached to a carboxyl group located at C-17.

The acidic character of some saponins is due to either a free carboxyl group on the sapogenin, one or more bound uronic acids, or acylation with dicarboxylic acids on the aglycone or sugar residues. In rare cases, sulfuric acid esters were also found.

The biogenesis of steroid sapogenins occurs starting from cholesterol, presumably beginning with hydroxylation of C atom 26 or 27 and linkage of the resulting hydroxyl group with glucose. Subsequently, hydroxylation occurs at C-16 and C-22 with subsequent formation of the furan ring under dehydrogenation [293]. The starting substance can also be cycloartenol. In this case, cycloartenol is probably first converted to cholesterol [77]. At which stage the sugar chain is attached to the C-3 is still unclear. Intermediate in the conversion of 3β-compounds to 3α-compounds is probably a 3-keto derivative and in the conversion of 5β- to 5α-compounds a 5-$\Delta^5$-sapogenin [260].

The biogenesis of triterpene sapogenins occurs by cyclization of squalene, initially to dammarane derivatives, which can be converted to ursane derivatives by ring openings and ring closures via lupane derivatives and oleanane derivatives [263].

The first steroidal saponin isolated was digitonin (Fig. 9.3), which Schmiedeberg obtained in 1875 [216] from commercial digitonin, a mixture of substances isolated from seeds of the red foxglove, *Digitalis purpurea*. The sapogenin from digitonin, digitogenin, was obtained by Kiliani in 1890 [140]. Its constitution was determined in 1939 by Marker and Rohrmann [170]. The final structural formula for the glycoside was established by Tschesche and Wulff in 1963 [268].

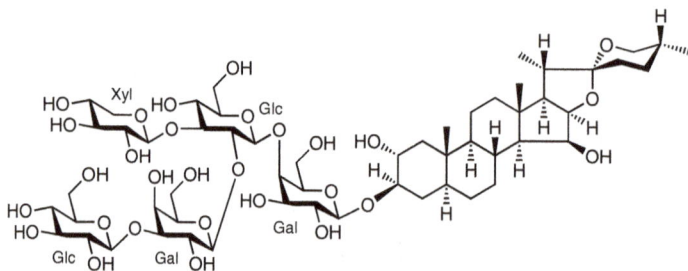

Digitonin

Fig. 9.3: Digitonin, a saponin of *Digitalis purpurea*.

The first triterpene saponin isolated was glycyrrhizic acid (Fig. 9.4). It was obtained from the root of the Spanish Licorice, *Glycyrrhiza glabra* L., as early as 1809 by

Robiquet [76]. The sapogenin of this compound, glycyrrhetinic acid, was obtained by Gorup-Besanez in 1861 [282]. Its structure was determined by Ruzicka and Jeger in 1942 [208], after the structural formula of the triterpene sapogenin oleanolic acid had already been established by Haworth in 1937 [93]. The structural elucidation of glycyrrhizic acid was completed in 1956 [164, 172].

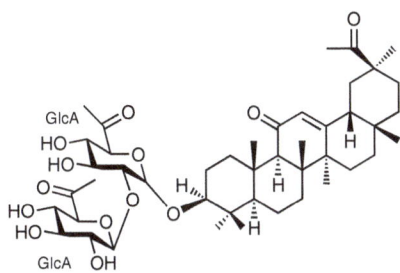

Glycyrrhizic acid

Fig. 9.4: Glycyrrhizic acid, a triterpene saponin of *Glycyrrhiza glabra*.

In more recent publications, especially by authors from the Asian region, the isolated substances usually are not given trivial names. Only the very complicated rational names or only structural formulae are published.

## 9.2 Distribution

Saponins are contained in more than 90 plant families comprising 500 genera. They are found preferentially in nutrient-rich tissues: roots, rhizomes, tubers, and seeds, but in leaves and flowers also. Many plant parts used as food and medicinal drugs contain saponins.

The distribution focus of steroid saponins is in monocotyledonous plants (monocots), namely in the families Amaryllidaceae, Araliaceae, Asparagaceae, Campanulaceae, Dioscoreaceae, Liliaceae, Melanthiaceae, and Smilacaceae. Steroid saponins are rarely found in dicotyledonous plants. There, they have been found in the Fabaceae, Solanaceae, and Scrophulariaceae, among others [281].

The distribution focus of triterpene saponins is in dicotyledonous plants (eudicots). About 60 families of this class contain triterpene saponins. They are particularly abundant in the families Amaranthaceae, Apiaceae, Araliaceae, Caryophyllaceae, Fabaceae, Primulaceae, Sapindaceae, Ranunculaceae, Sapotaceae, and Theaceae.

Completely free of saponins appear to be gymnosperms (Gymnospermae). In the Pteridophyta (vascular spore plants), they have been found in some genera.

The significance of saponins in living plants consists in their antimycotic [102, 187, 294], antiprotozoal [202], insecticidal [272], molluscicidal [105] properties, and in their mucous irritant effects in the mouth of predators. The very high concentrations of saponins in some storage organs, e.g., in some roots up to 30% of DW, suggest that saponins may also serve as energy stores.

The hebevinosides found in Hebeloma species (Basidiomycetes) are assigned to saponins by some authors [168, 219].

Steroid and triterpene derivatives with amphipathic character similar to saponins are also found in animals. The lipophilic part of these molecules is a steroid or a triterpene basic body. The polar part can be a monosaccharide or an oligosaccharide. Additionally, other polar groups at the parent body, e.g., sulfate groups, or coupling with amines are responsible for the lyobipolarity of the compounds.

These compounds have been found preferentially in marine animals, in sponges, cnidarians, corall like animals, echinoderms, hairy feather stars, sea stars, brittle stars, sea urchins, and sea cucumbers. Saponin-like substances have also been detected in the class of bony fishes and cartilaginous fishes and in some animals of the order of beetles [6].

## 9.3 Pharmacology

Saponins exert direct effects on cell membranes. After absorption of saponins they or their aglycons, the sapogenins, exert systemic effects by reacting with cell nuclei receptors. It is often unclear which mechanism caused an observed effect.

The **direct action of saponins** is exerted by their membrane activity. Due to their lyobipolarity they interact with sterols, phospholipids, and lipoproteins of cell membranes and form complexes with them. These complexes remain in the membranes or are released into the environment. Solutions with high concentrations of saponins can dissolve membranes (cytolysis [177, 296]). The direct effects are exerted on membranes attainable to saponins, e.g., on membranes of mucosa cells, isolated erythrocytes, epithelial cells of the gills of fishes, isolated cells in culture, spermatozoa, intestinal flora, and intestinal parasites. These effects are unspecific, triggered by all saponins and differ only in the strength of action.

An example of the pore-forming or membrane-destroying action of saponins is their hemolytic action, i.e., the damage of the membranes of erythrocytes resulting in leakage of hemoglobin by direct contact with saponins.

Saponins are able to react with membrane-integrated active components. Even in noncytolytic concentrations they can affect enzymes, carriers, ion channels, receptors, or transport processes [8, 80, 175].

Saponins can influence the absorption of substances. By solubilizing lipophilic substances and/or increasing the permeability of the membranes of the cells of the intestinal mucosa, they promote the absorption of compounds, including their own absorption, and, for example, that of nutrients and xenobiotics. They can also promote the absorption of drugs or toxins, e.g., cardioactive steroid glycosides, thereby enhancing the systemic effects of these substances. Saponins also facilitate the absorption of macromolecules, e.g., of the ribosome-inactivating protein (RIP) agrostin (see Section 9.13) or of receptor-specific protein chimeras used for therapy [149].

In this context it is worth mentioning that in Africa juices or extracts from saponin-containing plants, e.g., *Asparagus africanus* Lam., *A. exuvialis* Burchell, and *Dracaena aethiopica* (Thunb.) Byang et Christenh., are used as additives to arrow poisons to promote solubilization and/or absorption of the toxic substances [112].

The promotion of allergen absorption by saponins may increase the risk for sensitization to dietary proteins [166].

On the other hand, complex formation of saponins with substances may impede the availability of nutrients and trace elements, e.g., iron ions [128, 197, 245]. This may account for the growth retardation observed in animals due to intake of saponin-containing plants in their diet, e.g., Medicago species.

The formation of complexes of saponins with cholesterol in the intestine reduces the absorption of cholesterol. In addition, saponins stimulate the formation of bile acids from cholesterol, which are excreted in complex with cholesterol. Thus, saponins reduce blood cholesterol levels [171, 174, 203].

Worth mentioning are the stimulation of interferon release [1], the inhibition of platelet aggregation [13, 148], the immunomodulation [20, 38, 209], hepatoprotective effects [185, 209], and the promotion of expectoration (see Section 9.1.10). The antioxidant effects of saponins are due to their interaction with membrane lipids too [158].

**Effects of saponins on tumor cells** in culture are based, at least partly, on direct effects, causing alteration and destruction of cell membranes leading to inhibition of proliferation, apoptosis, and cytolysis. The effective concentrations of saponins ranged from 1 to 100 µmol/L culture media [120, 180]. Antitumoral activity postulated from in vitro results is only meaningful if its specificity for tumor cells is confirmed and is not only a case of common cytotoxicity. For some saponins, anticarcinogenic effects could be shown in several solid tumor models. For instance, triterpene saponins like ginsenosides and saikosaponins, and steroid saponins like dioscin, platycodin D, polyphyllin, and timosaponin were tested. The compounds lead to decrease in tumor volume, prevention of metastasis, and inhibition of tumor vascularization. These effects are possibly caused by influence on expression of specific genes leading to activation or inhibition of the production of specific proteins [125, 297].

Well-studied is the anticarcinogenic effect of platycodin D (Fig. 9.5), a triterpene saponin from *Platycodon grandiflorus* (Jaqc.) A. DC., ballon flower, Campanulaceae (Ph. 9.1). The plant occurs in northeast Asia and is a popular balcony and garden plant in temperate latitudes. Its root (Platycodi radix) is used in TCM to treat numerous diseases, including malignant tumors. Platycodin D leads to arrest of cell division, to apoptosis, and to autophagy in numerous tumor cell lines. Applied in vivo, it induces significant growth retardation of solid tumors. With a $LD_{50}$ of more than 2,000 mg/kg BW, mouse, p.o., applied for more than 14 days, the toxicity is very low [153]. The use of platycodin D for chemotherapy of tumors, combined with other chemotherapeutic agents, has been suggested [139].

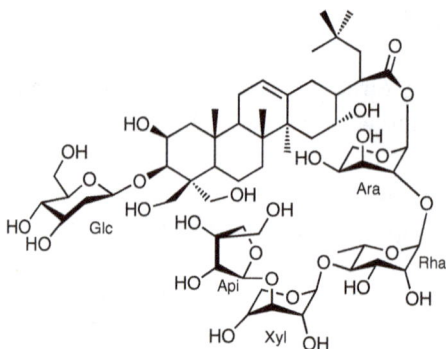

Platycodin D

**Fig. 9.5:** Platycodin D, a triterpene saponin of *Platycodon grandiflorus.*

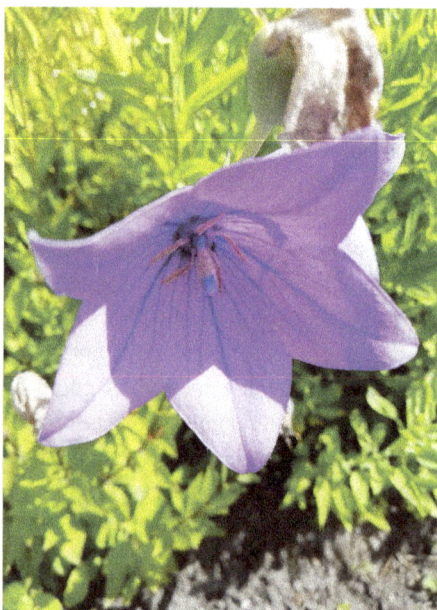

**Ph. 9.1:** *Platycodon grandiflorus* (source: Ulrike Lindequist).

Saponins of our nutrition are thought to be involved in prophylaxis of cancer and of degenerative aging processes, which result from permanent inflammation and oxidation processes in human bodies [113].

The **virostatic effect of saponins**, observed in vitro, is probably triggered by changes in the membrane properties of host cells. Saponins prevent the adsorption and penetration of viruses. However, because of the high concentrations required for this effect (>10 µmol/L), saponins are unlikely to be useful for the treatment of viral infections. But they seem to be suitable as adjuvants, for example, in COVID-19 vaccines [226]. The antiviral effect of some saponins could also be through stimulation of host immune defenses.

The **ichthyotoxicity of saponins** is the result of membrane disruption. External exposure of fish gills to saponins leads to an increase in the permeability of the gill epithelium. Small-molecule components are washed out of the animals' blood by the circulating water. The consequence is the death of the fishes. This enables the use of extracts from plants containing saponins as fish poisons [32].

The **systemic effects of saponins** are specific and essentially determined by the chemical structure of the sapogenin. They are based on the selective modulation of receptors in or on the cell nucleus by absorbed saponins or sapogenins. During absorption, the number of bound monosaccharide residues inversely correlates with the resorption rate, i.e., the fewer monosaccharide residues the lipophilic parent carries, the higher is the resorption rate.

The sapogenins or short-chain saponins modulate ligand-activated transcription factors controlled by nuclear receptors. These receptors include estrogen, glucocorticoid, mineralocorticoid, androgen receptors, and receptors, whose ligands are still unknown (orphan receptors [310]).

In **phytotherapy**, saponins and saponin-containing plants are used as expectorants, diuretics, antiexudatives, geriatrics, and dermatics.

The use of saponin-containing drugs as expectorants is mainly based on the mucosal irritant effect of saponins, which penetrate deep into the bronchial branches by spreading from the pharynx when applied perorally and on the promotion of secretion transport by their surface-active effect. The irritant effect in the upper part of the digestive tract may also lead to secretolysis by reflex promotion of the formation of thin mucus [287].

The diuretic effect of saponins is probably induced by affecting mineralocorticoid receptors [310]. The antiexsudative and anti-inflammatory effects of some saponins, e.g., those of licorice (see Section 9.16) and horse chestnut (see Section 9.11), are linked to the integrity of the adrenal cortex. They are presumably a consequence of a saponin-induced increase in tissue concentrations of adrenocortical hormones and an influence on glucocorticoid receptors.

Future fields of application of saponins may be their use in the therapy of obesity and diabetes. The antiadipose effect has been demonstrated, e.g., for high doses of glycyrrhizin, for fruits of *Momordica charantia* L., seeds of rice melon, *Chenopodium quinoa* Willd., flower buds of black tea, *Camellia chinensis* L. Kuntze, and even in low concentrations, for some isolated saponins, e.g., platytodine D, escins, soyasapogenol B$_7$, and dioscin. The antiadipose effect is due to, e.g., inhibition of pancreatic lipase and suppression of the appetite signal emanating from the hypothalamus [123, 171].

The antidiabetic effects of some saponins result from their influence on the dysregulated adipokine secretion via cell nuclear receptors [59, 60].

In pharmaceutical technology, saponins serve as detergents and emulsifiers [**149**].

Triterpene saponins have a great potential for the nonviral gene delivery [43]. Reviews on the effects of saponins: [98, 168, 219].

## 9.4 Toxicology

In our daily diet saponins are present in small amounts. The intake by humans averages 0.03 g daily. These amounts are harmless even when ingested continuously. On the contrary, they even have positive effects (see Section 9.3).

> ☠ After p.o. intake of larger amounts of saponin-containing plant parts, irritation of the mucous membranes of the digestive organs leads to nausea and vomiting, and in the case of very large amounts, which are absorbed partially, to pain during urinary bladder emptying and to kidney damage. Miosis has been observed. Symptoms following ingestion of large amounts of licorice extracts are a special case (see Section 9.16). Inhalation of saponin-containing dusts results in sneezing and coughing irritation. When saponin-containing dusts enter the eye, they may cause lacrimation and conjunctivitis.

Injection of saponins into the bloodstream causes severe symptoms of poisoning as a result of hemolysis [280]. Sinusoidal and endothelial cells of the liver and bone marrow are also damaged. Mice died within 6 h after i.v. administration of 0.1 mg of an unspecified saponin. Hemorrhages, clumping of the blood, and hepatic necrosis were observed [182].

> 🩹 Therapy of poisoning by saponins consists of toxin removal, e.g., by activated charcoal, symptomatic measures, and administration of copious fluids. In cases of hypokalemia, cautious administration of potassium chloride solution may be tried.

🐏 Some plants, e.g., *Narthecium ossifragum* (L.) Huds., bog asphodel (Ph. 9.2, see Section 6.2) and *N. asiaticum*, Maxim., Nartheciaceae, contain in the flowering tops saponins with the aglycon sarsapogenin. The annual plants are native to the Mediterranean region and widespread worldwide. They may cause kidney damage and hepatotoxic effects in grazing animals, mainly lambs, and lead to jaundice and, presumably due to their hepatotoxic effects, to secondary photosensitization [49]. In North-West Europe the photosensitization disease of lambs is known as 'plocheach', 'yellowses', 'saut', or 'alveld'. It is doubted that bog asphodel is the only cause of the disease. Toxins from fungi or cyanobacteria may also play a role [196].

🐏 Disease of sheep caused by *Tribulus terrestris* L., cat's head, Zygophyllaceae (Ph. 9.3, see Section 6.2), which has been observed preferentially in South Africa, is known as 'geeldikkop' [248, **159**]. Hepatogenous photosensitization caused by *T. terrestris* has also been observed in weaner sheep in New South Wales in Australia [76].

In addition to the steroid saponins (furostanol and spirostanol sapogenins), indole alkaloids occurring in *T. terrestris* are remarkable, e.g., harmine and norharman. The use of preparations from the plant as aphrodisiacs or muscle builders is strongly discouraged [**49**].

Poisonings of sheep by the signal or Surinam grass, *Brachiaria decumbens* Stapf, and its hybrids, Poaceae, native to tropical Africa, widespread in tropical and subtropical regions, are also due to the hepatotoxic action of saponins, esp. to protodioscin [55, 162].

Ph. 9.2: *Narthecium ossifragum* (source: AndyRoland/iStock/Getty Images Plus).

Ph. 9.3: *Tribulus terrestris* (source: Dackycards/iStock/Getty Images Plus).

## 9.5 Steroid Saponins of Herb Paris (*Paris quadrifolia*)

The genus Paris, Melanthiaceae, includes 33 species. *Paris quadrifolia* L., herb Paris, true lover's knot, occurs only in Central Europe. The other Paris species are native to western Asia, eastern Siberia, and the Kamchatka Peninsula. *P. polyphylla* Sm. thrives in the Himalayan region.

*Paris quadrifolia* (Ph. 9.4) is a perennial, growing to 40 cm tall, with an underground creeping basal axis. The flower stalk is glabrous and bears four leaves mostly arranged in a whorl to 10 cm long at the top and a terminal flower with light green petals. The fruit is a black-blue berry up to 1 cm thick. The plant occurs scattered in deciduous and mixed forests, in shrubbery and on moist shady rocks up to the upper tree line in almost all of Europe, Asia Minor, and Siberia.

Ph. 9.4: *Paris quadrifolia* (source: Petrovval/iStock/Getty Images Plus).

About 0.1% saponins have been isolated from the fresh herb of *P. quadrifolia* (about 1% of DW).

The main saponin is the pennogenin tetraglycoside, accompanied by further pennogenin glycosides, and 1-dehydrotrillenogenin (Fig. 9.6 [71, 184]). They have cytotoxic effects in vitro [71, 249].

Pennogenin tetraglycoside exhibits an antihypertensive effect at doses of 1–10 mg/kg BW, i.v., in mice, bullfrogs, and rabbits. Amplitude and tone of cardiac activity are increased [184]. It showed potent cardiotoxicity in isolated heart cells from neonatal rats [124].

In folk medicine, the fresh, crushed herb is used to treat wounds ('plague berry').

The rhizomes of *P. polyphylla* var. *yunnanensis* (Franch) Hand, Mazz, and *P. polyphylla* var. *chinensis* (Franch) H. Hara provide the drug Rhizoma Paridis, used in TCM.

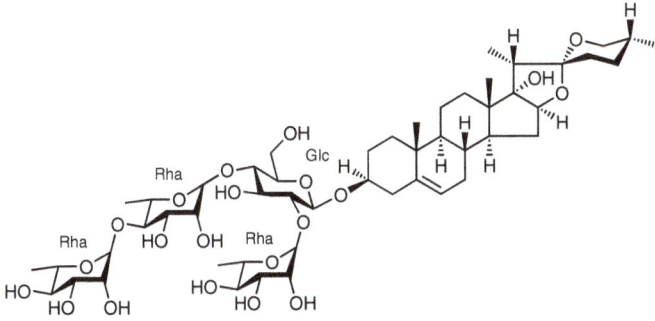

Pennogenintetraglycoside

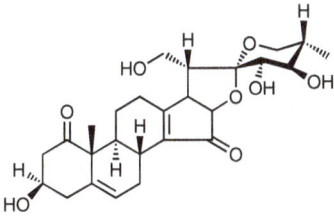

1-Dehydrotrillenogenin

Fig. 9.6: Steroid saponin and sapogenin of *Paris quadrifolia*.

Well over 100 saponins have been detected in them, including the diosgenin glycosides polyphyllin A and polyphyllin D [165, 199, 242, 316].

> ☠ Poisoning can result from confusion of the fruits of *P. quadrifolia* with blueberries. Nausea, dizziness, vomiting, headache, and miosis have been observed as symptoms of poisoning [**133**]. Deaths in humans after consumption of the fruits have not been described in recent decades.

🐕 Among animals, poultry are particularly at risk from fruits of *P. quadrifolia*.

## 9.6 Steroid Saponins of Asparagus Species

The genus Asparagus, Asparagaceae, includes about 300 species, native to low-rainfall areas of Europe, Asia, and Africa. In Central Europe are found *Asparagus officinalis* L., garden asparagus, and *A. tenuifolius* Lam., tender-leaved asparagus. Used as houseplants and cut plants to add to bouquets are *A. setaceus* (Kunth) Jessop (*A. plumosus* Baker), asparagus fern, and *A. densiflorus* (Kunth) Jessop 'Sprengeri', Sprenger's ornamental asparagus and less commonly other species.

The root of *A. officinalis* is used in traditional medicine for osteoporosis, cardiovascular diseases, rheumatism, diabetes, as galactogogum, and as diuretic. *A. officinalis* and *A. racemosus* Willd., Shatvari, are used in Ayurvedic medicine as 'rejuvenators' for women, *A. curillus* Buch.-Ham. ex Roxb., asparagus cyrillus, is used as a galactogogue and to induce abortion. The effects are due to the estrogenic effects of the saponins [183, 241]. *A. filicinus* Buch.-Ham. ex D. Don., fern asparagus, is used in East Asia as tranquilizer, sedative, and for its anti-inflammatory action [152] and in traditional Tibetan medicine as an expectorant and diuretic [111, 157].

Extracts of the roots and stems of *A. cochinchinensis* (Lour.). Merr., Chinese asparagus, known as tianmendong, are used in TCM and in Ayurveda medicine for treating lung and spleen-related diseases. Recent studies indicate anti-inflammatory and neuroprotective properties and antiproliferative effects on human tumor cells [41, 141, 160, 255, 309]. Twenty-seven steroid saponins have been isolated so far [160].

*Asparagus officinalis* (Ph. 9.5) is a dioecious perennial up to 1.5 m tall with a thick, woody basal axis. The stem bears small, scale-like leaflets with elongated branches or three to six needle-like green flat shoots, arranged in clusters in the axils. The fruits are about pea-sized, brick-red berries. The plant occurs scattered in Central Europe, Western Siberia, and the Caucasus region on sandy fields, grassy slopes, dunes, rubble areas, and hedgerows.

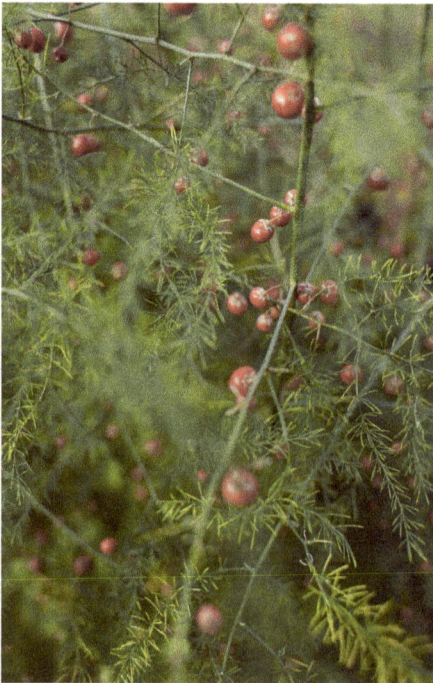

Ph. 9.5: *Asparagus officinalis* (source: Irina Pislar/iStock/Getty Images Plus).

Cultivars of the vegetable asparagus are grown to obtain the young stem shoots, sprouting from buried root pieces as vegetables. These are harvested either before they break through the mounds of mounded soil about 1 m high above the root pieces – white asparagus – or when they have reached a height of 20–30 cm above the ground-green asparagus [194]. The seeds of the vegetable asparagus were formerly used in a roasted state as a coffee substitute [223].

A lot of different saponins have been identified in *A. officinalis*. The spirostane derivatives sarsapogenin, diosgenin, its diastereomer yamogenin (neodiscogenin), and three furostane derivatives (Tab. 9.1, Fig. 9.7) were identified as sapogenins.

The main saponins of the plant are protodioscin (furostanol I), asparanin A, and yamogenin II (Fig. 9.7 [85, 194]). Asparagosides A, B, D, E (Fig. 9.8), F, G, H, and I and officinalisnin I and II, among others, could be isolated from the root, and asparaginosides I and II from the shoots. The bitter saponins asparasaponin I and officinalisnin I and II are found especially in the shoot heads of white asparagus eaten as a vegetable. The seeds contain glycosides of (25 *R*)-α-22-methoxyfurost-5-en-3β,26-diol and (25 *R*)-α-22-hydroxyfurost-5-en-3β,26-diol (Tab. 9.1 [81, 111, 135, 136, 151, 190, 194, 223]).

The saponins found in the underground organs of *A. tenuifolius* contain sarsapogenin as an aglycone. Two of the isolated compounds are identical to asparagoside C and D [189].

Of *A. setaceus*, native to South Africa, the structure of 3-spirostanol and 2-furostanol saponins has been elucidated. Aglycones are yamogenin, (25S)-furost-5-en-3β-22,26-triol, and (25S)-22a-methoxy-furost-5-en-3β,26-diol [211].

In the roots of *A. densiflorus*, native to South Africa (Natal), sprengerins A, B, C, and D occur. Their aglycone is diosgenin [230].

In addition to sarsapogenin, diosgenin, and yamogenin, in other Asparagus species the sapogenins pennogenin, ruscogenin, and the corresponding furostanol precursors have been detected. Worth mentioning is the occurrence of the above-mentioned 22-methoxyfurostanol and ((25R)-furosta-5,20-diene-3β,26-diol [159, 228]), both of which do not allow spiroketal formation after cleavage of the glucose residue bound at position 26. These sapogenins are also present in the fruits of the Asparagus species studied [159, 228–230, 232, 281].

As accompanying substances are noteworthy the amino acid L-asparagine and a number of sulfur-containing constituents, among them the nematicidal aspartic acid (1,2-dithiolane-4-carboxylic acid), its methyl and ethyl esters and *S*-oxides, 3-mercaptoisobutyric acid, *S*-(2-carboxy-3-mercaptopropyl)-L-cysteine, the disulfides of these compounds, and *N*-cysteinyl-L-proline (captopril). Some of the compounds are significantly involved in the odor and taste of asparagus [47, 119, 133, 134, 257, 266]. The characteristic odorant occurring in urine after consumption of vegetable asparagus is *S*-methyl-3 (methylthio)-thiopropionate [194, 289].

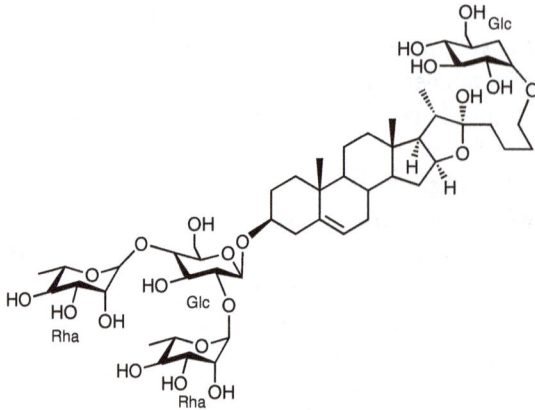

Protodioscin

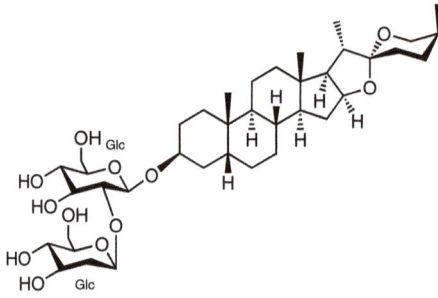

Asparanin A

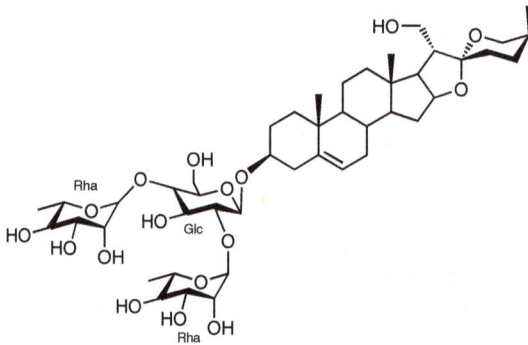

Yamogenin II

**Fig. 9.7:** Main saponins of *Asparagus officinalis*.

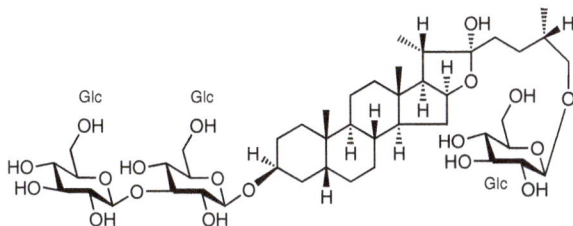

Asparagoside E

Fig. 9.8: Asparagoside E, a steroid saponin of *Asparagus officinalis*.

Tab. 9.1: Steroid saponins of *Asparagus officinalis*.

| Saponine | Structure |
|---|---|
| Asparagoside A | S-Sar-3←1βGlc |
| Asparagoside B | S-Sar-26←1βGlc |
| Asparagoside C | S-Sar-3←1βGlc3←1βGlc |
| Asparagoside D | S-Sar-3←βGlc(3←1βGlc)4←1βGlc |
| Asparagoside E | F-Sar $\begin{cases} -3\leftarrow 1\beta Glc3\leftarrow 1\beta Glc \\ -26\leftarrow 1\beta Glc \end{cases}$ |
| Asparagoside F | S-Sar-3←1βGlc(3←1βGlc)4←1βGlc←1βXyl |
| Asparagoside G | F-Sar $\begin{cases} -3\leftarrow 1\beta Glc(3\leftarrow 1\beta Glc)4\leftarrow 1\beta Glc \\ -26\leftarrow 1\beta Glc \end{cases}$ |
| Unnamed | S-Sar-3-Glc(2←1βGlc and 4←1αRha) |
| Asparasaponin I | F-Yam $\begin{cases} -3\leftarrow 1\beta Glc(2\leftarrow 1\alpha Rha\ and\ 4\leftarrow 1\alpha Rha) \\ -26\leftarrow 1\beta Glc \end{cases}$ |
| Asparasaponin II | F-Yam $\begin{cases} -3\leftarrow 1\beta Glc4\leftarrow 1\alpha Rha \\ -26\leftarrow 1\beta Glc \end{cases}$ |
| Officinalisnin I | F-Sar $\begin{cases} -3\leftarrow 1\beta Glc2\leftarrow 1\beta Glc \\ -26\leftarrow 1\beta Glc \end{cases}$ |
| Officinalisnin II | F-Sar $\begin{cases} -3\leftarrow 1\beta Glc(4\leftarrow 1\beta Xyl)2\leftarrow 1\beta Glc \\ -26\leftarrow 1\beta Glc \end{cases}$ |

Tab. 9.1 (continued)

| Saponine | Structure |
|---|---|
| Unnamed | F-Dios $\begin{cases} -3\leftarrow 1\alpha Rha \begin{cases} (1\rightarrow 2)\alpha Rha \\ (1\rightarrow 4)\beta Glc \end{cases} \\ -26\leftarrow 1\beta Glc \end{cases}$ |
| Unnamed | F-MDios $\begin{cases} -3\leftarrow 1\alpha Rha \begin{cases} (1\rightarrow 2)\alpha Rha \\ (1\rightarrow 4)\beta Glc \end{cases} \\ -26\leftarrow 1\beta Glc \end{cases}$ |

Abbreviations: S-Sar, sarsapogenin; F-Sar, (25S)-5α-furostan-3α,22α,26-triol; F-Yam, (25S)-furost-5-en-3α,22α,26-triol; F-Dios, (25 R)-22α-hydroxyfurost-5-en-3β, 26-diol; FMDios, (25 R)-22α-methoxyfurost-5-en-3β, 26-diol; Glc, D-glucose; Xyl, D-xylose; Rha, L-rhamnose

⚠ The berries of vegetable asparagus are considered toxic. While no symptoms of poisoning have been observed after ingestion of two to seven berries, gastrointestinal irritation occurred only after ingestion of larger quantities.

🔖 Ingestion of more than six asparagus berries should be treated by copious administration of fluids and activated charcoal, and for more than 10 berries by gastric emptying and administration of activated charcoal [161].

✱ Fresh asparagus may cause type I or type II allergic reactions. In persons who handle asparagus occupationally, such as workers in canneries, cooks, greengrocers, and asparagus growers, and rarely in amateur gardeners, allergic contact dermatitis develops after contact with young shoots, especially on the hands. The face may also be affected. In some cases, asthmatic complaints occurred [206, 65, 133]. So-called 'asparagus scabies' was observed in thousands of workers at a canning factory after World War I. One of the allergens that can be inactivated by cooking is the 1,2,3-trithiane-5-carboxylic acid contained in the plant at the beginning of the asparagus season [195, 200, 106]. In addition, so-called lipid transfer proteins, profilin and glycoproteins rich in asparagine, can trigger allergies [50, 256].

Saponins from edible spears of wild asparagus inhibit different signaling pathways and induce apoptosis in human colon cancer cells [120].

## 9.7 Steroid Saponins of Solomon's Seal Species (Polygonatum Species)

The genus Polygonatum, Asparagaceae, includes about 70 species that occur in the temperate zone of the northern hemisphere. In Central Europe are found *Polygonatum odoratum* (Mill.) Druce var. *odoratum* (*P. officinale* (L.) All.), lesser Solomon's seal, *P. multiflorum* (L.) All., many-flowered Solomon's seal, ladder-to-heaven, *P. verticillatum* (L.) All., whorled Solomon's seal, and *P. hirtum* (Poir.) Pursh, Crug's wide-leaf Solomon's seal.

Polygonatum species are perennial herbs with a thick, fleshy underground basal axis. The stem bears numerous alternate or whorled, broadly elliptic to lanceolate stem leaves. The cylindrical perianth consists of fused white petals and has a short greenish fringe. *P. verticillatum* has leaf whorls of six to seven leaves and occurs scattered in shady places, especially in forests, throughout Central Europe, the Caucasus region, northern Iran. The other species have alternate leaves. *P. multiflorum* has a round stem, and flowers are two to five in leaf axils. The plant is found in shady deciduous forests in Europe, the Caucasus and Himalayan regions, and Japan. In *P. hirtum* and *P. odoratum* (Ph. 9.6) the stems are angular. Main distribution area of *P. hirtum* is Southeastern Europe. *P. odoratum*, in contrast to the other species, prefers dry, sunny sites. It is widespread in all parts of Europe and from there reaches China and Korea via Siberia. In China, 32 species of Polygonatum are found.

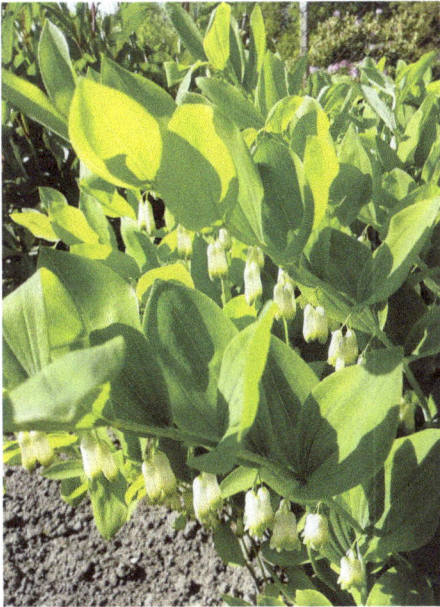

Ph. 9.6: *Polygonatum odoratum* (source: Ulrike Lindequist).

*P. multiflorum* and *P. odoratum* are cultivated as ornamental plants in gardens, along with the North American species *P. biflorum* (Walter) Elliot, two-flowered Solomon's seal.

Thirty-seven Polygonatum species, used in TCM, Japan, India, and Iran, have been intensively studied in recent years because of their potential antitumor, antiviral, and postulated antiaging effects [315]. They are used mainly as tonics for age-related complaints such as symptoms of fatigue, weakness, digestive disorders, diabetes, and pulmonary diseases. Pharmacological studies found antiosteoporotic, neuroprotective, cardiotonic [44], and immunomodulatory effects and confirmed some traditional uses. The lectins and polysaccharides may be involved in the effects in addition to the saponins [313].

All species studied to date contain steroid and/or triterpene saponins [138, 315]. The toxic azetidine-2-carboxylic acid (ca. 0.2%), common in the order Asparagaceae, is worth mentioning as an accompanying substance [67].

Particularly high concentrations of saponins were found in the seeds of *P. odoratum* [68]. Odoroside (Fig. 9.9 [79]), among others, was detected in the rhizome. It possesses strong hemolytic activity. Of note is a monoglycoside glycosylated only at C-26 [198].

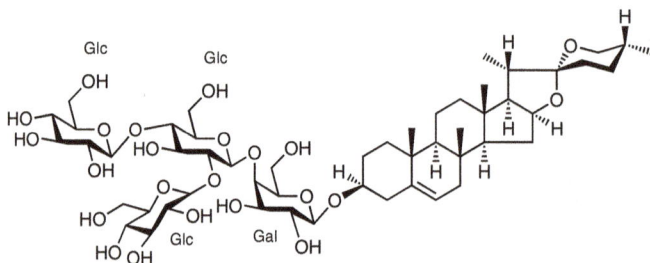

Odoroside

Fig. 9.9: Odoroside, a steroid saponin of *Polygonatum odoratum*.

The saponin content of the rhizome of *P. multiflorum* is reported to be 2.5%, that of the roots is higher, and that of the leaves is lower [117, 118, 222]. Two ribosome-inactivating proteins were isolated from the leaves [278].

Saponosides A, B, C, and D were detected in *P. verticillatum*. Saponoside D was identified as dioscin [117].

From *P. hirtum*, eight saponins could be obtained [143].

In other species, not native in Central Europe, smilagenin (*P. polyanthum* Bl., [84]), pennogenin (*P. stenophyllum* Meisner [115]), neopracerigenin A ((25$S$)-spirost-5-en-3β,14α-diol), (25$S$)-22α-methoxy-furost-5-en-3β,14α,26-triol, their (25 $R$)-analogs, and yamogenin (*P. odoratum* Lour. var. *pluriflorum* [254]) were found as sapogenins.

---

☠ Poisoning can occur after consumption of unripe berries of Polygonatum species in children. Symptoms of poisoning include vomiting, diarrhea, headache, and dizziness.

🐕 A dog exhibited vomiting persisting for 3 days after eating the leaves of *P. multiflorum* [18].

## 9.8 Steroid Saponins of May Lily (*Maianthemum bifolium*)

*Maianthemum bifolium* (L.) F.W. Schmidt, May lily, Asparagaceae (Ph. 9.7), is a perennial plant with a thin, creeping rhizome, a stem with two heart-ovate, pointed leaves, and a terminal, spike-like inflorescence composed of small, two- to three-flowered umbels with four-toothed, 2–3 mm-long white flowers. The ripe pea-sized berries are cherry red. It is found widely in deciduous and coniferous forests poor in lime from Europe to northern Japan.

Ph. 9.7: *Maianthemum bifolium* (source: Liudmyla Liudmyla/iStock/Getty Images Plus).

Saponins are present in the red berries, which have been reported to cause poisonings [236]. The structure of the saponins is unknown.

Saponins were also isolated from *M. canadense* Desf. and *M. dilatatum* Nels. et Macbr. [**68**].

After consumption of the berries, irritation of the gastrointestinal tract is to be expected. However, poisonings caused by fruits of the May lily are not known to us.

In folk medicine, *M. bifolium* is used as diuretic.

## 9.9 Steroid Saponins of Dragon Tree Species (Dracaena Species)

The genus Dracaena, Asparagaceae, includes about 150 species. In 2018, the genus *Sansevieria* was placed in the genus Dracaena. The plants are predominantly succulent shrubs and tree-like shrubs occurring in Africa, Australia, and Southeast India. Some species are cultivated as popular ornamental plants indoors as effective air cleaners or outdoors for their beautiful foliage. Especially *Dracaena marginata* Lam., Madagascar dragon tree, and *D. fragrans* (L.) Ker Gawl., cornstalk dracaena, serve as indoor plants [264].

One of the most famous representatives is *Dracaena draco* (L.) L., the Canary Islands dragon tree (Ph. 9.8).

Ph. 9.8: *Dracaena draco* (source: Horst Pilgrim).

From the representatives in genus Dracaena, a total of about 200 steroid saponins (spirostanol- and furastanol-di, tri-, and tetraglycosides) were isolated [264].

*Dracaena cinnabari* Balf. F., Socotra dragon tree, dragon blood tree, is a close relative of *D. draco*. It is an umbrella-shaped low tree, found endemically on the Socotra archipelago and at the southern coast of Yemen. It produces, like some other trees of the genus Dracaena, a deep red resin, dragon blood. Extracts of the resin serve as a dye and are used folk medicinally for traumatic, dermal, dental, and eye injuries (Two Brother's Blood, Damm Alakhwain). Nothing is known about its saponins. Methanol extracts of the deep red resin exerted only mild toxic effects in rats when administered by gavage. Up to 2,000 mg/kg BW was tolerated without changes in pathological and hematological parameters [3, 4].

*Dracaena cochinchinensis* (Lour.) S. C. Chen, dragon blood tree, is a tree-like shrub common in China. Its resin accelerates blood circulation and is used in traumatic injuries, blood stasis, and pain. It contains triglycosides of furostanol-5,25(27)-dien-1β,3β,22β,26-tetraol and of furostanol-5,20 [22, 25(27-)-trien-1β,3β-26-triol]. Application of 3 g of the resin/kg BW/day, p.o., for 90 days to rabbits was tolerated without symptoms. Pathological examinations showed no changes in organs and blood [62].

*Dracaena trifasciata* (Prain) Mabb. (*Sansevieria trifasciata* Prain), snake plant, nicknamed mother-in-law's tongue, serves as a popular pot plant in several cultivars, e.g., 'Laurentii' (with golden-yellow leaf margins), 'Craigii' (with white longitudinal stripes) and 'Hahnii' (of rosette-shaped habit, banded with silver-gray [264]). It contains saponins in the leaves with the aglycones (25S)-ruscogenin, sansevierigenin, and the structurally unusual abamagenin (23S or 24S)-dichloro-spirost-5-ene-1β,3β-diol [78]. Another interesting ingredient isolated was *n*-butyl-4-ol-*n*-propyl phthalate. Its pharmacology is not investigated [192]. The $LD_{50}$ of an ethanolic leaf extract in rats is 774.60 mg/kg BW, i.p. Oral administration of the extract at 18,000 mg/kg BW did not cause any behavioral changes in the animals [109].

☠ We are not aware of poisoning of humans by Dracaena species.

🐾 Due to their saponins all Dracaena species are toxic to cats, dogs, and pets. Symptoms of poisoning are loss of appetite, vomiting, diarrhea, abdominal pain, excessive drooling, lethargy, and swelling of tongue and lips. Allergic reactions are potentially life-threatening [107]. A dog showed ataxia and foaming at the mouth after eating a whole *D. trifasciata* plant [121].

## 9.10 Steroid Saponins of Century Plant Species (Agave Species)

Agave species, Asparagaceae, include about 400 representatives. They are xerophytic, perennial, or hapaxanthic plants with a short stem bearing a rosette of mostly succulent leaves. In some species, the inflorescence stem can grow to 10 m tall. They are distributed from the central states of the United States to the northern areas of South America. The center of origin is Mexico. They have been naturalized in many subtropical and tropical areas. *Agave americana* L., century plant (Ph. 9.9), has become the character plant of the Mediterranean region. Some Agave species are grown for the production of fibers, especially sisal hemp, pulque (agave wine), and tequila (agave brandy) [237, **68**].

Saponins were detected in about 60% of the Agave species studied. The main sapogenin is hecogenin. In addition, agavogenin, chlorogenin, dehydrohecogenin, diosgenin, gitogenin, mexogenin, sarsapogenin, sisalagenin, smilagenin, tigogenin, yuccagenin, and others were found [**68**].

From the leaves of *A. americana*, 1.6–2.6 g saponins/kg DW were obtained [126, 127]. Sapogenins that have been reported include, among others, hecogenin and chlorogenin.

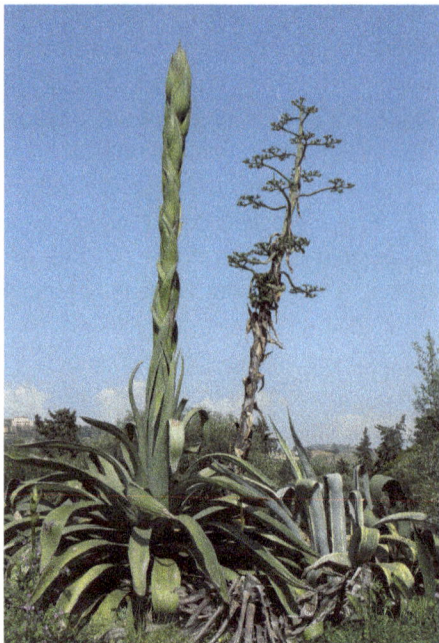

Ph. 9.9: *Agave americana* (source: Peter Schönfelder).

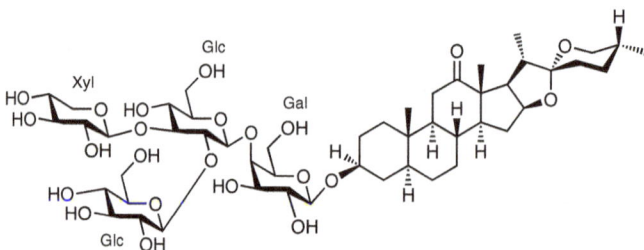

Hecogenintetraglycoside

Fig. 9.10: Hecogenintetraglycoside C, a steroid saponin of *Agave americana*.

In the older literature, also 3β-hydroxy-spirostan-12-one (gekgonin) and 12-hydroxy-tigogenin (rocogenin) are mentioned [286, 291]. Glycosides include hecogenintetraglycoside C (Fig. 9.10) and other [304]. The naming of the glycosides is handled very differently by the individual authors: agavasaponins, Agave saponins, and agamenosides [237, 291].

*Agave sisalana* Perrine, sisal agave, yields the sisal hemp and is used to isolate steroids for the semisynthesis of corticosteroids. It contains glycosides of the sapogenins tigogenin, hecogenin, barbourgenin, rockogenin, and chlorogenin, among them the dongnosides A–E with the aglycone tigogenin [24, 51].

*Agave lechuguilla* Torr., shin dagger, ixtle fiber, cultivated as fiber plant [234] and found also feral, contains oligosaccharides of the sapogenin smilagenin in the leaves [286].

☠ Contact with the sap of leaves of *A. americana* and some other Agave species causes skin irritation. Contact with the eyes results in conjunctivitis [26, 37, 137, 205, 210]. Injuries with the abundant oxalate needles in the plant enable the saponins to entry into the tissues [**49**]. The irritation symptoms occur in about 80% of workers in tequila factories processing the sap of *A. tequilana* Weber, tequila agave. One milliliter of the sap contains about 3,000–5,000 oxalate needles, 30–500 μm long, with very sharp tips. Eye inflammation is also very common among workers in Agave plantations [37].

🐕 Agave species are toxic to dogs, goats, sheep, and other animals, when ingested. Symptoms are irritation, burning, redness, swelling in the throat, and breathing difficulties. The animals may experience vomiting or diarrhea, e.g., *A. lecheguilla* triggered irritation of the mouth, nose, ears, and mucous membranes, dogs may develop blisters [28, 176]. It is thought that these symptoms are the result of photosensitization caused by the saponins and of the oxalate needles [31].

## 9.11  Triterpene Saponins of Horse Chestnut Species (Aesculus Species)

The genus Aesculus, Sapindaceae, includes about 15 species with several varieties, cultivars, or hybrids. The representatives are distributed in North America, North China, from Near East to India, Japan, and Southern Europe.

The most important representative of the genus is *Aesculus hippocastanum* L., common horse chestnut (Ph. 9.10), which is native to the Balkans. It is cultivated throughout Europe and in North America in several forms. It also occurs feral. Other cultivated species

Ph. 9.10: *Aesculus hippocastanum* (source: WALA Heilmittel GmbH).

are, e.g., *Ae. pavia* L., red-flowered horse chestnut, pavia, *Ae. glabra* Willd., bald horse chestnut, Ohio horse chestnut, Ohio buckeye, *Ae. flava* Sol, yellow horse chestnut, *Ae. turbinata* Blume, Japanese horse chestnut, and *Ae. parviflora* Walter, shrub horse chestnut.

The starch-rich seeds of *Ae. hippocastanum* contain 3–10% (up to 28%?), the immature seeds in August about 12% saponins [132].

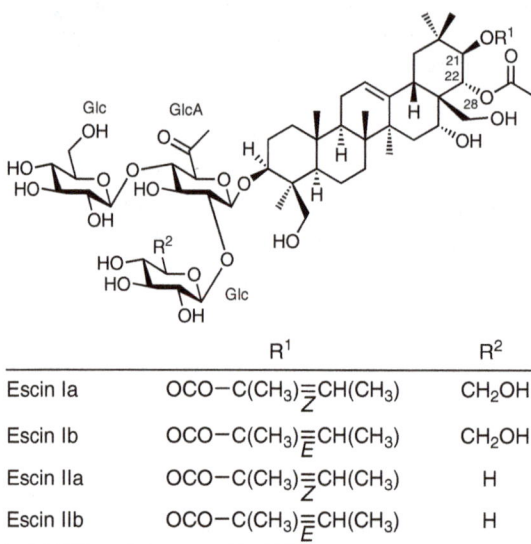

| | R¹ | R² |
|---|---|---|
| Escin Ia | OCO−C(CH₃)=CH(CH₃) Z | CH₂OH |
| Escin Ib | OCO−C(CH₃)=CH(CH₃) E | CH₂OH |
| Escin IIa | OCO−C(CH₃)=CH(CH₃) Z | H |
| Escin IIb | OCO−C(CH₃)=CH(CH₃) E | H |

Fig. 9.11: Triterpene saponins of *Aesculus hippocastanum*.

From escin (aescin) of commerce, obtained from the seeds, as main saponins the escins (aescins) Ia and Ib could be isolated first. These were followed by further escins (Fig. 9.11) as well as isoescins (presumably isolation artifacts), monoacylescins, desacylescins, aesculiosides, aesculisides, and assamicins.

About 80 different saponins have been isolated from Aesculus species to date. The sapogenins are mono- or diesters of protoaescigenin and barringtogenol C. The saponins carry a trisaccharide residue on the OH group at C-3. The OH group at C-21 is esterified with angelic acid (*E*-form) or tiglic acid (*Z*-form), that at C-22 usually carries an acetate residue or is free. In the isoescins the acetate residue is located at C-28. More rarely, isobutyric acid or 2-methylbutyric acid occur as ester components [54, 305, 306, 311].

Based on the solubility, three fractions of saponins can be distinguished: β-escins, cryptoescins, and α-escins. The representatives of the β-escin fraction are sparingly soluble in water and hemolytically active (H.I. about 40,000). This fraction consists predominantly of escins [83]. Spontaneous migration of acetyl groups from C-22 to C-28 results in the formation of cryptoescins, which consist predominantly of isoescins. The members of this group are highly water soluble and hemolytically inactive.

The α-escins are the equilibrium mixture of the C-22 and the C-28 acetates. Because of their cytotoxic properties, the 21-*O*-22-*O*-tigloyl-angeloyl-$R_1$-barrigenol l, called hippoesculin, and the 21-*O*-angeloyl-barringtogenol C deserve special mention [103, 146, 259, 285, 305].

The fruit shells of the capsule fruits of horse chestnut contain a mixture of saponins known as hippocastanoside with barringtogenol C and $R_1$-barringenol as sapogenins [277].

In the seeds of *Ae. punduana* Wall. (Himalaya) barringtogenol C, protoaescigenin, and oleanolic acid were found [134]. In those of *Ae. turbinata* (Japan) aescigenin, barringtogenol C and 16-deoxy-barringtogenol C [189], in the seeds of *Ae. glabra* barringtogenol C and R1-barringenol [11], in those of *Ae. indica* L. (Afghanistan) protoaescigenin [212, 213, 239, 240], and in those of *Ae. chinensis* Bunge (Northern China) protoaescigenin and barringtogenol C [290, 312] were found as aglycons.

Triterpene saponins of seeds of other Aesculus species include aesculiosides G1–G16, aesculiosides Ia, 1d, IIa-, IIc, Iva–IVc, vaccarosides A and B, and pavosides A–H [54].

Small amounts of saponins can occur in other organs of Aesculus species: foliage leaves up to 0.038% and petals 0.12% [131].

Accompanying substances include 0.2–0.3% flavonol glycosides (biosides and triosides of quercetin and kaempferol) in the seeds, in the bark, and in small amounts in the leaves, besides hydroxycoumarins and their glycosides (aesculin, fraxin, and scopolin [108]).

Due to their saponin content, medicines from horse chestnut and other Aesculus species possess antiedematous, anti-inflammatory, hypoglycemic, vein tonic, and antiviral effects. The action of saponins is supported by the flavonoids. The mentioned effects are caused in various ways, e.g., by inhibition of lysosomal enzymes like hyaluronidase, that degrade proteoglycans of the blood vessel walls, by promotion of release of $PGF_2$, and by antagonism to histamine [36, 243]. Some of the effects, esp. the anti-inflammatory effect, are glucocorticoid-like. They are based, at least in part, on the increased expression of the glucocorticoid receptor [74, 314].

Partial absorption of chestnut saponins after p.o. application is proven. In animal experiments 13–16% of the ingested saponins is absorbed [58, 72].

$LD_{50}$ values after p.o. administration in mice were 320 mg/kg BW for α-escins and 134 mg/kg BW for β-escins and in rats 720 mg/kg BW for α-escins and 400 mg/kg BW for β-escins [86]. Poisonings of chickens and hamsters resulted in muscle incoordination, paralysis, coma, and death [292]. Acute toxicity is much higher after i.v. than after p.o. administration. In animals, i.v. given escins lower blood pressure and accelerate respiration and pulse rate [150]. When given i.v., the $LD_{50}$ values for mice are 3.2 mg/kg BW for α-escins, 1.4 mg/kg BW for β-escins, and 21 mg/kg BW for cryptoescins and for rats 5.4 mg/kg BW for α-escins, 2.0 mg/kg BW for β-escins, and 34 mg/kg BW for cryptoescins [86]. Lethal doses cause severe hemolysis and consequent hypoxia of vital tissues [174]. Death occurs by uremia or respiratory paralysis [58].

☠ Symptoms of poisonings after ingestion of horse chestnut seeds or the green horse chestnut fruit peels are predominantly gastrointestinal. After ingestion of ¼ to one seed by an adult, redness of the skin, edema, in some cases lowering of blood pressure with loss of consciousness or collapse have been observed after a short time. With p.o. medicinal use of horse chestnut preparations, mucosal irritation of the gastrointestinal tract occasionally occurred, associated with nausea, and rarely itching, skin redness, skin scaling, and skin dryness [101]. Escin-induced renal failure in overdose of a horse chestnut preparation has been described [217, 283].

☦ A case of poisoning with fatal outcome, described in the older literature [221], is attributed to mucosal damage caused by repeated consumption of horse chestnut and the resulting favored absorption.

🛟 Treatment of poisoning should be symptomatic after primary toxin removal.

✱ Allergic reactions, including anaphylactic shock, sometimes occurred with medicinal use of horse chestnut preparations [145, 284]. The pollen can cause inhalant allergy. Chestnut pollen is responsible for about 10% of spring pollinosis.

Therapeutically, horse chestnut preparations or escins are used in chronic insufficiency of the leg veins [72, 243]. The usability of the antiviral [142] or the hypoglycemic [224] effects may become possible. Escin could become an add-on therapy in acute lung injury related to COVID 19 infection [73].

## 9.12 Triterpene Saponins of Ivy Species (Hedera Species)

The genus Hedera, Araliaceae, ivy, includes about 15 species of which only *Hedera helix* L., common ivy, thrives in western, central, and eastern Europe and is cultivated in many varieties as ornamental plant. In North America, common ivy is naturalized. Feral from cultivation, *H. colchica* (K. Koch) K. Koch, colchic ivy, native to the Caucasus region, is sometimes found in Europe and North America. Other species occur in southern and western Europe, northern Africa, the Himalayan region, and Japan.

Common ivy (Ph. 9.11) is a creeping or clinging-rooted woody plant that can reach heights of up to 20 m by climbing. The wintergreen, leathery leaves of the flowering shoots, hairy when young and glabrous when old, are ovate-lanceolate, those of the nonflowering ones three- to five-lobed (heterophyllous). In cultivated forms the leaves may also be variegated. The five-digit flowers have greenish, inconspicuous petals. The berries are pea-sized, first reddish-purple, then dark brown, finally black. They contain three to five kidney-shaped seeds.

Saponins are found in all organs of Hedera species. Leaves have the highest content (2.5–6% of DW in *H. helix*), but berries and wood also contain significant amounts of saponins.

Ph. 9.11: *Hedera helix* (source: Nadin Schultze).

Oleanolic acid, hederagenin, bayogenin, and echinocystic acid were found as sapogenins, and D-glucose, L-rhamnose, and L-arabinose as monosaccharide components. The main saponin, accounting for about 80% of the saponin mixture, is the bisdesmosidic hederasaponin C (hederacoside C, hederasaponoside C, kalopanaxsaponin, Fig. 9.12). It readily converts, e.g., upon alkaline hydrolysis, to the monodesmosidic α-hederin by loss of the oligosaccharide residue at C-28 [12, 104, 271]. Helixosides A and B were isolated from the fruits along with four other saponins ([19], Table 9.2).

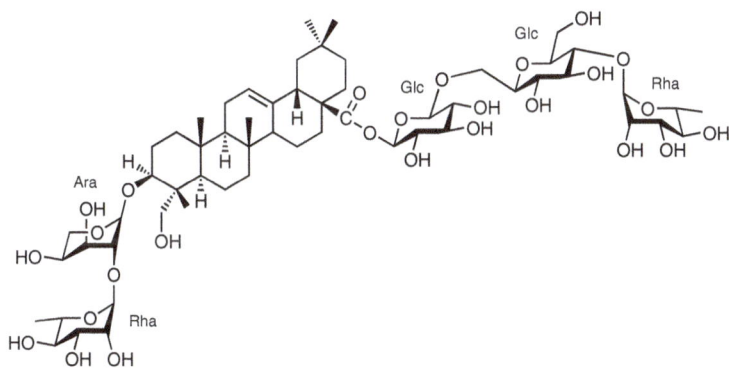

Hederasaponin C

Fig. 9.12: Hederasaponin C, a triterpene saponin of *Hedera helix*.

Other triterpene saponins occur in small amounts, including 28-*O*-acylglycosides, esters of sulfate, of oleanolic acid, and echinocystic acid [61].

Tab. 9.2: Triterpene saponins of *Hedera helix*.

| Saponin | Structure |
|---|---|
| Hederasaponin C (hederacoside C) | Hed-Sap$\begin{cases} -3 \leftarrow 1\alpha Ara2 \leftarrow 1\alpha Rha \\ -28 \leftarrow 1\beta Glc6 \leftarrow 1\beta Glc4 \leftarrow 1\alpha Rha \end{cases}$ |
| α-Hederin | Hed-Sap-3←1αAra2←1αRha |
| Hederasaponin D (hederasaponin K10) | Hed-Sap$\begin{cases} -3 \leftarrow 1\alpha Ara \\ -28 \leftarrow 1\beta Glc6 \leftarrow 1\beta Glc4 \leftarrow 1\alpha Rha \end{cases}$ |
| Hederasaponin A (hederacoside A) | Hed-Sap-3←1αAra4←1βGlc |
| Saponin 2 | Hed-Sap-3←1βGlc |
| Saponin 4 | Hed-Sap-3←1βGlc2←1βGlc |
| Unnamed | Hed-Sap-3←1αAra |
| Unnamed | Hed-Sap-3←1αRha2←1αAra |
| Helixoside A | Hed-Sap$\begin{cases} -3 \leftarrow 1\beta Glc2 \leftarrow 1\beta Glc \\ -28 \leftarrow 1\beta Glc6 \leftarrow 1\beta Glc \end{cases}$ |
| Helixoside B | Ole-Sap$\begin{cases} -3 \leftarrow 1\beta Glc2 \leftarrow 1\beta Glc \\ -28 \leftarrow 1\beta Glc6 \leftarrow 1\beta Glc \end{cases}$ |
| Hederasaponin B (hederacoside B, eleuthroside B) | Ole-Sap$\begin{cases} -3 \leftarrow 1\alpha Ara2 \leftarrow 1\alpha Rha \\ -28 \leftarrow 1\beta Glc6 \leftarrow 1\beta Glc4 \leftarrow 1\alpha Rha \end{cases}$ |
| β-Hederin | Ole-Sap3 ← 1αAra2 ← 1αRha |

Abbreviations: Ole-Sap, oleanolic acid; Hed-Sap, hederagenin; Glc, D-glucose; Ara, L-arabinose; Rha, L-rhamnose

The polyynes falcarinol and 11,12-didehydrofalcarinol, among others, have been detected in the leaves, falcarinol, falcarinone, and (Z)-9,10-epoxy-heptadec-1-ene-4,6-diin-3-one in the fruits of *H. helix*.

*Hedera colchica* contains hederacolchisides A–F in its leaves [45]. Hederacolchiside D is identical to hederasaponin C. Hederacolchiside E has oleanolic acid as aglycone and hederacolchiside F hederagenin. Both are bisdesmosidic hexaosides [46]. Colchiside A (hederagenin-3-*O*-β-D-xylopyranoside) and colchiside B, a bisdesmosidic oleanolic acid glycoside containing D-glucuronic acid, L-rhamnose, and D-glucose residues, were among the compounds isolated from berries [181].

Sapogenins from *H. rhombea* (Miq.) Bean [144] and *H. taurica* Carr [161, 233] include hederagenin, (20*S*)-dammar-24-en-3β,6α,20,26-tetraol, 3-oxo-(20*S*)-dammar-24-en-6α,20,26-triol, (20*S*)-dammar-24-en-3β,6α,20,26-tetraol, and (20*S*)-dammar-24-en-3β,20,26-triol. In the leaf galls of *H. rhombea* Siebold et Zucc, the antimicrobial active $C_{17}$-diacetylen-hederin A was found [27, 42, 69].

The monodesmosidic $\alpha$-hederin shows much stronger hemolytic activity (H.I. 150,000) than hederasaponin C (H.I. 400) and is much more toxic than the bisdesmosidic glycosides. The free carboxyl group is important for the hemolytic, the terminal rhamnose residue for the antimycotic activity. The aglycone hederagenin is not membrane-active [258]. Effects directed against parasites, e.g., nematodes and protozoa, e.g., *Trichomonas*, *Leishmanania*, and snails, e.g., *Biophalaria glabrata*, have been demonstrated [104].

Nepalins, triterpene saponins from *H. nepalensis* K. Koch, immobilize human spermatozoa [191].

The absorption rate of ivy saponins is relatively low.

The $LD_{50}$ of $\alpha$-hederin is greater than 4,000 mg/kg BW after p.o. ingestion in mice and 1,800 m g/kg BW when applied i.p [265].

The promotion of expectoration by ivy saponins occurs by influencing membrane properties of the lung epithelium directly. Inhibition of endocytosis of membrane areas of the lung epithelium with high density of $\beta_2$-receptors is thought to increase the effect of the neurotransmitter epinephrine and thus the secretion of surfactant II, which facilitates expectoration [214].

In addition to the direct effects, saponins of ivy have spasmolytic, analgesic, anti-inflammatory, and antimutagenic activity [163, 169].

> ♟ Consumption of common ivy fruits, more than five, may cause nausea, vomiting, colic, diarrhea, facial flushing, and drowsiness, especially in children. The risk of ingestion of larger quantities of the blue–black berries is low. The berries taste very bitter. Information on severe, sometimes fatal poisoning can be found in the older literature. Their truthfulness is questionable.

† A 37-year-old male was found dead approximately 1 week after death in a public garden in the state of advanced putrefaction. At the autopsy an incredible quantity of leaves of common ivy was discovered in the mouth, throat, and esophagus, but no debris in the stomach. Toxicological analysis found hederacoside C (857 ng/ml) in the gastric juice. It was concluded that the man committed suicide and that the death was caused by suffocation (!) by leaves of common ivy [70].

When extracts of ivy leaves were used therapeutically, stomach complaints occasionally occurred. In infants and children up to 2 years of age, laryngospasm may be caused by drugs made from ivy leaves [155, 275, **75**, **133**].

> 🛠 If more than five ivy fruits are ingested, primary toxin removal (induced vomiting, activated charcoal) and abundant liquid intake are necessary, possibly symptomatic treatment must follow.

✹ Of toxicological importance is the skin-irritating and allergenic effect of ivy leaves. Responsible allergens, in addition to the saponins, are the polyynes falcarinol and didehydrofalcarinol. Allergic contact dermatitis can develop during horticultural handling of the plants or during medicinal or cosmetic application of ivy-containing preparations. Cross-reactivity with other Araliaceae has been described [92, 186, 188, 301, **65**].

🐕 Poisoning of dogs, cats, pets, rodents, and horses after eating ivy leaves has been observed. The symptoms are diarrhea, vomiting, increase in pulse rates, cramps up to respiratory failure. Bird can eat the berries without harming them. For them, the berries are a valuable source of foods during the winter [**60**].

In medicine, extracts of ivy leaves are used for their expectorant, spasmolytic, and anti-inflammatory effects, especially in respiratory diseases [**20**, **149**].

## 9.13 Triterpene Saponins of Corn Cockle (*Agrostemma githago*)

*Agrostemma githago* L., common corn cockle, is the only representative of the genus Agrostemma, Caryophyllaceae, which comprises only four species, native to Central Europe. *A. linicola* Terechov is native to Eastern Europe, *A. gracilis* Boiss to Greece and Turkey.

Ph. 9.12: *Agrostemma githago* (source: Wolfgang Kiefer).

*Agrostemma githago* (Ph. 9.12) is an annual to 1 m tall plant with linear-lanceolate, downy-hairy leaves. The inflorescence, a two- to three-flowered coil, bears five-petaled flowers with long-pointed sepals that excel the purple corolla. The 2.5–3.5 mm seeds are pointed black.

Corn cockle was a common cereal weed in Europe in earlier times. It has completely disappeared today due to modern farming techniques, especially seed cleaning and the use of herbicides and occurs now only as an adventitious plant.

The entire plant contains saponins. The seeds have a particularly high content of saponins (6–7%). The main saponin is the monodesmosidic githagoside with the aglycon gypsogenin [267]. Further saponins with gypsogenin (githagenin) or quillaic acid as aglycons were detected [43, 238].

Concomitants of interest are the compounds orcialanin (2,4-dihydroxy-6-methyl-phenylalanine [218]) and agrostin, a ribosome-inactivating protein (RIP 1, Mr 27 kDa, 8–400 mg/100 g plant material [250, 288]), found in the seeds.

Ribosome-inactivating proteins (RIPs) are *N*-glycosylases that inactivate ribosomal RNA of bacteria, fungi, plants, and animals. Thereby they irreversibly inhibit protein synthesis and lead to cell death. Supercoiled DNA, e.g., in plasmids, is also attacked. RIPs have also been detected in some other plants, e.g., in the seeds of *Asparagus officinalis* (see Section 9.6), the latex *of Hura crepitans* (see Section 5.3.3), the roots of Phytolacca species (see Section 9.14), and the leaves of *Polygonatum multiflorum* (see Section 9.7). Ricin, a highly toxic lectin from *Ricinus communis*, is also a RIP. RIPs have been used to treat viral and tumor diseases and to produce transgenic virus-resistant bacteria, fungi, and insects [110].

The toxicity of Agrostemma saponins to mice is low ($LD_{50}$ 750 mg/kg BW, p.o. [173]). It is assumed that agrostin is involved in the poisoning symptoms caused by *A. githago* and that the saponins only enable the uptake of agrostin into the cells. This is supported by studies on cell cultures in which it was shown that a combination of agrostin, which alone is not cytotoxic, with Agrostemma saponins exerts strong cytotoxic effects. Potentiation of the action of agrostin requires certain structures of the saponins. An aldehyde group at C-4 supports the action [96, 176].

In humans, 3–4 g of the seeds are reported to cause severe symptoms of poisoning and may be letal [**133**]. It is likely that the sensitivity of individuals varies widely and may depend on preexisting mucosal damage.

Poisonings of humans, sometimes with fatal outcome, by bread cereal flours, oatmeal, or grain coffee, which in the nineteenth century often contained seeds of corn cockle in large quantities, were frequently reported in the older literature.

☠ Symptoms of poisoning included scratching in the mouth and throat, nausea, belching, vomiting, diarrhea, as well as headache, dizziness, delirium, circulatory disorders, possibly convulsions, and in severe cases death by respiratory paralysis.

🚑 Treatment of poisoning is symptomatic after primary toxin removal.

🐗 Animal poisoning from grain contaminated with the seeds of *A. githago* or from eating the plant is described. For pigs, the LD is 2–5 g of seeds/kg BW and for poultry

80 g/kg BW [250]. 2.5–5 g of seeds/kg BW were lethal to cattle [244]. Death of goats, pigs, and chickens and severe poisoning of horses have also been reported [**163**].

Since corn cockle hardly occurs in Europe anymore and the grains are thoroughly cleaned from seeds of other plants, poisonings by corn cockle seeds should not play a role today in this region.

In folk medicine, corn cockle seeds were used for skin blemishes and gastritis.

Agrostemmoside E, an acetylated triterpene saponin with quillaic acid as aglycon, has great potential for delivery of gene-loaded nanoplexes and is therefore a new tool for gene transfections [43].

## 9.14 Triterpene Saponins of Pokeweed Species (Phytolacca Species)

The genus Phytolacca, pokeweed, Phytolaccaceae, includes about 35 species native to subtropical and tropical areas.

*Phytolacca americana* L. (*Ph. decandra* L.), American pokeweed, makeup berry (Ph. 9.13), is distributed from Canada through the United States to Mexico and is grown as an ornamental in many areas of the world. It has gone wild in many places from gardens and from vineyards where it was cultivated for the ability of its berries to darken light red wines.

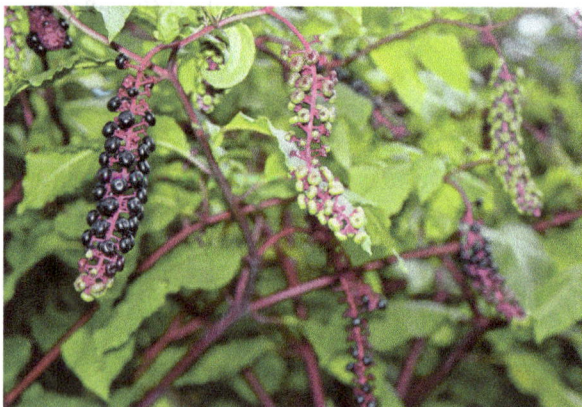

Ph. 9.13: *Phytolacca americana* (source: Eberhard Teuscher).

*Ph. americana* is a perennial with a turnip-shaped rootstock. The erect, branched stem grows to 3 m tall. The dark red, later almost black fruits are berry-like. The young shoots are often eaten, before the reddish discoloration begins, as an asparagus-like vegetable.

In *Ph. americana* 28,30-dicarboxy-olean-12-enes occur as sapogenins. Esculentic acid, jagilonic acid, phytolaccagenin (jagilonic acid-30 methyl ester), and phytolaccagenic acid (esculentic acid-30 methyl ester) were detected (Tab. 9.3). In addition, small amounts of acinosolic acid-30 methyl ester (pokeberrygenin) were found in the berries [39, 130].

Tab. 9.3: Triterpene saponins of *Phytolacca americana*.

| Name | Structure |
|---|---|
| Phytolaccoside A | Ec-30-M-Sap-3←1βXyl |
| Phytolaccoside B | Ja-30-M-Sap-3←1βXyl |
| Phytolaccoside C | Ec-Sap-3←1βGlc |
| Phytolaccoside D | Ec-30-M-Sap-3←1βXyl4←1βGlc |
| Phytolaccoside E | Ja-30-M-Sap-3←1βXyl4←1βGlc |
| Phytolaccoside G | Ja-Sap-3←1βXyl |
| Phytolaccasaponin B | Ja-30-M $\begin{cases} -3\leftarrow1\beta Xyl4\leftarrow1\beta Glc \\ -28\leftarrow1\beta Glc \end{cases}$ |

Abbreviations: Ec-30 M-Sap, esculentic-30-methylester;
Ja-30 M-Sap, jaligonic acid-30-methylester;
Ec-Sap, esculentic acid; Ja-Sap, jaligonic acid;
Xyl, D-xylose; Glc, D-glucose

*Phytolacca dodecandra* L' Hérit, inkweed, gopo berry, native to southern Africa and Madagascar, is used there and in many areas of Africa for treatment of malaria, rabies, ascariasis, and skin diseases, as a drastic laxative, as an emetic, and as a diuretic [21, 48]. In addition to phytolaccagenin and phytolaccagenic acid, the plant contains as sapogenins bayogenin, hederagenin, oleanolic acid, 2-hydroxy-oleanolic acid, and hederagenin [21, 48, 262].

*Phytolacca esculenta* Van Houtte, Asian pokeweed, is originally native to China, Bhutan, India, Japan, and Korea and was introduced into the Mediterranean area and to South America. The plant contains the esculentosides H, I, J, K, L, M, N, O, P, and Q with the sapogenins jagilonic acid, phytolaccagenin, and phytolaccagenic acid [21, 39, 303]. The berries were used as a dye. They are not edible!

*Phytolacca acinosa* Roxb., Indian pokeweed, is native to southeast Asia. The saponins found are phytolaccacinoside A, phytolaccosides B, E, and F, esculentosides G and U, as sapogenins jagilonic acid, 11β-methoxyjagilonic acid-30-methyl ester, 2,23-dihydroxyoleanane-11, and 13(18)-diene-28,29-dioic acid 29-methyl ester [56, 75].

Other sapogenins detected in the above species were e.g., 28,30-dicarboxy-olean-12-ene derivatives, such as acinosolic acid A, phytolaccagenin A (3β-acetyl-jagilonic acid-30-methyl ester), and spergulagenic acid [89, 193, 246, 298].

The fruits are colored by betalains. Notable accessory compounds occurring in fruits are caffeic aldehyde and dimers or trimers of these compounds with free aldehyde groups, including americanins A, B, and D and isoamericanin A [91, 295]. Histamine (0.13–0.16%) and GABA are present in the roots [68].

Representatives of the genus Phytolacca have been used as therapeutics for thousands of years in traditional medicine worldwide, esp. in TCM and Tibetan medicine. Mostly used are extracts from the roots of *Ph. acinosa* and *Ph. esculenta*, today also of *Ph. americana*. Anti-inflammatory effects have been exploited in arthritis, nephritis, and rheumatism. The plants have also been used in cancer treatment. Histamine and GABA are thought to be responsible for their hypotensive effects [68]. The antiviral and abortive properties are probably due to peptides (PAP, antiviral proteins [15, 276, 302]). *Ph. americana* also contains mitogenic lectins (pokeweed mitogen) and ribosome-inactivating proteins (Mr 29 kDa [220]), which are involved in the toxicity of the plants [16]). Extracts from the plants also exert potent antimycotic [302], spermicidal [251], and molluscicidal [52, 154, 262] effects. Therefore, they serve as antiparasitics, molluscicides, and insecticides [14].

In spring and early summer shoots and leaves, not the roots, are edible with proper cooking by boiling two times with water drained and replaced each time (poke sallet). But later they become poisonous. Birds are unaffected so that the berries are a good food source for them.

Poisoning by Phytolacca species is relatively common in North America, the home of pokeweed [82, 114, 156, **49, 144**]. Causes of poisoning are consumption of the raw berries, esp. by children, in adults by eating improperly prepared leaves and shoots and by mistaking the root for an edible tuber, by folk medicinal use of infusions from the roots or leaves [114], or confusion of the roots with carrots, horseradish, or other root vegetables. Toxins of the juice of the pokeberries can be absorbed through the skin. Do not contact the plant with bare skin! The plant can cause dermatitis.

🦂 Symptoms of poisonings after consumption of the leaves, roots, and fruits of American pokeweed and of extracts from this plant are irritation of the mucous membranes of the mouth and of the gastrointestinal tract, characterized by burning, salivation, in severe cases abdominal cramping, vomiting, diarrhea, sometimes bloody, persistent thirst, furthermore urticaria, headache, dizziness, somnolence, drop in blood pressure, tachycardia, convulsions, and, especially in children, death by respiratory paralysis [**49, 169**].

🚑 First measures in case of ingestion of up to three pokeweed berries or some flowers should be abundant supply of liquids, in case of ingestion of more than three berries in young children and more than 10 in school children and adults, primary removal of poison by administration of activated charcoal. Subsequently symptomatic treatment may be required [**161**].

🐾 Most animals avoid the plants. Nevertheless, poisoning is possible with contaminated hay. Horses, sheep, and cattle have also been poisoned by eating fresh leaves, e.g., in green fodder, pigs by eating the roots. In case of poisoning of grazing animals and horses by the fruits of *Ph. americana*, severe gastrointestinal irritation, bloody diarrhea, convulsions, purging, and death by respiratory paralysis have been observed and abortions in cows too [17, 88, 300]. Fatal poisonings of grazing animals and chickens by the herb and fruits of *P. dioica* have been reported from Australia [252]. Berries of *Ph. americana* in feed of turkey poults caused weight loss, inability to walk, ataxia, and sometimes death [17].

## 9.15 Triterpene Saponins of Cyclamen Species

The genus Cyclamen, Primulaceae, includes 22 species, which are preferentially native in the eastern part, some also in the western part of the Mediterranean region. Central Europe is reached only by *Cyclamen purpurascens* Mill. (*C. europaeum* L.), European cyclamen, wild cyclamen, summer alpine violet (Ph. 9.14) and *C. hederifolium* Aiton (*C. neapolitanum* Ten.), ivy-leaved cyclamen, autumn alpine violet, Neapolitan cyclamen. *C. mirabile* Hildebr. sowbread, is endemic to Turkey. These species, *C. coum* Mill., eastern cyclamen and several subspecies distributed in southeastern Europe and the Caspian Sea, are used for outdoor cultivation in central Europe and the United States. Various cultivated forms of *C. persicum* Mill, indoor cyclamen, are used as potted plants. Its wild form is native to southwestern Asia and North Africa and was introduced to Europe in the seventeenth century.

Ph. 9.14: *Cyclamen purpurascens* (source: Eberhard Teuscher).

*Cyclamen purpurascens* is a perennial plant with 1.5–5 cm thick bulb, loosely rooted over the entire surface of the bulb. The foliage leaves are evergreen, kidney to heart-shaped. The upper side of the leaf is dark green and has a silvery-spotted zone. The flowers are crimson and strongly fragrant. The plant occurs in mountain forests in Central Europe on calcareous soils. The tubers of *C. hederifolium* are rooted in the middle of the underside. The corolla lobes of the flowers bear auricles at the base. The leaves have an elongated-heart shape. It occurs sporadically in Switzerland.

*Cyclamen purpurascens* has been studied most intensively. The main component of the saponin mixture of the tubers is the monodesmosidic cyclamin with the aglycon cyclamiretin A (Fig. 9.13 [273, 274]). It is accompanied by desglucocyclamin I (glucose residue 1 absent) and desglucocyclamin II (glucose residue 2 absent). Furthermore, the presumably genuine sapogenins cyclamigenin $A_2$ and cyclamigenin B were obtained after hydrolysis [53].

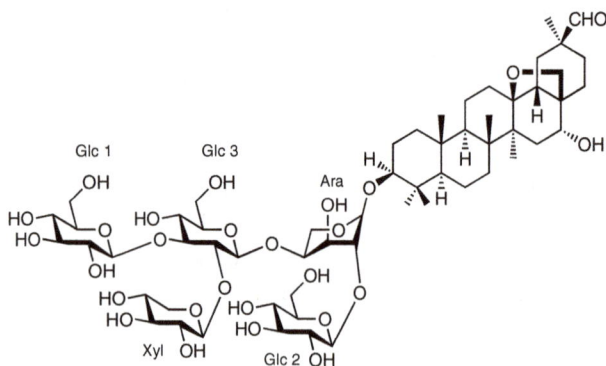

Cyclamin

Fig. 9.13: Cyclamin, a triterpene saponin of *Cyclamen purpurascens*.

Similar saponins occur in the tubers of other Cyclamen species [5, 7, 29, 204]. The leaves of Cyclamen species also contain saponins [204].

*C. hederifolium* contains, among others, hederofoliosides A, B, C, D, and E. Sapogenins are 3β,16α-dihydroxy-13β,28-epoxyoleanan-30-oic acid, 3β,16α-dihydroxy-13β,25-epoxy-oleanan-30-al, and 3β,16α,30-trihydroxy-13β,25,30-trihydroxy-olean-12-en-28-al [5].

Numerous monodesmosidic saponins have been isolated from *C. coum* tubers, including deglucocyclamin, cyclaminorin, cyclacoumin, mirabilin lactone, coumoside A, and coumoside B. Their aglycones are cyclamiretin A, 3β,16-dihydroxy-olean-12-ene -30,28-lactone, and 3β,16,23-trihydroxy-13β,28-epoxyolean-30-al [30, 299, 300].

*C. persicum* tubers yielded, among others, the saponins cyclamin I, cyclamorin I, and lysicokianoside I. The sapogenin isolated was 16-hydroxy-28-epoxy-oeleanane [178].

Cyclamin has the highest hemolytic index measured to date: H.I. 390,000 [269]. It is partially absorbed after peroral administration. Like the other saponins, cyclamin has antibiotic and cytostatic activity [215, 274]. The $LD_{50}$ in rats is 13 mg/kg BW, i.v [280]. In Sicily, extracts from the tuber are used for fishing. Because pigs are largely insensitive to cyclamen saponins, the tubers produced in nurseries are used steamed as pig feed (soebread! [60]).

The peeled tubers of *C. coum* are used in Turkish folk medicine like globules to treat female infertility [30]. In nontoxic doses lyophilized extracts from the tubers of *C. purpurascens* were effective in the treatment of rhinosinusitis because of its muco-lytic and possibly also of its virostatic activity [65, 129].

Poisonings may occur when the very bitter tuber of the European cyclamen was used as a drastic and emmenagogue. The tuber (0.2 g) was said to be poisonous, and 8 g caused death [247, **169**].

Despite the popularity of cyclamen as a houseplant, only a few cases of poisoning have been reported in recent times [178, 247].

♦ A 7-month-old infant who had eaten two to three leaves of *C. persicum* was admitted to a hospital 2 h later in a comatose state [261].

Some Cyclamen flowers were consumed by a child without causing symptoms of poisoning. In other cases, children ate flowers, leaves, and flower buds and chewed on stem parts. Nausea and vomiting occurred [**161**].

♨ Ingestion of parts of tubers and other parts of European cyclamen may lead to local irritation, nausea, vomiting, stomach pain, and diarrhea. After ingestion of large amounts, dizziness, sweating, convulsions, and paralysis can occur [**161**].

🚑 First measures in case of ingestion of large portions of a Cyclamen tuber or several leaves should be gastric emptying, administration of activated charcoal, and abundant supply of liquids [**161**].

✳ Pollen of Cyclamen species can have a sensitizing effect.

♦ A 46-year-old female flower cultivator had cultivated cyclamens for 7 years. After 2 years exposure, she developed allergenic rhinitis and asthma with cough, wheeze, and chest tightness during the blooming season [10].

🐾 Cyclamen species, esp. tubers, are toxic to dogs, cats, and horses. Clinical symptoms are drooling, vomiting, and diarrhea if leaves or flowers have been ingested. In severe cases, if tubers or parts of tubers were ingested, heart rhythm abnormalities, seizures and death may occur.

## 9.16 Triterpene Saponins of Sweetwood Species (Glycyrrhiza Species)

About 30 species of the genus Glycyrrhiza, Fabaceae, are known. Of medicinal and toxicological importance is especially *Glycyrrhiza glabra* L., liquorice, Spanish licorice (Ph. 9.15). The plant is a perennial up to 2 m tall with a strong taproot, thick secondary roots, and underground stolones that grow many meters long.

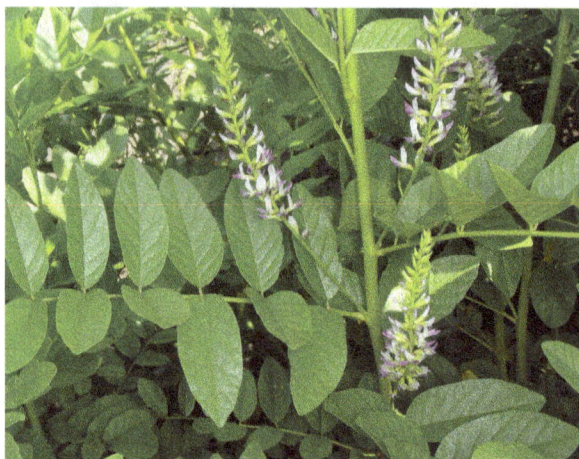

Ph. 9.15: *Glycyrrhiza glabra* (source: Peter Schönfelder).

The unpeeled or peeled, dried roots, and underground stolones of *G. glabra* are used medicinally, as well as licorice, their thickened extract.

The plant is native in southern Europe and the Caucasus region. Cultivation and collection from wild occurrences take place in several countries in southern Europe. The drug is commercially available unpeeled or peeled. It is probably also derived from *G. uralensis* Fisch. ex DC., Chinese licorice.

Besides its broad medicinal use (see below) and the use as solubilizer of coexisting bioactive constituents in herbal extracts, licorice is used as a sweetener in food products, as functional food, confectionery, as spice, in cosmetics, and to flavor tobacco [64].

Roots and stolones of *G. glabra* contain 3–9% triterpene saponins. In the leaves and stems, the saponin content is about 0.9 and 1.7% resp. [19].

Until 2020, 77 triterpene saponins have been isolated from Glycyrrhiza species: 38 from *G. glabra*, 50 from *G. uralensis*, and 13 from *G. inflata* Batal [64].

The main component of the saponin fraction of the mentioned species is the strongly sweet-tasting, hardly hemolytic active glycyrrhizic acid. Its ammonium–calcium–sodium–potassium salt is called glycyrrhizin. This term is often used to refer to glycyrrhizic acid too. The aglycone is 18β-glycyrrhetinic acid, an oleanane-type triterpenoic acid. A β-D-glucuronyl-(1→2)-β-D-glucuronyl residue is attached to its OH group at C-3 (Fig. 9.4). Further components of the saponin mixture are, among others, glycosides of hydroxy derivatives of glycyrrhetinic acid, its lactones, and 11-deoxy derivatives, e.g., 24-hydroxy-glycyrrhizin, which is also highly sweet [64, 90, 94, 225, **19**, **143**].

Glycyrrhizic acid is metabolized in the intestine by the intestinal flora predominantly to 18-glycyrrhetinic acid, which is well absorbed. It enters the enterohepatic circulation, is bound to lipoproteins of the blood, and is excreted very slowly.

Accompanying substances include flavonoid glycosides, e.g., with the aglycones liquiritigenin and its chalcone isoliquiritigenin, isoflavonoide glycosides, e.g., with the aglycones formononetin and licoricidin, 6-prenylated coumestane derivatives, including glycyrol and isoglycyrol, hydroxycoumarins, including herniarin, umbelliferon, and liqcumarin, and phytosterols [**19**, **20**, **143**].

Glycyrrhizic acid and its metabolite 3β-D-glucuronyl-18β-glycyrrhetinic acid, formed in the liver, inhibit steroid dehydrogenases (SDR, NADPH-dependent short-chain dehydrogenases/reductases), including 4,5-reductase and 11β-hydroxysteroid dehydrogenase, which are responsible for the degradation of steroid hormones: glucocorticoids, mineralocorticoids, and estrogens. The prevention of the degradation of these steroids, complemented by the modulatory intrinsic action of glycyrrhizic acid on glucocorticoid receptors, leads to several systemic effects. Very important is the anti-inflammatory activity. Inhibition of mineralocorticoid degradation results in hypokalemia. Inhibition of the breakdown of estrogens in the placenta causes undesirable effects on the fetus.

Numerous studies are made on the effects of extracts from Glycyrrhiza species. The effects are determined mainly, but not solely by glycyrrhizic acid. The accompanying substances in the extract, mentioned above, contribute to the overall effects. Worth mentioning are anti-inflammatory, antibacterial, antiviral, cytostatic, antitumoral, hepatoprotective, and neuroprotective activities [64, 90, 179, 225].

The flavonoids may contribute to the anti-inflammatory effect by suppressing the formation of TNF and IL-1 RNA. The hepatoprotective and estrogenic properties of glycyrrhizic acid are supported by isoflavones and 6-prenylated coumestane derivatives [2] and the spasmolytic effects by flavones, isoflavones, and hydroxycoumarins [185, 235]. The antimicrobial activity of glycyrrhizic acid against *Helicobacter pylori* is, additionally to anti-inflammatory effects, useful in the treatment of gastric ulcers [179].

According to the broad spectrum of activities licorice root or extracts prepared from it are used as expectorants in bronchial catharses, as spasmolytic components of laxatives and as components of finished medicinal products for the treatment of gastrointestinal inflammation [**20**, **149**].

Traditionally, licorice serves, especially in East Asia, as a tonic and 'rejuvenating' medicine. In Ayurvedic medicine it is known as 'Rasayana'. TCM uses *G. glabra, G. uralensis* and *G. inflata* for dyspnea, cough, influenza, gastric ulcers, pain, and hepatitis. Kampo medicine uses the neuroprotective effect of licorice [179].

Glycyrrhizin is well tolerated when used in low doses. Higher doses are toxic. However, even low doses can cause severe side effects in sensitive individuals. Simultaneous administration of other drugs, e.g., cardioactive steroid glycosides, may cause severe symptoms of intoxication because the saponins promote the solubilization and absorption of other substances and inhibit their degradation.

> ♀ Inhibition of 11β-hydroxysteroid dehydrogenase type 2 activity by glycyrrhizinic acid suppresses the degradation of cortisol. Cortisol acts like the mineralocorticoid aldosterone on mineralocorticoid receptors. Pseudohyperaldosteronism occurs, characterized by $Na^+$ ion retention in cells, and increased $K^+$ ion excretion from cells. Consequences are hypokalemia and hypernatremia. This leads to hypertension, congestive heart failures, bradycardia, renal failures, formation of edema, asthma cardiale, intestinal colic, headache, visual disturbances, metabolic alkalosis, muscle weakness, partial paralysis, rhabdomyolysis, somnolence, and paralysis. In severe untreated cases, dysfunction of kidneys, salivary glands, colon, and placenta may occur. In premenstrual syndrome, licorice may cause water retention and flatulence. Deterioration of glucose tolerance may cause failure in hypoglycemic therapy in diabetic patients [22, 34, 57, 87, 147, 225].

Especially remarkable is the harmful estrogenic effect of licorice extract on the fetus in pregnant women. When pregnant women ingest large amounts of glycyrrhizin (more than 100 mg/day and about 50 g/day licorice), placental 11β-hydroxysteroid dehydrogenase type 2 and thus the degradation of maternal estrogens are strongly inhibited. The fetus is exposed to high concentrations of estrogens in the placenta. Presumably, the estrogenic action of glycyrrhetinic acid itself is also involved in this effect.

In Finland, studies were conducted with pregnant women who had a high consumption of licorice extract caused by the therapeutic use as expectorant or by excessive snacking on licorice candies. External hemorrhages and hematomas occurred in the pregnant women themselves. The gestational period was shortened (<37 weeks). The number of preterm births was doubled, and miscarriages occurred more frequently [40]. The birth weight of the children was not affected [253]. The girls of these mothers, had, determined at the ages of 8 and 12 years, poorer cognitive abilities, poorer memory performance, attention deficits, and externalizing symptoms. IQ was lower than that of peers. They were taller and had higher body weight for their age than normal children. Physical and psychological maturity (Tanner staging score) occurred earlier [201].

Symptoms of intoxication may occur with continuous use of licorice also, e.g., as a tea, and as preparations from licorice, e.g., confectioneries containing licorice (more than 500 mg glycyrrhizin/week). The content of such confectioneries may be 30–200 mg of glycyrrhizinic acid per 100 g. Consumption of southern French pastis (containing 1.13 g/L glycyrrhizin [33]), of chewing gum or chewing tobacco containing glycyrrhizic acid for several months may also cause adverse health effects [22, 207].

In the USA, the FDA has set the following maximum limits for glycyrrhizin: Baked goods 500 mg/kg, alcoholic beverages 1,000 mg/kg, nonalcoholic beverages 1,500 mg/kg, chewing gum 1,100 mg/kg, hard candy 160,000 mg/kg, soft candy 31,000 mg/kg, herbs and spices 1,500 mg/kg, allowable concentration in dietary supplements 0.05–0.15% [63]. FEMA (Flavor and Extract Manufactures Association of the US) has set lower limits [112]. The safety limit glycyrrhizin was set at 100 mg/day [63]. However, as little as 60–80 mg daily was already capable to raise blood pressure in susceptible individuals [122]. For strong licorice content (>200 mg glycyrrhizin/100 g) labeling is mandatory in Germany [22].

The FDA warns against excessive consumption of licorice. 'Persons over 40 years of age who consumed more than 2 ounces (about 60 mg), or about 60 g, of Licorice daily over a 2 week period are at risk of being hospitalized with cardiac arrhythmias' [9]. The no-effect level is reported as 2 mg glycyrrhizin/kg BW [23]. The acceptable daily intake of glycyrrhizic acid is 0.2 mg/kg BW. Because of their side effects, the duration of therapeutic use of licorice extracts without medical advice should not exceed 4–6 weeks.

Particularly at risk are hypertensives, diabetics, people with cardiovascular diseases, and pregnant women. It seems that licorice and confectionaries with it are not well suited for children. Adults should control the ingested amounts. Pregnant women should avoid licorice.

There are no objections to the use of the drug as a flavor corrigendum and as an ingredient in confectionery up to a daily dose equivalent to 100 mg glycyrrhizin [22].

🐾 Dogs and cats can eat low quantities of licorice, e.g., licorice candy. But it is not healthy for them. Higher doses are toxic. Gastrointestinal problems including vomiting and diarrhoe can occur.

## References

(for numbers in bold see cross chapter literature)

[1]     Abe N et al. (1982) Microbiol Immunol 26: 535
[2]     Akao T et al. (1992) Chem Pharm Bull 40: 1208
[3]     Al-Afifi NA et al. (2018) BMC Compl Altern Med 18: 50
[4]     Al-Awthan YS, Bahattab OS (2021) Bio Med Res Int 2021: 8561696
[5]     Altunkeyik H et al. (2012) Phytochemistry 73: 127
[6]     Anderson L et al. (1989) Toxicon 27: 179
[7]     Anil H (1977) Doga 1: 148
[8]     Anisimov MM, Cirva JV (1980) Pharmacy 35. 731
[9]     Anonymous (2017) Dtsch Apoth Ztg Online, 11/ 23/2017
[10]    Ariano R et al. (2006) Eur Ann Allerg Clin Immunol 38(3): 90
[11]    Aurada E et al. (1984) Planta Med 50(5): 391
[12]    Babadjamian A et al. (1988) Spectrosc Lett 21: 565
[13]    Bader G, Hiller K (1987) Pharmacology 42: 577
[14]    Bailly C (2021) Digital Chin Med 4(3): 159
[15]    Barbieri L et al. (1989) Biochem j 257(3): 801
[16]    Barker BE et al. (1966) Pediatrics 38: 490

[17]    Barnett BD (1975) Poultr sci 54(4): 1215
[18]    Baxter CP (1983) Vet Rec 113: 247
[19]    Bedir E et al. (2000) Phytochemistry 53(8): 905
[20]    Behl T et al. (2021) Biomed Pharmacother 133: 110959
[21]    Beressa TB et al. (2020) Infect Dis (auckl) 13(1). 10.1177/11786337209443509
[22]    Bielenberg J (1999) Dtsch Apoth Ztg 139(35): 3282 see also 139 (37):3461
[23]    Bijlsma J et al. (1996) Rijksinstitut for Volksgezondheit en Milieu Bilthoven: Report No 348801004
[24]    Blunden G, Patel AV (1986) J Nat Prod 49(4): 68
[25]    Boll PM, Hansen L (1987) Phytochemistry 26(1): 2955
[26]    Brazelli V et al. (1995) Contact Dermat 33(1): 60
[27]    Bruhn G et al. (1987) Z Naturforsch b Chem Sci 42: 1328
[28]    Burrows GE, Stairs EL (1990) Vet Hum Toxicol 32(3): 259
[29]    Calis I et al. (1997) J Nat Prod 60(3): 315
[30]    Calis I et al. (1997) Planta Med 63(2): 166
[31]    Camp BJ et al. (1988) Vet Hum Toxicol 30: 533
[32]    Cannon JG et al. (2004) J Chem Educ 2004 81(10): 1457
[33]    Cerada JM et al. (1983) Lancet i 8339: 1442
[34]    Chamberlain TJ (1970) J Am Med Assoc 213(8): 1343
[35]    Chandel RS, Rastogil RP (1980) Phytochemistry 19(9): 1889
[36]    Cheong DHJ et al. (2018) Cancer Lett 422: 1
[37]    Cherpelis BS, Fenske NA (2000) Cutis 66(4): 287
[38]    Chiu LCM et al. (2001) Int J Oncol 19(1): 137
[39]    Choe S et al. (2020) J Chromatogr B 1149: 122123
[40]    Choi JS et al. (2013) Planta Med 29(2): 97
[41]    Choi JY et al. (2018) J Clin Med 7(10): 377
[42]    Christensen LP et al. (1991) Phytochemistry 30(12): 4151
[43]    Clochard J et al. (2020) Int J Pharm 589: 119822
[44]    Constantinescu E, Sommer L (1968) Herba Hung 7: 59
[45]    Dekanoside GE et al. (1971) Soobschenja AN TCCP 61: 609
[46]    Dekanoside GE et al. (1984) Khim Prir Soedin 747
[47]    Demirkol O et al. (2004) J Agric Food Chem 52(26): 8151
[48]    Desta KT, AMA E-A (2021) Mini Rev Med Chem 21(1): 23
[49]    Diaz AEC et al. (2019) Phytochemistry 164: 67
[50]    Diaz-Perales A et al. (2002) J Allergy Clin Immunol 110(5): 790
[51]    Ding Y et al (1993) Chem Pharm Bull 41(3): 557
[52]    Domon B, Hostettmann K (1984) Helv Chim Acta 67: 1310
[53]    Dorchai RP et al. (1968) Tetrahedron 24: 5649
[54]    Draghici LR (2020) J Agroaliment Proc Technol 26(4): 422
[55]    Driermeier D et al. (2002) Toxicon 40(7): 1027
[56]    Du L et al. (2018) China J Chin Materia Med 43(12): 2552 in Chinese
[57]    Duax WL, Ghosh D (1998) Endocr Res 24(3/4): 521
[58]    Eisenburger R et al. (1974) Arzneim Forsch 24: 1621
[59]    Elekofehinti OO et al. (2015) Pathophysiology 22(2): 95
[60]    Elekofehinti OO et al. (2017) Word J Diabetes 8(7): 337
[61]    Elias R et al. (1991) J Nat Prod 54(1): 98
[62]    Fan JY et al. (2014) Molecules 19(7): 10650
[63]    FDA, Licorice and Licorice Derivatives, US Food and Drug Administration
[64]    Feifei L et al. (2020) Molecules 25(17): 3904
[65]    Fernández-Campos F et al. (2019) Pharmaceutics 11(9): 426

[66]  Findlay JA et al. (1992) J Nat Prod 55(1): 93
[67]  Fowden L, Bryant M (1958) Biochem J 70(4): 626
[68]  Funayama S, Hikino H (1979) J Nat Prod 42(6): 672
[69]  Gafner F et al. (1989) Phytochemistry 28(4): 1256
[70]  Gaillard Y et al. (2003) J Anal Toxicol 27(4): 257
[71]  Gajdus J et al. (2014) Pharmacogn Mag 10(suppl 2): S324
[72]  Gallelli L (2019) Drug Des Devel Ther 13: 3425
[73]  Gallelli L (2020) J Clin Pharmacol 60(7): 815
[74]  Gallelli L et al. (2021) Drug Des Devel Ther 15: 699
[75]  Gao M et al. (2009) J Asian Nat Prod Res 11(5): 433
[76]  Glastonbury JR et al. (1984) Aust Vet J 61(10): 314
[77]  Gonzalez AG (1977) Tetrahedron Lett 2959
[78]  Gonzalez AG et al. (1972) Tetrahedron 28: 1289
[79]  Goodfellow RM, Goad LJ (1983) Comp Biochem Physiol B 76 B: 575
[80]  Gorshkova IA et al. (1989) Toxicon 27: 927
[81]  Goryana G et al. (1976) Khim Prir Soedin 6: 823+400
[82]  Gossel TA, Bricker JD (1990) Principles of Clinical Toxicology, 2 nd ed. Raven Press, New York
[83]  Griffini A et al. (1997) Fitotherapia 68: 520
[84]  Groof, de, RC, Narahashi T (1976) Eur J Pharmacol 36: 337
[85]  Guo Q et al. (2020) J Functional Food 103727
[86]  Hampel H et al. (1970) Arzneim Forsch 20: 209
[87]  Hardens H, Rausch-Stroomann IG (1953) Muench Med Wochenschr 95: 580
[88]  Hardin JW (1973) North Carolina Agr Exp Sta Bull 414
[89]  Harkar S et al. (1984) Phytochemistry 23(12): 2893
[90]  Hasan K et al. (2021) Heliyon 7(6): e07240
[91]  Hasegawa T et al. (1987) Chem Lett 329
[92]  Hausen BM et al. (1987) Contact Dermat 17(1): 1
[93]  Haworth RD (1938) Helv Chim Acta 21: 1378
[94]  Hayashi H et al. (1993) Planta Med 59(4): 351
[95]  Hebestreit P et al. (2003) Z Phytother 24(5): 249
[96]  Hebestreit P, Melzig MF (2003) Planta Med 69(10): 921
[97]  Heftmann E (1967) J Nat Prod 30: 209
[98]  Hiller K (1996) Dtsch Apoth Ztg 136(9): 723
[99]  Hiller K, Adler C (1982) Pharmazie 37: 619
[100] Hiller K, Voigt G (1977) Pharmazie 32: 365
[101] Hiller K et al. (1966) Pharmazie 21: 713
[102] Hiller K et al. (1981) Pharmazie 36: 133
[103] Hoppe W et al. (1968) Angew Chem 80: 563
[104] Hostettmann K (1980) Helv Chim Acta 63: 606
[105] Hostettmann K et al. (1982) Planta Med 44(1): 34
[106] https://petcareadvisor.com/cats/isdracaena-toxic-to-cats/
[107] https://simplifygardening.com/are-snake-plants-toxic-to-pets/
[108] Hübner G et al. (1999) Planta Med 65(7): 636
[109] Ighodaro OM et al. (2017) J Intercult Ethnopharmacol 6(2): 234
[110] Iglesias R et al. (2016) Biochem Biopys Acta (bba)-General Subjects 1860(6): 1256
[111] Iqbal M et al. (2017) J Plant Biochem Physiol 5(1): 1
[112] Isbrucker RA, Burdock GA (2006) Regul Toxicol Pharmacol 46(3): 167
[113] Islam MA et al. (2016) Oxid Med Cell Longev 2016: 5137431
[114] Jaeckle K, Freeman FR (1981) South Med J 74: 639

[115]  Janeczko Z (1978) Planta Med 36: 266

[116]  Janeczko Z (1980) Acta Polon Pharm 37: 559

[117]  Janeczko Z et al. (1987) Planta Med 53(1): 52

[118]  Janeczko Z, Sibiga A (1982) Herba Polon 28: 115

[119]  Jang DS et al. (2004) Agric Food Chem 52(8): 2218

[120]  Jaramillo S et al. (2016) J Funct Foods 26: 1

[121]  Jaspersen-Schib R (1987) Schweiz Apoth Ztg 125: 129

[122]  JECFA/IPSC, Joint FAO/WHO Expert Committee on Food Additives (JECFA). International Program on Chemical Safety (IPSC) WHO Geneva, 2006

[123]  Jeepipalli SPK et al. (2020) Food Chem 318: 126474

[124]  Jenett-Siems K et al. (2012) Z Naturforsch C 67(11-12): 565

[125]  Ji Y et al. (2012) Curr Oncol 19(suppl 2): eS1). 10.3747/co.19.1139

[126]  Jin JM et al. (2003) J Asian Nat Prod Res 5(2): 95

[127]  Jin JM et al. (2004) Chem Pharm Bull 52(6): 654

[128]  Johnson IT et al. (1986) J Nutr 116: 227015

[129]  Jurkiewicz D et al. (2016) Otolaryngol Pol 70: 1

[130]  Kang SS, Woo WS (1980) J Nat Prod 43(4): 510

[131]  Kartnig T, Herbst R, Graune FJ (1965) Planta Med 13(1): 39

[132]  Karuza-Stojakovic L et al. (1991) Pharmazie 46: 303

[133]  Kasai T (1981) J Agric Biol Chem 45(2): 433

[134]  Kasai T et al. (1981) Phytochemistry 20(9): 2209

[135]  Kawano K et al. (1975) Agric Biol Chem 39: 1999

[136]  Kawano K et al. (1977) Agric Biol Chem 41: 1

[137]  Kerner J et al. (1973) Arch Dermatol (USA) 108: 102

[138]  Khan H, Rauf A (2015) Am J Biomed Life Sci 3(2-1): 5

[139]  Khan M et al. (2016) J Cell Mol Med 20(3): 389

[140]  Kiliani H (1890) Ber Dtsch Chem Ges 23: 1555

[141]  Kim HR et al. (2020) BMC Compl Med Ther 20(1): 325

[142]  Kim JW et al. (2017) Bioorg Med Chem Lett 27(13): 3019

[143]  Kintya PK et al. (1976), Khim Prir Soedin: 670

[144]  Kizu H et al. (1985) Chem Pharm Bull 33: 1400+3473

[145]  Klose P, Pistor K (1976) Münch Med Wochenschr 118: 719

[146]  Konoshima T, Lee KH (1986) J Nat Prod 49(4): 650

[147]  Koster M, David GK (1968) N Engl J Med 278(25): 1381

[148]  Kubo M et al. (1984) Yakugaku Zasshi 104: 757

[149]  Kumar A et al. (2014) World J Pharm Res 3(6): 2099

[150]  Kurt SM, Meljutschuk OP (1960) Farmakologija Tokssikologija 23: 61

[151]  Lazurevskii GV et al. (1976) Dokl Akad Nauk SSSR 231: 1479

[152]  Lee DY et al. (2009) J Ethnopharmacol 121(1): 28

[153]  Lee WH et al. (2011) Toxicol Res 27(4): 217

[154]  Lemma A, Yau P (1974) Ethiop Med J 12: 115

[155]  Leschke E (1932) Münch Med Wochenschr 1837

[156]  Lewis WH, Smith PR (1979) J Am Med Assoc 242: 2759

[157]  Li Q et al. (2019) Mitochondrial DNA B 4(2): 3135

[158]  Li YH et al. (2015) World Gastroenterol 21(31): 9262

[159]  Liang ZZ et al. (1988) Planta Med 54(4): 344

[160]  Liu B et al. (2021) Bioorganic Chem 115: 105237

[161]  Loloiko AA et al. (1988) Khim Prir Soedin 379

[162]  Lozano MCet al. (2017) Toxins (Basel) 9(7): 220

[163]  Lutsenko Y et al. (2010) Herba Polonica 56(1): 83

[164]  Lythgoe B, Trippett S (1950) J Chem Soc 1983

[165]  Ma JCN, Lau FW (1985) Phytochemistry 24(7): 1561

[166]  Maharaj I et al. (1986) Can J Microbiol 32: 414

[167]  Mahato SB et al. (1982) Phytochemistry 21(5): 959

[168]  Mahato SB et al. (1988) Phytochemistry 27(10): 3037

[169]  Majester-Savornin B et al. (1991) Planta Med 57(3): 260

[170]  Marker RE, Rohrmann E (1939) J Am Chem Soc 61: 2724

[171]  Marrelli M et al. (2016) Molecules 21(10): 1404

[172]  Marsh CA, Levvy GA (1956) Biochem J 63: 9

[173]  Matsumoto M et al. (2000) Zygote 8 Suppl 1: S62

[174]  Mebel H, Patt P (1960) Arzneim Forsch 10: 280

[175]  Melzig MF et al. (2001) Planta Med 67(1): 43

[176]  Melzig MF et al. (2005) Planta Med 71(11): 1088

[177]  Meyer-Bertenrath J, Kaffanik H (1970) Arzneim Forsch 20: 147

[178]  Mihci-Gaidi G et al. (2010) Nat Prod Commun 5(7): 1023

[179]  Ming LJ et al. (2013) Nat Prod Commun 8(3): 415

[180]  Moses T et al. (2014) Crit Rev Biochem Mol Biol 49(6): 439

[181]  Mshvildadze V et al. (2001) Chem Pharm Bull 49(6): 752

[182]  Nakamura S, Mori KJ (1984) Toxicology 29: 235

[183]  Negi JS et al. (2010) Pharmacogn Rev 4(8): 215

[184]  Nohara T et al. (1982) Chem Pharm Bull 30: 1851

[185]  Nose M et al. (1994) Planta Med 60(2): 136

[186]  Oka K (1999) Contact Dermat 40(4): 209

[187]  Oleszek W et al. (1990) J Agric Food Chem 38: 1810

[188]  Özdemir C et al. (2003) Dermatologist 54(10): 966

[189]  Panova D, et al. (1982) Farmatsiya 32: 42

[190]  Pant G et al. (1988) Phytochemistry 27(10): 3324

[191]  Pant G et al. (1988) Pharmazie 43: 294

[192]  Pare JRJ et al. (1981) J Nat Prod 44(4): 490

[193]  Parkhurst RM et al. (1990) Phytochemistry 29(4): 1171

[194]  Pegiou E et al. (2020) Metabolites 10(1): 17

[195]  Pilz J et al. (1992) Curr Derm 18: 361

[196]  Pollock MI et al. (2015) Vet J 206(3): 275

[197]  Price KR et al. (1987) CRC Crit Rev Food Sci Nutr 26: 27

[198]  Qin HL et al. (2004) Zhongguo Zhong Yao Za Zhi 29(1): 42

[199]  Quin X-J et al. (2016) Phytochemistry vol. 121: 20

[200]  Rademaker M, Yung A (2000) Aust J Dermatol 41(4): 262

[201]  Räikkönen K et al. (2017) Am J Epidemiol 185(5): 317

[202]  Ramos-Morales E et al. (2019) FEMS Microbiol Lett 366(13): fnz 144

[203]  Rao AV, Kendall CW (1986) Food Chem Toxicol 24: 441

[204]  Reznicek G et al. (1989) Phytochemistry 28(3): 825

[205]  Ricks MR et al. (1999) J Am Acad Dermatol 40(2Pt 2): 356

[206]  Rieker J et al. (2004) Dermatologist 55(4): 397

[207]  Rosseel M, Schoors D (1993) Lancet 341: 175

[208]  Rucizka L, Jeger O (1942) Helv Chim Acta 25: 775

[209]  Safayhi H, Saller ER (1997) Planta Med 63(6): 487

[210]  Salinas ML et al. (2001) Contact Dermat 44(2): 94

[211]  Sati OP, Pant G (1985) J Nat Prod 48(3): 390

[212]   Sati OP, Rana U (1987) Int J Crude Drug Res 25: 158

[213]   Sati OP, Rana U (1987) Pharmazie 42: 141

[214]   Schlenger R (2003) Dtsch Apoth Ztg 143(49): 6278

[215]   Schlösser E (1971) Acta Phytopathol 6: 85

[216]   Schmiedeberg O (1875) Arch Exp Path Pharm 3: 16

[217]   Schmitt M, Cremer W (1983) Kidney Hypertens Dis 12(8): 306

[218]   Schneider G (1958) Biochem Z 330: 428

[219]   Schöpke T, Hiller K (1990) Pharmacology 45: 313

[220]   Schrot J et al. (2015) Toxins 7(5): 1556

[221]   Schweitzer H (1952) Med Clin 47: 683

[222]   Sendrai J, Janeczko Z (1981) Herba Hung 20: 187

[223]   Shao Y et al. (1997) Planta Med 63(3): 258

[224]   Sharipov A et al. (2021) Molecules 26(13): 3784

[225]   Sharivi-Rad J et al. (2021) Oxid Med Cell Longev 2021: 7571132

[226]   Sharma R et al. (2020) Hum Vaccine Immunother 16(12): 2944

[227]   Sharma SC (1981) Pharmazie 36: 709

[228]   Sharma SC, Sharma HC (1984) Phytochemistry 23(3): 645

[229]   Sharma SC et al. (1980) Pharmazie 35: 711

[230]   Sharma SC et al (1981) Phytochemistry 22(10): 2259

[231]   Sharma SC et al. (1982) Phytochemistry 21(8): 2075

[232]   Sharma SC et al. (1983) Planta Med 47(2): 117

[233]   Shaskov AS et al (1987) Khim Prir Soedin 363

[234]   Sheldon S (1980) Econ Bot 34: 376

[235]   Shim SB et al. (2000) Planta Med 66(1): 40

[236]   Sibiga A et al. (1986) Herba Polon 31(1): 21

[237]   Sidana J et al. (2016) Phytochemistry 130: 22

[238]   Siepmann C et al (1998) Planta Med 64(2): 159

[239]   Singh B et al. (1986) Planta Med 52(5): 409

[240]   Singh B et al. (1987) J Nat Prod 50(5): 781

[241]   Singh R Geetanjali (2016) Nat Prod Res 30(17): 1896

[242]   Singh SB et al. (1982) Phytochemistry 21(8): 2079

[243]   Sirtori CR (2001) Pharmacol Res 44(3): 183

[244]   Smith RA et al. (1997) Vet Hum Toxicol 39(4): 250

[245]   Southon S et al. (1988) Br J Nutr 59: 389

[246]   Spengel SM et al. (1995) Planta Med 61(4): 385

[247]   Spoerke DG et al. (1987) Vet Hum Toxicol 29(3): 250

[248]   Ştefănescu R et al. (2020) Biomolecules 10(5): 752

[249]   Stefanowicz-Hajduk J et al. (2015) PLoS ONE 10(8): e0135993

[250]   Stirpe F et al. (1983) Biochem J 216(3): 617

[251]   Stolzenberg SJ, Parkhurst RM (1974) Contraception 10: 135

[252]   Storie GJ et al. (1990) Aust Vet J 69(1): 21

[253]   Strandberg TE et al. (2002) Am J Epidemiol 156(9): 803

[254]   Sugiyama M et al. (1984) Chem Pharm Bull 32: 1365

[255]   Sung JE et al. (2017) Int J Mol Med 40(5): 1365

[256]   Tabar AI et al. (2004) Clin Exp Allergy 34(1): 131

[257]   Takasugi M et al. (1975) Chem Lett 1: 43

[258]   Takechi M, Tanaka Y (1990) Phytochemistry 29(2): 451

[259]   Tetenyi P (1977) Plant Med Phytother 11: 158

[260]   Thakur M et al. (2011) Botanics: Targets Ther 1: 19

[261] Theus L (1994) Dissertation, University of Basel

[262] Thilborg ST et al. (1993) Phytochemistry 32(5): 1167

[263] Thimmappa R et al. (2014) Ann Rev Plant Biol 65: 225

[264] Thu ZM et al. (2021) Molecules 26(7)): 1916

[265] Timon-David P et al. (1980) Ann Pharm Fr 38: 545

[266] Tressl R et al. (1977) J Agric Food Chem 25(3): 455

[267] Tschesche R, Schulze H (1974) Chem Ber 107: 2710

[268] Tschesche R, Wulff G (1963) Tetrahedron 19: 621

[269] Tschesche R, Wulff G (1964) Planta Med 12(3): 272

[270] Tschesche R, Wulff G (1973) Fortschr Chem Org Naturst 30: 461

[271] Tschesche R et al. (1965) Z Naturforsch 20b 708

[272] Tschesche R et al. (1970) Z Naturforsch 25b 999

[273] Tschesche R et al. (1966) Liebigs Ann Chem 691: 165

[274] Tschesche R et al. (1969) Liebigs Ann Chem 721: 194

[275] Turton (1925) Brit Med J 298

[276] Ussery MA et al (1977) Ann NY Acad Sci 284: 431

[277] Vadkerti A et al (1989) Chem Papers 43: 783

[278] Van Damme EJ et al (2000) Eur J Biochem 267(9): 2746

[279] Vincken JP et al. (2007) Phytochemistry 68(3): 275

[280] Vogel G (1963) Pharmazie 11: 362

[281] Voigt G, Hiller K (1987) Sci Pharm 55: 201

[282] von Gorup-Besanez E (1861) Liebig's Ann Chem 118: 236

[283] von Mühlendahl KE et al. (1978) Intensive Care Med 15(2): 65

[284] von Sethe CH, Berning H (1964) Dtsch Med Wochenschr 89: 1555

[285] Wagner J et al. (1971) Arch Pharm 304: 804

[286] Wall ME, Fenske CS (1961) Econ Bot 15: 131

[287] Wang D et al. (2009) Nat Prod Res 23(10)

[288] Weise Ch et al (2020) Sci Rep 10(1): 15377

[289] White RH (1975) Science 189: 810

[290] Wie F et al. (2004) Chem Pharm Bull (tokyo) 52(10): 1246

[291] Wilkomirski B et al. (1975) Phytochemistry 14(12): 2657

[292] Williams MC, Olsen JD (1984) Am J Vet Res 45(3): 539b

[293] Woitke HD et al. (1970) Pharmazie 25: 133+213

[294] Wolters B (1966) Planta Med 14(4): 392

[295] Woo WS et al. (1980) Tetrahedron Lett 4255

[296] Wulff G, Tschesche R (1973) Fortschr Chem Org Naturst 30: 461

[297] Xu XH et al. (2016) Molecules 21(10): 1326

[298] Yang-Hua YI (1991) Phytochemistry 30(12): 4179

[299] Yayli N et al. (1998) Planta Med 64(4): 382

[300] Yayli N et al. (1998) Phytochemistry 48(5): 881

[301] Yesudian PD, Franks A (2002) Contact Dermat 46(2): 125

[302] Yeung HW et al. (1987) J Ethnopharmacol 21: 31

[303] Yi YH et al. (1990) Pharm Sinica 25(10): 745

[304] Yokosuka A et al. (2000) Planta Med 66(4): 393

[305] Yoshikawa M, Matsuda H (2000) In: Olezek W, Marston H Eds. Saponins in Food, Feedstuff and Medicinal Plant. Springer, Berlin

[306] Yoshikawa M et al. (1994) Chem Pharm Bull 42(6): 1357

[307] Yoshikawa M et al. (1996) Chem Pharm Bull 44(8): 1454

[308] Yosioka I, Kitayama I (1967) Tetrahedron Lett 2577

[309]  Zhang RS et al. (2021) Nat Prod Bioprospect 11(6): 651
[310]  Zhang T et al. (2020) Critical Rev Food Sci Nutr 60(1): 94
[311]  Zhang Z et al. (2010) Pharm Crops 1: 24
[312]  Zhao J, Yang XW (2003) J Asian Nat Prod Res 5(3): 197
[313]  Zhao P et al. (2018) J Ethnopharmacol 214: 274
[314]  Zhao SQ et al. (2018) Chin J Nat Med 16(2): 105
[315]  Zhao X, Li J (2015) Nat Prod Commun 10(4): 683
[316]  Zhou N et al. (2021) Front Bioeng Biotechnol 713860

# Cross-Chapter Literature and Further Reading

(Numbers inside the chapters in bold)

[1]   Aktories K et al., eds (2017) Allgemeine und Spezielle Pharmakologie und Toxikologie, 12th ed, Urban & Fischer, München, Jena and Elsevier, Amsterdam. Formerly: Forth W et al. (2001) Allgemeine und spezielle Pharmakologie und Toxikologie. Urban & Fischer, München

[2]   Ali SI, Venugopalan V (2019) Poisonous Plants. LAP Lambert Academic Publishers, Chisinau, Moldova

[3]   Alonso-Castro AJ et al. (2017) Medicinal plants from North and Central America and the Caribbean considered toxic for humans: the other side of the coin.eCAM 2017 9439868

[4]   Altmann H (1997) Giftpflanzen – Gifttiere: Die wichtigsten Arten–Erkennen, Giftwirkung, Therapie, 3rd ed. BLV, München

[5]   Ammon Hpt ed (2001) Arzneimittelneben- und Wechselwirkungen, 4th ed. WVG Stuttgart

[6]   Armstrong J, Pascu O (2015) Toxicology Handbook, 3rd ed. Elsevier, Amsterdam

[7]   Aronson JK (2014) Plant Poisons and Traditional Medicine. In: Farrar J, Hotez PJ, Junghanss T et al. (eds.) Manson's Tropical Infectious Diseases, 23rd ed. W.B. Sauders, London. p 1128-1150

[8]   Aronson JK ed (2009) Meyler's side effects of herbal drugs. The International Encyclopedia of Adverse Drug Reactions and Interactions. Elsevier, Amsterdam

[9]   Avigan MI et al. (2016) Scientific and regulatory perspectives in herbal and dietary supplement associated hepatotoxicity in the united states. Int J Mol Sci 17(3): 331

[10]  Barceloux DG (2008) Medical Toxicology of Natural Substance, Foods, Fungi, Medicinal Herbs, Plants, and Venomous Animals, John Wiley & Sons Inc., Hoboken, New Jersey

[11]  Barile FA (2010) Clinical Toxicology, 2nd ed. CRC Press, Boca Raton

[12]  Barton S, Nakanishi K eds (1999) Comprehensive Natural Products Chemistry, Vol. 1–9. Elsevier, Amsterdam

[13]  Beasley V (2004) Veterinary toxicology. Int Vet Information Service www.ivis.org/, Arden Hills

[14]  Belitz HD, Grosch W (1992) Lehrbuch der Lebensmittelchemie, 4th ed. Springer, Berlin

[15]  Bentz H (1969) Nutztiervergiftung. Erkennung und Verhütung, Fischer-Verlag, Jena

[16]  BfR (2000) Bekanntmachung einer Liste giftiger Pflanzen. Bundesanzeiger 52(86): 8517

[17]  BfR (2017) Risiko Vergiftungsunfälle bei Kindern. www.bfr.bund.de/cm/350/risiko-vergiftungsunfaelle-bei-kindern.pdf

[18]  Bhakuni DS, Rawat DS (2005) Bioactive Marine Natural Products. Springer, New York

[19]  Blaschek W et al. (edition as CD ROM since 2015) Hager ROM/Hagers Enzyklopädie der Arzneistoffe und Drogen, Springer-Verlag Berlin, Heidelberg, New York, 2019, formerly Hagers Handbuch der Pharmazeutischen Praxis. Last printed edition Hänsel, R et al. eds. (1992-1998), Drogen und Folgebände Drogen, 5th ed.

[20]  Blaschek W ed (2016) Wichtl – Teedrogen und Phytopharmaka. Ein Handbuch für die Praxis, 6th edition, WVG, Stuttgart. Formerly: Wichtl M ed Teedrogen und Phytopharmaka, WVG Stuttgart

[21]  Blunt JW, Munro HGM eds (2007) Dictionary of Marine Natural Products. Chapman & Hall/CRC, London

[22]  Bös B, Giftpflanzencompendium. www.giftpflanzen.com

[23]  Brugsch H, Klimmer OR (1966) Vergiftungen im Kindesalter. Ferdinand Enke, Stuttgart

[24]  Buckingham J ed (2020) Dictionary of Natural Products, Vol 1-6, Suppl 1. Taylor and Francis Ltd., London

[25]  Caius JF (2003) The Medicinal and Poisonous Plants of India. Scientific Publishers, Jodhpur

[26]  Calitz C et al. (2015) Herbal hepatotoxicity: current status, examples, and challenges. Expert Opin Drug Metab Toxicol 11: 1551

https://doi.org/10.1515/9783110724738-010

[27] Canadian Biodiversity Information Facility (2015) Poisonous plants information system. http://www.cbif.gc.ca/eng/species-bank/canadian-poisonous-plants-information-system/poisonous-plants-sites/?id=1370403265037

[28] Centers for Disease Control and Prevention. Poisonous plants. http://www.cdc.gov/niosh/topics/plants/

[29] Chen Y et al. (2019) The etiology, prevalence and morbidity of outbreaks of photosensitisation in livestock: a review. PLoS ONE 14(2): e0211625

[30] CliniTox: www.clinitox.ch / www.giftpflanzen.ch

[31] Collett MG (2019) Photosensitisation diseases in animals: classification and a weight of evidence approach to primary causes. Toxicon: X 3: 100012

[32] Dabrowski WM, Sikorski ZE eds (2004) Toxins in Food. CRC Press, Boca Raton

[33] Dauncey E, Lawson S (2018) Plants that Kill: A Natural History of the World's Most Poisonous Plants. Princeton University Press

[34] Daunderer M (1990) Drogenhandbuch für Klinik und Praxis. Diagnose, Therapie, Nachweis, Prophylaxe, Recht, Drogenprofile, Vol. III. Pharmakologie, ecomed, Landsberg/Lech

[35] Daunderer M (1995) Lexikon der Pflanzen- und Tiergifte, Diagnostik Und Therapie, Nikol Verlagsges. Hamburg

[36] D'Mello JFP ed (1997) Handbook of Plant and Fungal Toxicants. CRC Press, Boca Raton

[37] Dingermann Th et al. (2004) Arzneidrogen, 5th ed. Spektrum Akademischer Verlag, München

[38] Dunkelberg H et al. eds (2007) Handbuch der Lebensmitteltoxikologie. Vol. 1–5, Wiley VCH, Weinheim

[39] Durrant M (2013) Handbook of Clinical Toxicology, Hayle medical, New York

[40] Efferth T (2006) Molekulare Pharmakologie und Toxikologie. Springer, Berlin

[41] Erhardt W et al. eds (2014) Zander–Handwörterbuch der Pflanzennamen. 19th and Former Eds. Ulmer, Stuttgart

[42] Eisenbrandt G et al. (2005) Toxikologie für Naturwissenschaftler und Mediziner, 3rd ed. Wiley-VCH, Weinheim

[43] Esser P (1910) Die Giftpflanzen Deutschlands, Edition Kramer, Reprint 2019. Rhenania Verlagsges, Koblenz

[44] Everist SL (1981), Poisonous Plants of Australia, 2nd ed. Angus and Robertson, Sydney

[45] Flora of North America. www.eFloras.org

[46] Franke G ed (1994/95) Nutzpflanzen der Tropen und Subtropen, Vols I, II und III. Eugen Ulmer, Stuttgart

[47] Frenzel C, Teschke R (2016) Herbal hepatotoxicity: clinical characteristics and listing compilation. Int J Mol Sci 17: 588

[48] Frohne D, Jensen U (1998) Systematik des Pflanzenreichs, unter besonderer Berücksichtigung chemischer Merkmale und pflanzlicher Drogen, 5th ed. Fischer-Verlag, Jena

[49] Frohne D, Pfänder HJ (2005) Poisonous Plants, 2nd ed. Manson Publishing LTD, London

[50] Fu PP et al. (2013) Phototoxicity of herbal plants and herbal products. J Environ Sci Health C Environ Carcinog Ecotoxicol Rev 31: 213

[51] Geisslinger G et al. (2020), Mutschler Arzneimittelwirkungen. Pharmakologie – Klinische Pharmakologie – Toxikologie, 11th Ed, WVG, Stuttgart. Formerly: Mutschler E Et Al.: Arzneimittelwirkungen. Lehrbuch der Pharmakologie und Toxikologie, WVG Stuttgart

[52] Geschwinde Th (2018) Rauschdrogen–Marktformen und Wirkungsweisen, 8th ed. Springer Verlag, Berlin

[53] Gessner O (1931: 1st ed, 1953: 2nd ed, 1974: 3rd ed, published and reedited by Orzechowski G), Gift- und Arzneipflanzen von Mitteleuropa. Carl-Winter-Universitäts-Verlag, Heidelberg

[54] Ghorani-Azam A et al. (2018) Plant toxins and acute medicinal plant poisoning in children: a systematic literature review. J Res Med Sci 23: 26

[55] Gonçalves J et al. (2021) Psychoactive substances of natural origin: toxicological aspects, therapeutic properties and analysis in biological samples. Molecules 26: 1397

[56] Gopalakrishnakone P et al. eds (2017) Plant Toxins. Springer, Berlin, Heidelberg, New York

[57] Gossel TA, Bricker JD (2019) Principles of Clinical Toxicology, 3rd ed. CRC Press, Boca Raton

[58] Grandjean P (2016) Paracelsus revisited: the dose concept in a complex world, Basic Clin Pharmacol Toxicol 119(2): 126

[59] Gupta RC ed (2012) Veterinary Toxicology: Basic and Clinical Principles, 2nd ed. Elsevier INC, San Diego

[60] Habermehl GG, Ziemer P (1999) Mitteleuropäische Giftpflanzen und ihre Wirkstoffe, 2nd ed. Springer-Verlag, Berlin, Heidelberg, New York

[61] Hapke HJ (1988) Toxikologie für Veterinärmedizin, 2nd ed. Ferdinand Enke, Stuttgart

[62] Harborne JB et al. (1977) Dictionary of Plant Toxins, Wiley, Chichester

[63] Hardin JW, Arena JM (1977) Human Poisoning from Native and Cultivated Plants. Duke University Press, Durham, North Carolina

[64] Hartwig A et al. (2020) Mode of action-based risk assessment of genotoxic carcinogens. Arch Toxicol 94: 1787

[65] Hausen BM, Vieluf IK (1997) Allergiepflanzen, Handbuch und Atlas, 2nd ed. Nikol, Hamburg

[66] He S et al. (2019) Herb-induced liver injury: phylogenetic relationship, structure-toxicity relationship, and herb-ingredient network analysis, Int J Mol Sci 20: 3633

[67] Hegi G (1909–1931: 1st ed since 1936, 2nd ed since 1966, 3rd ed) Illustrierte Flora von Mitteleuropa. Publisher of the 1. ed: Carl Hauser Verlag, München

[68] Hegnauer R (1962–1990) Chemotaxonomie der Pflanzen I – X, XIa, XIb-1, XIb-2. Birkhäuser, Basel

[69] Helfrich W, Winter CK eds (2000) Food Toxicology. Taylor and Francis Ltd., London

[70] Herrmanns-Clausen M et al. (2014) Neubewertung von Giftpflanzen. www.bfr.bund.de/cm/343/neu bewertung-von-giftpflanzen.pdf

[71] Herrmanns-Clausen M et al. (2019) Akzidentelle Vergiftungen mit Gartenpflanzen und Pflanzen in der freien Natur. Bundesgesundheitsbl 62: 73–83. https://doi.org/10.1007/s00103-018-2853-5

[72] Herrmanns-Clausen M et al. (2019) Risiko Pflanze–ein neuer Ansatz zur Einschätzung des Vergiftungsrisikos für Kleinkinder. Bundesgesundheitsbl https://doi.org/10.1007/s00103-019-03034-5

[73] Hiller K, Melzig FM (2010) Lexikon der Arzneipflanzen und Drogen, 2nd ed. Spektrum Akad Verlag, Heidelberg

[74] Hodgson E, Roe M (2014) Dictionary of Toxicology, 3rd ed. Elsevier, Amsterdam

[75] Jaspersen-Schib R et al. (1996) Wichtige Pflanzenvergiftungen in der Schweiz 1966-1994. Schweiz Med Wochenschr 126: 1085–1098

[76] Karrer W et al. (1958) Konstitution und Vorkommen der organischen Pflanzenstoffe (exclusive Alkaloide), Suppl vol 1 (1977), Suppl vol 2, part 1 (1981), part 2 (1985), Birkhäuser, Basel

[77] Keeler RF, Tu AT eds (1983 and 1991), Handbook of Natural Toxins, Vol I: Toxicology of Plant and Fungal Compounds, Marcel Dekker New York and CRC Press, Boca Raton

[78] Keeler RF et al. (1978) Effects of Poisonous Plants on Livestock. Academic Press, New York

[79] Kingsbury JB (1964) Poisonous Plants of the United States and Canada, Prentice Hall Inc. Englewood Cliffs, New Jersey

[80] Klaassen CD ed (2008) Casarett and Doull's Toxicology–The Basic Science of Poisons. 7th ed. McGraw-Hill, New York

[81] Krenzelok EP, Mrvos R (2011) Friends and foes in the plant world: a profile of plant ingestions and fatalities. Clin Toxicol (Phila) 49(3): 142

[82] Krienke EG et al. (1986) Vergiftungen im Kindesalter. Ferdinand Enke, Stuttgart

[83] Kupper J, Demuth D (2010) Giftige Pflanzen für Klein- und Haustiere. Ferdinand Enke, Stuttgart

[84] Lander DG (2017) Veterinary Toxicology. Agri Bio Vet Press, Delhi

[85] Lendac Data Systems Ltdex System. www.pharmaceuticalonline.com/doc/toxicology-database-0001

[86] Leung AY (1996) Encyclopedia of Common Natural Ingredients Used in Food, Drugs and Cosmetics. Wiley, Cichester

[87] Lewin L (1992) Gifte und Vergiftungen–Lehrbuch der Toxikologie, 6th ed. Karl F Haug Verlag, Heidelberg, reprint of the original edition from 1929

[88] Lewin L (2007) Die Gifte in der Weltgeschichte, Updated and Expanded New Edition of Sorgegfrey C, by Tosa in the Publishing House of Carl Ueberreuter, Wien

[89] Lewis WH, Elvi-Lewis MPF (2003) Medical Botany–Plants Affecting Man's Health, Wiley, New York

[90] Liebenow H, Liebenow K (1988) Giftpflanzen: Ein Vademekum für Tierärzte, Humanmediziner, Biologen, Landwirte. Fischer-Verlag, Jena

[91] Lindner E (1990) Toxikologie der Nahrungsmittel. Thieme, Stuttgart, New York

[92] Linger RS, Flaherty DK (2020) A Guide to the Toxicology of Selected Medicinal Plants and Herbs of Easter North America, Publ Rebecca Linger, Charleston

[93] List PH, Hörhammer L (1967-1980) Hagers Handbuch der Pharmazeutischen Praxis, 4. Neuausgabe Band 1-8. Springer, Berlin

[94] LiverTox: Clinical and Research Information on Drug-Induced Liver Injury [Internet]. Bethesda (MD): National Institute of Diabetes and Digestive and Kidney Diseases (2012) https://www.ncbi.nlm.nih.gov/books/

[95] Luch A ed, Molecular, Clinical and Environmental Toxicology. Vol 1: Molecular Toxicology (2009), Vol 2: Clinical Toxicology (2010), Vol. 3: Environmental Toxicology (2012), Birkhäuser, Basel

[96] Luckner M (1984) Secondary Metabolism in Microorganisms, Plants and Animals. Fischer-Verlag, Jena

[97] Ludewig R, Regenthal R (2015) Akute Vergiftungen und Arzneimittelüberdosierungen, 11th ed. WVG Stuttgart

[98] Lüllmann H et al. (2016) Pharmakologie und Toxikologie, 18th ed. Thieme, Stuttgart

[99] Mabberley DJ (2017) Mabberley's Plant-Book, a Portable Dictionary of Plants, Their Classification and Uses, 4th ed. Cambridge University Press, Cambridge

[100] Machholz R, Lewerenz HJ (1989) Lebensmitteltoxikologie. Akademie Verlag, Berlin

[101] Mahady GB et al. (2001) Botanical Dietary Supplements: Quality, Safety and Efficacy. Swets & Zeitlinger Publ, Lisse

[102] Mansfelds R (1986) Verzeichnis landwirtschaftlicher und gärtnerischer Kulturpflanzen (ohne Zierpflanzen), 2nd ed (ed Schultze-Motel I). Akademie-Verlag, Berlin

[103] Marquardt H et al. (2019) Toxicology, 4th ed. WVG, Stuttgart

[104] McFarland SE et al. (2017) Systematic account of animal poisonings in Germany, 2012-2015. Vet Rec 180(13): 327

[105] McKenzie R (2012) Australia's Poisonous Plants, Fungi and Cyanobacteria: A Guide to Species of Medical and Veterinary Importance. CSIRO Publishing, Collingwood

[106] Mitchell J, Rook A (1979) Botanical Dermatology, Plants and Plant Products Injurious to the Skin. Greensgrass Ltd., Vancouver

[107] Moeschlin S (1986) Klinik und Therapie der Vergiftungen. Thieme, Stuttgart

[108] Moshobane MC et al. (2020) Plants and mushrooms associated with animal poisoning incidents in south africa. Vet Rec Open 7: e000402

[109] Mtewa AG et al. eds (2021) Poisonous Plants and Phytochemicals in Drug Development, 1st ed. Wiley & Sons, Inc., Hoboken, New Jersey

[110] National Poison Data System. https://www://aapcc.org//national-poison-data-system (NPDS)

[111] Nelson LS et al. (2007) Handbook of Poisonous and Injurious Plants. New York Botanical Garden. Springer, New York

[112] Neuwinger HD (1994) Afrikanische Arzneipflanzen und Jagdgifte. WVG, Stuttgart

[113] Nuhn P (2006) Naturstoffchemie: Mikrobielle, pflanzliche und tierische Naturstoffe. Hirzel, Stuttgart

[114] Oberdisse E, Hackenthal E eds (2002) Pharmakologie und Toxikologie, 3rd ed. Springer, Berlin

[115]  Petejova N et al. (2019) Acute toxic kidney injury. Renal Failure 41(1): 576

[116]  Peterson ME et al. (2006) Small Animal Toxicology. W.B. Saunders, Philadelphia

[117]  Plenert B et al. (2012) Plant exposures reported to the poison information centre erfurt from 2001–2010. Planta Med 78: 401

[118]  Plumlee KH (2004) Clinical Veterinary Toxicology. Mosby, St. Louis

[119]  POISINDEX®System, IBM Micromedex, access via WVG, Stuttgart

[120]  Quattrocchi U (2012) CRC World Dictionary of Medicinal and Poisonous Plants. Routledge, Abingdon-on-Thames-Oxfordshire

[121]  Queensland Health, Poison Information Centre (2015) Plants and mushrooms (fungi) poisonous to people in Queensland. https://www.childrens.health.qld.gov.au/chq/our-services/queensland-poisons-information-centre/plants-mushrooms/

[122]  Quinn JC et al. (2014) Secondary plant products causing photosensitization in grazing herbivores: their structure, activity and regulation. Int J Mol Sci 15: 1441.

[123]  Rätsch Ch (1998) Enzyklopädie der psychoaktiven Pflanzen – Botanik, Ethnopharmakologie und Anwendungen. AT-Verlag, Aarau, WVG, Stuttgart

[124]  Rätsch Ch (2006) Räucherstoffe. Der Atem des Drachen. WVG, Stuttgart

[125]  Rätsch Ch, Müller-Ebeling C (2006) Lexikon der Liebesmittel, AT Verlag Aarau 2003. licensed ed of WVG, Stuttgart

[126]  Riet-Correa F et al. (2011) Poisoning by Plants, Mycotoxins and Related Toxins. Cabi, Wallingford

[127]  Roth L et al. (2008) Giftpflanzen, Pflanzengifte, 5th ed. Nikol, Hamburg

[128]  Schade F, Jokusch H (2016) Giftpflanzen in unserer Umgebung. Springer Spektrum

[129]  Schäfer C, Marschall-Kunz B (2014) Gifte und Vergiftungen in Haushalt, Garten, Freizeit. WVG, Stuttgart

[130]  Schmidbauer W, Scheidt, vom, J (2004) Handbuch der Rauschdrogen. 11. Ed, Nymphenburger in der FA Herbig Verlagsbuchhandlung. München

[131]  Schmidt RJ, Botanical Dermatology Database. http://www.botanical-dermatology-database.info

[132]  Schultes RE, Hofmann A (1981) Pflanzen der Götter–Die magischen Kräfte der Rausch- und Giftgewächse. Hallwag Verlag, Bern

[133]  Seeger R, Neumann HG (1990) Giftlexikon. Ein Handbuch für Ärzte, Apotheker und Naturwissenschaftler, last actualization 2015 by Schrenk D et al., Dtsch Apoth-Verl, Stuttgart

[134]  Shepherd RCH (2004) Pretty but Poisonous. Plants Poisonous to People. An Illustrated Guide for Australia. RG and FJ Richardson, Melbourne

[135]  Singh BR, Tu AT eds (2011) Natural Toxins 2: Structures, Mechanism of Action and Detection. Springer

[136]  Singh YD (2014), Pathology of Plant Poisonings in Animals–A Guide to Practitioners and Farmer. LAP Lambert Academic Publishers, Chisinau (Moldova)

[137]  Soyka M (2010) Drogennotfälle: Diagnostik, Klinisches Erscheinungsbild, Therapie. Schattauer, Stuttgart

[138]  Stary F, Berger Z (1983) Giftpflanzen. Artia-Verlag Prag

[139]  Stewart A (2011) Wicked Plants, the Weed That Killed Lincoln's Mother & Other Botanical Atrocities, Algoquin Books of Chapel Hill, in German, BV, Berlin

[140]  Sticher O et al. eds (2015), Hänsel/Sticher Pharmakognosie–Phytopharmazie, 10th ed. WVG, Stuttgart

[141]  Stickel F, Shouval D (2015) Hepatotoxicity of herbal and dietary supplements: an update. Arch Toxicol 89(6): 851

[142]  Usda Agricultural Res Service Poisonous Plant Res. www.ars.usda.gov

[143]  Tang W, Eisenbrand G (2011) Chinese Drugs of Plant Origin, Chemistry, Pharmacology, and Use in Traditional and Modern Medicine. Springer, Berlin

[144] Tedeschi CG et al. (1977), Forensic Medicine – A Study in Trauma and Environmental Hazards, III, Environmental Hazards. WB Saunders Company, Philadelphia

[145] Teschke R, Eickhoff A (2015) Herbal hepatotoxicity in traditional and modern medicine: actual key issues and new encouraging steps. Front Pharmacol 6: 72

[146] Teuscher E (2006) Medicinal Spices, A Handbook of Culinary Herbs, Spices, Spice Mixtures and Their Essential Oils. Medpharm Scientific Publishers, Stuttgart

[147] Teuscher E, Lindequist U (2010) Biogene Gifte, Biologie–Chemie–Pharmakologie–Toxikologie, 3rd ed. WVG, Stuttgart

[148] Teuscher E ed (2018) Gewürze und Küchenkräuter. WVG, Stuttgart

[149] Teuscher E et al. (2020) Biogene Arzneimittel, 8th ed. WVG, Stuttgart

[150] The Merck Veterinary Manual (2008) http://www.merckvetmanual.com/mvm/index.jsp?cfile=htm/bc/210800.htm&word=prussic%2cacid

[151] Tox Info Suisse Zürich (2019) Giftige Garten- und Wildpflanzen. https://toxinfo.ch/customer/files/28/Giftige-Garten-und-Wildpflanzen-fuer-Tox-Info-Suisse.pdf

[152] TOXLINE (National Library of Medicine bibliographic database for toxicology). www.nlm.nih.gov/databases/download/toxlinesubset.html

[153] Tu AT ed (2019) Handbook of Natural Toxins, Vol. 7. Food Poisoning, Marcel Dekker New York

[154] Tubes G (2017) Giftigste Pflanzen Deutschlands, Kennzeichen–Standorte–Wirkung. Quelle & Meyer Verlag, Wiebelsheim

[155] University of California. Safe and poisonous garden plants. http://ucanr.edu/sites/poisonous_safe_plants/Toxic_Plants_by_Scientific_Name_685/

[156] Uter W (2020) Contact allergy–emerging allergens and public health impact. Int J Environ Res Public Health 17: 2404

[157] U.S. National Plant Germplasm System. GRIN-Global

[158] Van Wyk BE (2005) Food Plants of the World. WVG Stuttgart

[159] Van Wyk BE et al. (2002) Poisonous Plants of South Africa. Briza Publ, Pretoria

[160] Vergiftungs-Informations-Zentrale Freiburg (2019) Liste ausgewählter Giftpflanzen. https://www.uniklinik-freiburg.de/giftberatung/liste-ausgewaehlter-giftpflanzen.html

[161] Von Mühlendahl KE et al. (2003) Vergiftungen im Kindesalter, 4th ed. Thieme, Stuttgart

[162] Wagstaff DJ (2008) International Poisonous Plant Checklist. CRC Press, Boca Raton

[163] Watt JM, Breyer-Brandwijk MG (1982) The Medicinal and Poisonous Plants of Southern and Eastern Afrika. E und S Livingstone LTD, Edinburgh

[164] Weilemann S et al. (2006) Giftberatung Pflanzen, 3rd ed. Govi Verlag, Eschborn

[165] Wexler P (2014) Encyclopedia of Toxicology, 3rd ed. Elsevier, Amsterdam

[166] WFO (2021) World Flora Online. www.worldfloraonline.org

[167] Wink M, van Wyk BE (2008) Mind-altering and Poisonous Plants of the World. Timber Press, Portland Oregon

[168] Wink M (2015) Modes of action of herbal medicines and plant secondary metabolites. Medicines 2: 251

[169] Wirth W, Gloxhuber C (1994) Toxikologie, 5th ed, Thieme, Stuttgart, New York

[170] Xu X et al. (2020) Nephrotoxicity of herbal medicine and its prevention. Front Pharmacol 11: 569551

# Poison Information Centers

(Selection of PI numbers below, without guarantee of correctness)

**Dial the emergency number of the country you are in (EN)**, numbers see https://www.taschenhirn.de/allgemeinbildung-reisen/internationale-notrufnummern) **or from abroad** (**PI**, see international dialing code in the internet file above), and you will be connected with the relevant poison information center in this country

**Worldwide**, PI: (+45) 82 12 12 12
**All European countries.** EN: 112
Austria, PI: +44(1) 406 43 43
Australia, PI: 131 126 or EN: 000
Belgium, PI: +32(70) 245 245
Brazil, EN: 192
Bulgary, PI: +359(2) 515 32
Denmark, PI: +45 (35) 316-060
Egypt, EN: 123
England, PI: +44 (171) 635 91 91, 999 works in England and in all form British colonies
Finland, PI: +358(9) 472 977
France, PI: +33 (3) 883 737 37
Germany, EN: 112, this number also works in India, Great Britain, and in all European countries
Greece, PI: +30 (1) 799 37 77
Hungaria, PI: +36 (1) 215 215
India, PI: 1800 345 033
Indonesia, EN: 112
Italy, PI: +39 (6) 490 663
Norway, PI: +47(22) 591 300
People's Republic of China, EN: 120, works also in Japan
Poland, PI: +48 (42) 657 99 00
Russian Federation, PI: +7 (95) 928 16 47
South Africa, EN: 10 177
Spain, PI: +34 (91) 562 84 69
Sweden, PI: +46 (8) 736 03 84
Switzerland, PI: +41(1) 251 51 51
The Netherlands, PI: +31 (39) 274 88 88
Turkey, PI: +90(312)433 70 01
USA, PI: 1-800-222-1222 or 11911 or EN: 911, works also for Philippines
Other numbers use EN, see above.

https://doi.org/10.1515/9783110724738-011

# Index

The pages with the photos of the corresponding plants and the structural formulas of the corresponding compounds are printed in bold.

https://doi.org/10.1515/9783110724738-012

www.ingramcontent.com/pod-product-compliance
Lightning Source LLC
Chambersburg PA
CBHW080912220326

41598CB00034B/5553